Research Progress in Groundwater Contamination and Treatment

Research Progress in Groundwater Contamination and Treatment

Guest Editors

Panagiotis Papazotos
Eleni Vasileiou
Eleni Gianni
Simeone Chianese

MDPI

Basel • Beijing • Wuhan • Barcelona • Belgrade • Novi Sad • Cluj • Manchester

Guest Editors
Panagiotis Papazotos
School of Mining and
Metallurgical Engineering
National Technical University
of Athens
Athens
Greece

Eleni Vasileiou
School of Mining and
Metallurgical Engineering
National Technical University
of Athens
Athens
Greece

Eleni Gianni
Department of Environment
Ionian University
Zakynthos
Greece

Simeone Chianese
Department of Engineering
University of Campania
Luigi Vanvitelli
Aversa
Italy

Editorial Office
MDPI AG
Grosspeteranlage 5
4052 Basel, Switzerland

This is a reprint of the Special Issue, published open access by the journal *Environments* (ISSN 2076-3298), freely accessible at: https://www.mdpi.com/journal/environments/special_issues/P0D0WHLK1X.

For citation purposes, cite each article independently as indicated on the article page online and as indicated below:

Lastname, A.A.; Lastname, B.B. Article Title. *Journal Name* **Year**, *Volume Number*, Page Range.

ISBN 978-3-7258-6508-6 (Hbk)
ISBN 978-3-7258-6509-3 (PDF)
https://doi.org/10.3390/books978-3-7258-6509-3

Cover image courtesy of Panagiotis Papazotos

Contents

About the Editors

Panagiotis Papazotos

Panagiotis Papazotos is currently a postdoctoral researcher at the School of Mining and Metallurgical Engineering (SMME) at the National Technical University of Athens (NTUA), an adjunct lecturer at the University of Western Macedonia (UOWM), and an external collaborator with the Hellenic Survey of Geology & Mineral Exploration (HSGME). His primary research interests are in environmental and applied geochemistry, hydrochemical modeling, mineralogy, water resource management, and soil science, stemming from his focus on the geochemical footprint of ultramafic rocks on groundwater quality. He has been actively involved in projects centered on water–rock interaction investigation, specifically examining the occurrence, mobility, fate, and transport of potentially toxic elements in the environment, with his doctoral research focusing on chromium and arsenic pollution . His achievements include obtaining a Ph.D. in 2020, two M.Sc. degrees, publishing over 50 articles in peer-reviewed journals and conference proceedings, receiving a scholarship from the NTUA Research Committee for his doctoral dissertation, and acting as a reviewer for more than 25 high-impact environmental science journals. He is also proficient in various analytical techniques and software tools for statistical analysis, spatial analysis, geochemical modeling, and hydrochemistry.

Eleni Vasileiou

Eleni Vasileiou has been working as Research and Teaching Staff at the School of Mining and Metallurgical Engineering of National Technical University of Athens (NTUA) since 2014. She teaches seven courses, all related to Environmental Hydrogeology, in two postgraduate programs of NTUA and Agricultural University of Athens and in two undergraduate programs of NTUA (Mining Engineering and Civil Engineering). She has participated in 46 national research projects and European projects all involved with water resources and geo-environment. She has authored numerous technical reports (>15) and environmental studies on water resources (>20). She has 100 publications in peer-reviewed journals, proceedings of conferences, and book chapters. Her main research interests are environmental hydrogeology, groundwater contamination, anthropogenic and geogenic sources of water pollution, vulnerability of aquifers (DRASTIC, GALDIT), heavy metal pollution, mine water management, environmental pressures and management, DPSIR and SWOT analysis, hydro-geo-chemical modeling, water resources management, risk assessment of water resources, rehabilitation of post mining areas, hydro-geochemistry, water rock–soil interaction, microplastics in groundwater, land use and water quality, climate change, and water resources. She acts as a reviewer in numerous journals concerning water environment and has given more than 15 invited lectures.

Eleni Gianni

Eleni Gianni, a Geologist, is currently a Laboratory Teaching Staff at the Department of Environment, Ionian University, Greece, specializing in Applied and Environmental Mineralogy with strong teaching skills. She obtained her BSc, MSc, and PhD degrees from the Department of Geology, University of Patras, Greece. Her thesis focused on the preparation of drug delivery systems based on halloysite clay minerals and their characterization via experimental data, molecular simulation methods (classical and quantum), and analytical techniques. During her studies, she also focused on clays, clay minerals, and their nanocomposites' applications in cosmetics, catalysts, and water or wastewater treatment materials. Her previous position at the Centre for Research & Technology Hellas enhanced her research skills and her interest in the decontamination of soil systems and the minerals' role in underground energy storage applications. Her research interests include the following: i. Mineralogical characterization. ii. Clay, clay minerals, and other natural materials' applications in cosmetics, pharmaceutical, energy, and environmental. iii. The synthesis and characterization of nanocomposites. iv. The decontamination of water bodies and soil systems. v. Geochemical modelling for underground H and CO_2 storage. vi. Molecular simulations (quantum and classical) in a variety of systems. She has more than 35 publications in journals of SCI, chapters, and full conference papers, and more than 30 contributions to conferences, and is a reviewer for more than 14 high-impact environmental and material science journals.

Simeone Chianese

Simeone Chianese, PhD, is an Assistant Professor of Chemical Plants at the Department of Engineering of the University of Campania Luigi Vanvitelli (Aversa, Italy). His research interests are mainly focused on environmental chemical engineering for the protection of environmental matrices and the removal of pollutants; innovative processes for value-added production from biomass/organic sources; and sustainable approaches for the recovery of valuable compounds from organic substrates.

Preface

The focus of this Reprint is the comprehensive study of groundwater quality and treatment, emphasizing the understanding, monitoring, and remediation of contamination.

The Reprint's content covers identifying contamination sources and pathways, evaluating the mobility, transport and fate of contaminants, and exploring innovative monitoring and remediation techniques. Specific contaminants addressed include Potentially Toxic Elements (PTEs), Pharmaceuticals and Personal Care Products (PPCPs), and benzene, toluene, ethylbenzene, and xylene (BTEX).

The aim of this Reprint is to advance knowledge in the field through the dissemination of novel research findings and technological innovations. The primary goal is to strengthen collective understanding and provide a solid foundation for informed decision making in the sustainable management and protection of groundwater resources.

Groundwater, as the primary source of freshwater worldwide, faces continuous threats from both human activities, including population growth and industrial development, and natural processes. This work is essential to safeguard and sustainably manage this critical resource, ensuring its quality and availability for current and future generations.

This work is addressed to researchers, policymakers, environmental scientists, geochemists, hydrogeologists, environmental engineers, materials' scientists, chemists, and public health professionals.

Panagiotis Papazotos, Eleni Vasileiou, Eleni Gianni, and Simeone Chianese
Guest Editors

environments MDPI

Editorial

Research Progress in Groundwater Contamination and Treatment

Eleni Gianni [1,*] and Panagiotis Papazotos [2,3,*]

1 Department of Environment, Ionian University, M. Minotou-Giannopoulou 26, 29100 Zakynthos, Greece
2 Division of Geo-Sciences, School of Mining and Metallurgical Engineering, National Technical University of Athens, 9 Heroon Polytechniou St., 15773 Zografou, Greece
3 Hellenic Survey of Geology and Mineral Exploration, 1 Sp. Louis St., 13677 Acharnae, Greece
* Correspondence: e.gianni@ionio.gr (E.G.); papazotos@metal.ntua.gr (P.P.)

Groundwater constitutes approximately 99% of the total freshwater volume circulating on Earth [1]. It is therefore the primary source of freshwater for the global population, supporting domestic, industrial, and agricultural applications [2]. Its importance extends beyond human and ecological systems to encompass hydrological processes as a whole [3]. However, the continuous increase in population and industrial activity has led to contamination or even pollution from multiple anthropogenic sources, while natural processes such as water–rock–soil interaction via mineralogical phases dissolution, salinization, and the occurrence, mobilization, transport, and environmental fate of potentially toxic elements (PTEs), also contribute to groundwater contamination, thereby threatening water quality. Researchers worldwide are developing innovative technologies for assessing and monitoring groundwater contamination or pollution, such as real-time Internet of Things (IoT) platforms, remote sensing techniques, artificial intelligence (AI) and machine learning (ML) approaches, and biosensors [4]. A detailed understanding of contaminant sources, pathways, and receptors enabled by these technologies is essential for the efficient design of treatment methods aimed at mitigating the impacts of groundwater contamination.

Innovative remediation methods have become a primary research focus to overcome the limitations of conventional technologies. To address these challenges, several in-situ and ex-situ approaches have been investigated, depending on the contaminant type, the distance from sensitive receptors, and the aquifer depth [4]. These approaches include advanced oxidation processes [5], membrane-based technologies [6], bioremediation [7], and hybrid systems [8].

This Special Issue, "Research Progress in Groundwater Contamination and Treatment," comprises eight research articles that collectively advance the understanding of groundwater quality and treatment across various regions worldwide. Relatively few studies address advanced remediation techniques, including the synthesis of titanium dioxide/tungsten trioxide (TiO_2/WO_3) photocatalysts for pharmaceutical pollutants using a novel hydrothermal synthesis method, which offers advantages over conventional physical mixing by enhancing photocatalytic performance through specific structural and morphological features (contribution 1).

The groundwater quality studies explored assessed contamination across diverse geographic and hydrogeological settings. In southeast Spain, the Solana aquifer was shown to exhibit stable water quality over time, dominated by calcium carbonate facies. Principal component analysis (PCA) and hierarchical cluster analysis (HCA) techniques verified salinity differentiations based on evaporitic dissolution and anthropogenic sources of nitrate (NO_3^-) relative to agricultural applications (contribution 2). In the Anaga Rural Park in Tenerife, Canary Islands, Spain, a UNESCO-protected area, the groundwater quality remained consistently high for over a decade due to effective soil protection and low anthropogenic pressure, highlighting the benefits of sustainable soil and groundwater management practices. In this study, the Free Residual Chlorine (FRC), in-situ chlorine (ISC), coliform bacteria, colony count at 22 °C, *Escherichia coli*, laboratory turbidity, in-situ turbidity, ammonium (NH_4^+), nitrate (NO_3^-), pH, electrical conductivity (EC), color, odor, and taste were examined in accordance with Spanish legislation and the European Union's (EU) Drinking Water Directive (DWD) (contribution 3). In northeastern Peloponnese, Greece, several areas were examined to assess the effectiveness of multiple geo-environmental indices for evaluating groundwater used for various purposes. Indices related to drinking and irrigation suitability, potentially toxic element (PTE) loadings, and ionic ratios were analyzed. Hydrogeochemical bivariate plots, correlation analyses, and multivariate statistical techniques (including factor analysis via PCA and hierarchical cluster analysis) were integrated to evaluate the robustness of commonly applied geo-environmental indices in groundwater quality assessments. The findings revealed lessons learned, important insights, and limitations about the successful application of the indices, highlighting the sources of PTEs in groundwater and demonstrating the need for the combined use of geo-environmental indices to ensure more dependable evaluations, as individual indices may yield misleading interpretations due to several factors (contribution 4).

In West–Central Florida, USA, two studies investigated the occurrence of fourteen pharmaceuticals and personal care products (PPCPs) in shallow lakes, revealing seasonal variations in compound concentrations and associated ecological risks, in two distinct environments; one influenced by septic tanks and the other unaffected. Liquid chromatography–tandem mass spectrometry (LC-MS) analysis consistently detected theophylline, caffeine, cotinine, N,N-diethyl-meta-toluamide (DEET), and testosterone, while additional compounds were observed during the wet season. These findings aligned with the risk quotients that exhibited higher risks for algae, crustaceans, and fishes during the same period, particularly in lakes surrounded by septic tanks, as was expected (contributions 5 & 6). In Limpopo Province, South Africa, groundwater chemistry was influenced by both natural and anthropogenic factors. Sodium adsorption ratio (SAR) analysis revealed high sodium (Na^+) concentrations that caused low water quality, while PCA identified the presence of PTEs, including vanadium (V) in very high concentrations. Microbial assessments indicated that water was suitable for both drinking and irrigation (contribution 7). In the Eastern Niger Delta, Nigeria, groundwater near petroleum infrastructure exhibited high concentrations of benzene, toluene, ethylbenzene, and xylene (BTEX), with increasing distance from the pipelines. Although the aquifer's natural aerobic attenuation potential provided partial mitigation of hydrocarbon contamination, complete decontamination would require a prolonged period, indicating that natural attenuation alone is insufficient (contribution 8).

Collectively, these studies illustrate the importance of integrating different technologies, including also conventional techniques and innovative tools, as well as hydrochemical characterization, advanced monitoring techniques, risk assessment, and innovative remediation strategies to safeguard groundwater resources globally. The active involvement of multiple stakeholders, including researchers, policymakers, and citizens, is crucial to promote the implementation of these approaches in large-scale applications. Such collaboration is key to ensuring equitable access to clean water and the sustainable management of groundwater resources in line with the Sustainable Development Goals (SDGs).

This Special Issue highlights the current research advancements in distinguishing the sources and pathways of groundwater contamination, as well as exploring remediation techniques and strategies, with a focus on evaluating the effectiveness and feasibility of various treatment methods. For future studies, it will be necessary to integrate fieldwork, laboratory experiments, modeling approaches, and innovative methodologies such as ML and multiple-criteria decision analysis (MCDA) to gain an in-depth understanding of the factors affecting groundwater contamination and treatment. Emerging contaminants such as per and polyfluoroalkyl substances (PFAS), micro/nanoplastics (e.g., polyethylene, polypropylene, polyvinyl chloride, polyethylene terephthalate, and polyamide), PPCPs, and pesticides, together with PTEs (e.g., arsenic—As; chromium—Cr; lead— Pb; mercury—Hg; cadmium—Cd; zinc—Zn; antimony—Sb, etc.) and the use of environmental isotopic signatures (e.g., δ^{53}Cr, ^{87}Sr/^{86}Sr, ^{206}Pb/^{204}Pb, ^{207}Pb/^{204}Pb, ^{208}Pb/^{204}Pb, δ^{11}B, $\delta^{15}N_{NO3}$, and $\delta^{18}O_{NO3}$), have become prominent research hotspots in environmental science.

We are deeply grateful to the authors for sharing their innovative ideas and pioneering methods in this Special Issue. Our sincere thanks also go to the reviewers, whose dedication and insightful evaluations have been invaluable throughout the process. It has been a genuine pleasure to collaborate with the *Environments* Editorial Office staff, whose professionalism and teamwork made this endeavor both rewarding and enjoyable. We would also like to extend our heartfelt thanks to Ms. Maria Chen, whose decisive contributions were essential to the successful completion of this Special Issue. Her support was consistently generous and dependable, and she played a key role in the overall success of this project. This collection aspires to foster sustained interdisciplinary collaboration and innovation at the nexus of environmental science, geochemistry, hydrogeology, environmental engineering, and public health, advancing knowledge and practical approaches for the protection and remediation of groundwater resources.

Conflicts of Interest: The authors declare no conflicts of interest.

List of Contributions:

1. Ghiloufi, M.; Schnabel, T.; Springer, C.; Mehling, S.; Wolfram, A.; Touati, F.; Kouass, S. Hydrothermal Versus Physical Mixing: Superior Photocatalytic Activity of TiO_2/WO_3 Nanocomposites for Water Treatment Applications. *Environments* **2025**, *12*, 359. https://doi.org/10.3390/environments12100359.
2. Sala-Sala, V.; Andreu, J.M.; Pérez-Gimeno, A.; Jordán, M.M.; Navarro-Pedreño, J.; Almendro-Candel, M.B. Spatial and Multivariate Analysis of Groundwater Hydrochemistry in the Solana Aquifer, SE Spain. *Environments* **2025**, *12*, 323. https://doi.org/10.3390/environments12090323.
3. Rodríguez-Alcántara, J.S.; Cruz-Pérez, N.; Rodríguez-Martín, J.; García-Gil, A.; Koritnik, J.; Santamarta, J.C. Improving Groundwater Quality Through Biosphere Reserve Management: Insights from the Anaga Reserve, Tenerife. *Environments* **2025**, *12*, 53. https://doi.org/10.3390/environments12020053.
4. Papazotos, P.; Vlachomitrou, M.; Psarraki, D.; Vasileiou, E.; Perraki, M. Coupling Advanced Geo-Environmental Indices for the Evaluation of Groundwater Quality: A Case Study in NE Peloponnese, Greece. *Environments* **2025**, *12*, 14. https://doi.org/10.3390/environments12010014.
5. Bialkowska-Jelinska, E.; van Beynen, P.; Calcul, L. Seasonality of Pharmaceuticals and Personal Care Products in Shallow Lakes, Florida, USA—Part A. *Environments* **2025**, *12*, 219. https://doi.org/10.3390/environments12070219.
6. Bialkowska-Jelinska, E.; van Beynen, P.; Calcul, L. Assessing Environmental Risk Posed by Pharmaceuticals and Personal Care Products in Shallow Lakes, Florida, USA—Part B. *Environments* **2025**, *12*, 231. https://doi.org/10.3390/environments12070231.
7. Shokoohi, E.; Moyo, N. Groundwater Quality in a Rural and Urbanized Region in Limpopo Province, South Africa. *Environments* **2025**, *12*, 174. https://doi.org/10.3390/environments12060174.
8. Aleku, D.L.; Biester, H.; Pichler, T. Pipeline-Related Residential Benzene Exposure and Groundwater Natural Attenuation Capacity in the Eastern Niger Delta, Nigeria. *Environments* **2024**, *11*, 221. https://doi.org/10.3390/environments11100221.

References

1. Younger, P.L. *Groundwater in the Environment: An Introduction*; John Wiley & Sons: New York, NY, USA, 2009; ISBN 1-4443-0904-8.
2. Li, P.; Karunanidhi, D.; Subramani, T.; Srinivasamoorthy, K. Sources and Consequences of Groundwater Contamination. *Arch. Environ. Contam. Toxicol.* **2021**, *80*, 1–10. [CrossRef] [PubMed]
3. Giordano, M. Global Groundwater? Issues and Solutions. *Annu. Rev. Environ. Resour.* **2009**, *34*, 153–178. [CrossRef]
4. Robles, K.P.V.; Monjardin, C.E.F. Assessment and Monitoring of Groundwater Contaminants in Heavily Urbanized Areas: A Review of Methods and Applications for Philippines. *Water* **2025**, *17*, 1903. [CrossRef]
5. Petala, A.; Arvaniti, O.S.; Travlou, G.; Mantzavinos, D.; Frontistis, Z. Solar Light Induced Photocatalytic Removal of Sulfamethoxazole from Water and Wastewater Using BiOCl Photocatalyst. *J. Environ. Sci. Health Part A* **2021**, *56*, 963–972. [CrossRef] [PubMed]
6. Foorginezhad, S.; Zerafat, M.M.; Ismail, A.F.; Goh, P.S. Emerging Membrane Technologies for Sustainable Water Treatment: A Review on Recent Advances. *Environ. Sci. Adv.* **2025**, *4*, 530–570. [CrossRef]
7. Romantschuk, M.; Lahti-Leikas, K.; Kontro, M.; Galitskaya, P.; Talvenmäki, H.; Simpanen, S.; Allen, J.A.; Sinkkonen, A. Bioremediation of Contaminated Soil and Groundwater by In Situ Biostimulation. *Front. Microbiol.* **2023**, *14*, 1258148. [CrossRef] [PubMed]
8. Anandan, S.; Kumar Ponnusamy, V.; Ashokkumar, M. A Review on Hybrid Techniques for the Degradation of Organic Pollutants in Aqueous Environment. *Ultrason. Sonochem.* **2020**, *67*, 105130. [CrossRef] [PubMed]

environments

MDPI

Article

Hydrothermal Versus Physical Mixing: Superior Photocatalytic Activity of TiO_2/WO_3 Nanocomposites for Water Treatment Applications

Mabrouka Ghiloufi [1,2], Tobias Schnabel [3,*], Christian Springer [4], Simon Mehling [3], Axel Wolfram [3], Fathi Touati [2] and Salah Kouass [1,2]

1 Faculty of Sciences of Bizerte, Carthage University, Tunis 1054, Tunisia; mabrouka.ghiloufi@fsb.ucar.tn (M.G.)
2 Laboratory of Materials: Treatment and Analysis (LMTA), National Institute of Research and Analysis Physicochemical (INRAP), Ariana 2020, Tunisia; fathi19612004@yahoo.fr
3 Research Group "Photonics and Water", Institute for Sustainable Water Systems, Hof University of Applied Sciences, 95028 Hof, Germany; simon.mehling@hof-university.de (S.M.); axel.wolfram@hof-university.de (A.W.)
4 Urban Water Management and Environmental Technology, Erfurt University of Applied Science, 99085 Erfurt, Germany; christian.springer@fh-erfurt.de
* Correspondence: tobias.schnabel@hof-university.de

Abstract

The photocatalytic efficiency of TiO_2 was significantly enhanced by coupling with WO_3 to form a TiO_2/WO_3 heterostructure, designed to operate effectively under UV-LED irradiation. The nanocomposites were synthesized via a hydrothermal route, and their activity was evaluated through the degradation of the pharmaceutical pollutant venlafaxine. Contaminants are rarely addressed in photocatalytic studies. Unlike a simple physical mixture of commercial TiO_2 and WO_3 powders, the hydrothermally synthesized TiO_2/WO_3 photocatalyst exhibited superior efficiency, attributable to its nanoscale dimensions achieved via the hydrothermal route, which promoted improved charge carrier separation, enhanced surface homogeneity, and the formation of an effective heterojunction interface. An optimization study varying the WO_3 content (5, 10, 20, and 30 wt.%) within the TiO_2 revealed that the 10 wt.% WO_3 composition achieved the highest performance, with ~52% venlafaxine degradation within 60 min. SEM, TEM, FTIR, Raman spectroscopy, XRD, and UV-Vis DRS revealed the successful incorporation of WO3 into the TiO_2 matrix, confirming phase purity and composition-dependent structural evolution of the nanocomposite, and evidencing extended light absorption and superior charge-transfer properties. Importantly, the optimized photocatalyst thin film retained excellent stability and reusability, maintaining high degradation efficiency over three consecutive cycles with negligible activity loss, which avoids slurry separation. These findings establish hydrothermally synthesized $TiO_2/10\%WO_3$ thin film heterostructures as effective and sustainable photocatalytic platforms for the removal of pharmaceutical pollutants in wastewater under UV-LED irradiation.

Keywords: TiO_2/WO_3 heterostructure; THIN film; photocatalysis; venlafaxine; hydrothermal synthesis; LED^{365}

1. Introduction

Water pollution has emerged as one of the most serious environmental challenges of the 21st century, largely driven by rapid urbanization, industrialization, and the overutilization of synthetic organic chemicals. Among the different classes of pollutants, pharmaceutical residues are of increasing concern owing to their continuous release into the environment via domestic, hospital, and industrial effluents. Classified as emerging pollutants, they are capable of exerting long-term ecological and health effects, even at low concentrations, and are not effectively eliminated by conventional groundwater purification methods [1–3]. Pharmaceuticals, such as anti-inflammatories, antibiotics, and antidepressants, are designed to be biologically active and stable, which makes them resistant to the natural degradation processes. Their presence in aquatic environments has been widely documented. One such compound is venlafaxine (Venla), a serotonin-norepinephrine reuptake inhibitor (SNRI) used extensively for the treatment of depression and anxiety disorders. Due to its widespread use, high water solubility, and poor biodegradability, Venla is frequently detected in surface waters, groundwater, and even drinking water supplies [3–5]. Researchers have reported its presence at different concentrations ranging from ng/L to several µg/L, raising serious concerns about its long-term ecotoxicological impacts, including behavioral and reproductive disorders in aquatic organisms [6].

The inefficiency of conventional groundwater purification in removing Venla and other micropollutants necessitates the use of more effective and advanced treatment methods [7,8]. Among these, heterogeneous photocatalysis has generated significant attention as a green and sustainable process capable of degrading a wide spectrum of organic pollutants into non-toxic end products (CO_2 and H_2O) [9–11]. Upon light irradiation, the semiconductor photocatalyst is photoexcited, generating electron–hole pairs. The photogenerated electrons can react with dissolved oxygen to produce reactive species such as superoxide radicals ($O_2\cdot^-$), while holes in the valence band can directly oxidize organic molecules or react with surface-adsorbed water to form hydroxyl radicals (HO·). These reactive species, together with direct hole oxidation, act synergistically to attack and degrade the organic pollutants. Titanium dioxide (TiO_2) is widely recognized as a highly active photocatalyst and is frequently employed as a benchmark for evaluating the performance of emerging photocatalysts due to its chemical stability, reduced cost, non-toxicity, and strong oxidative power. TiO_2 has two major disadvantages: its wide band gap (3.2 eV), which restricts activation to UV light, and the high electron–hole recombination rate, thus limiting its quantum efficiency [12–14]. To overcome these limitations, heterojunction engineering has been proposed as an improved method for boosting photocatalytic activity. Associating TiO_2 with a suitable narrow band gap semiconductor, such as tungsten trioxide (WO_3), has shown promising results. WO_3, with a band gap of 2.6–2.8 eV, is photoactive under UV and near visible light (440–480 nm), and can act as an effective electron acceptor co-coupled with TiO_2. The formation of TiO_2/WO_3 heterojunction facilitates spatial separation of charge carriers, prolongs their lifetime, and enhances the generation of reactive species [15,16]. Furthermore, WO_3 possesses high chemical stability and good compatibility with TiO_2, making the nanocomposite a promising candidate for photocatalytic applications.

In addition to material optimization, the choice of the light source plays a crucial role in the efficiency and sustainability of the photocatalytic process. Traditional UV mercury lamps are energy-intensive and environmentally hazardous due to mercury content. In contrast, UV-A light-emitting diodes (LEDs) offer several advantages, including low energy consumption, long operational life, low heat generation, and wavelength specificity, with a narrow spectral band near 365 nm. These features make LEDs more suitable for practical, scalable photocatalytic systems for water purification applications [17,18].

Given this context, the present study focuses on the photocatalytic degradation of Venla using TiO_2/WO_3 thin film under LED^{365} irradiation, with a particular emphasis on ecological applications. It explores how the preparation method of the photocatalyst, specifically hydrothermal synthesis versus physical mixing, significantly influences its photocatalytic efficiency. Furthermore, the systematic variation of WO_3 loading provides new insights into the optimal composition required for enhanced activity. To strengthen this understanding, a comprehensive set of characterization techniques (Raman spectroscopy, FTIR, SEM, and UV-Vis spectroscopy) was employed to establish a clear correlation between the structural features, morphological properties, and photocatalytic performance.

2. Materials and Methods

2.1. Materials

The precursors employed in this study were titanium dioxide (TiO_2 Aeroxide® P25, Evonik industries AG, Essen, Germany), tungsten (VI) oxide (WO_3, Sigma Aldrich, St. Louis, MO, USA), sodium tungstate dihydrate ($Na_2WO_4{\cdot}2H_2O$, Loba Chemie, Mumbai, India, 99, hydrochloric acid (HCl, Carl Roth, Karlsruhe, Germany), deionized water, isopropanol (C_3H_8O, Chem solute, Renningen, Germany), acetylacetone ($C_5H_8O_2$, Carl Roth, Karlsruhe, Germany), nitric acid (HNO_3, Carl Roth, Karlsruhe, Germany), titanium isopropoxide ($Ti(O\text{-}iPr)_4$, Sigma Aldrich, St. Louis, MO, USA), and polyethylene glycol (PEG, Sigma Aldrich, Overijse, Belgium).

2.2. Synthesis of Nanocomposites

In this study, two sorts of nanocomposites will be treated as photocatalysts and will be compared with an unmodified sample of nano-TiO_2. Nano-TiO_2 P25 was used as the reference photocatalyst. P25 consists of 70–80% anatase and 20–30% rutile, and its mixed-phase composition provides higher photocatalytic activity compared to pure anatase, owing to the synergistic interaction between the two crystalline phases [19]. The first type of nanocomposite was achieved by mixing two commercial oxides (C.O), TiO_2 and WO_3, through simple physical mixture. The oxides were ground using a clean mortar and a pestle for 30 min. The second sort of nanocomposite preparation will be based on synthesized oxides (S.O) by the hydrothermal method. The TiO_2/WO_3 nanocomposites with different WO_3 loading (5, 10, 20, and 30%) values were synthesized using 0.3 g of TiO_2 as the base and $Na_2WO_4{\cdot}2H_2O$ as the tungsten source. In a typical preparation, the required stoichiometric amount of $Na_2WO_4{\cdot}2H_2O$ corresponding to the desired WO_3 percentages was dissolved in 20 mL of deionized water under continuous magnetic stirring (300 rpm) for 30 min. The pH of the solution was carefully adjusted by the dropwise addition of a concentrated HCl (1 M) until the pH reached 2, thereby promoting the formation of WO_3 species. Subsequently, the predetermined amount of P25 TiO_2 powder was dispersed into the mixture under vigorous stirring (500 rpm) for one hour to ensure homogeneous mixing and intimate contact between TiO_2 and WO_3 precursors. The resulting suspension was transferred into a 45 mL Teflon-lined stainless-steel autoclave and subjected to hydrothermal treatment at 150 °C and autogenous pressure for 12 h. After cooling to room temperature, the solid products were washed with deionized water and ethanol to remove residual ions, and then dried at 100 °C for 3 h. The as-prepared nanocomposites, denoted as $TiO_2/x\%WO_3$ (x = 5, 10, 20, and 30%), were subsequently stored for further physicochemical and photocatalytic characterization.

2.3. Preparation of Thin Film

All the thin films were deposited onto square fluorine-doped tin oxide (FTO) glass substrates with the same thickness (5 cm × 5 cm, 3 mm thick, Sigma Aldrich). As a first step, a base solution was prepared via the sol–gel route. Initially, the deionized water was placed in a beaker under magnetic stirring (300 rpm) and mixed with isopropanol. Subsequently, acetylacetone was introduced and stirred for 5 min, followed by acidification with nitric acid. The mixture was stirred further before increasing the speed to 500 rpm, at which point titanium isopropoxide was added dropwise. The resulting solution was stirred for 2 h at ambient temperature until it had a transparent yellow appearance. The precise sequence of addition was essential to ensure controlled hydrolysis and to avoid premature precipitation.

As a second step, for each sample, 1 mL of the base sol–gel solution was mixed with 0.1 g of the prepared nanocomposites in a reactor under magnetic stirring for 1 h/300 rpm. Afterwards, the mixture was placed in an ultrasonic bath. Then, 0.25 mL of polyethylene glycol, 0.5 mL of 2-propanol, and 0.2 mL of acetylacetone were added sequentially to the reactor, ensuring that the components were added in the correct order. The coagulated mixture was coated on the surface of FTO glass to enable heating in an oven at 300 °C for 2 h.

2.4. Experimental Setup and Procedure

The experimental reactor was specifically designed to enable measurements of pollutant degradation. The unit consists of a 145 mL reactor vessel, in which the photocatalyst coated onto FTO glass was mounted on one side, oriented towards the interior. A magnetic stirrer was integrated at the base of the vessel to ensure effective turbulence. Illumination is provided by an LED array comprising nine UV-A diodes (365 nm, 1 W, Seoul, Republic of Korea, CUN66A1B). The array is positioned parallel to the catalyst at a distance of 5.8 cm, allowing the radiation to pass through the FTO substrate and reach the photocatalyst surface.

For the investigation of the photocatalytic performance of the nanocomposites, Venla was chosen as a pharmaceutical pollutant. The UV-Vis measurements were performed with Mettler Toledo UV5 Bio (Gießen, Germany) at 365 nm. The photocatalytic degradation generates reactive species that break down organic pollutants into less harmful substances like CO_2 and H_2O.

$$\text{Photocatalyst } (HO\cdot/O_2\cdot^-) + \text{pollutant} \rightarrow CO_2 + H_2O$$

This reaction involves the excitation of electrons, formation of electron–hole pairs, and subsequent redox reactions with the pollutant. The degradation efficiency is calculated using the equation:

$$\text{Degradation } (\%) = \left(\frac{C_0 - C_t}{C_0}\right) \times 100 \tag{1}$$

where C_0 is the initial concentration, C_t is the concentration at time t.

Assuming that the photocatalytic degradation of Venla follows a first-order kinetic model, the reaction can be described by the following equation:

$$r = -\frac{dC}{dt} \text{ and } \ln\left(\frac{C}{c_0}\right) = -Kt \tag{2}$$

In the expression, r represents the reaction rate, C is the concentration of contaminant at a given irradiation time t (min), C_0 is the initial concentration (mg/L), K is the rate constant (min^{-1}), and t is the irradiation duration.

The structural, morphological, and optical properties of the synthesized samples were characterized using Fourier-transform infrared spectroscopy (FTIR) (PerkinElmer FT-IR spectrometer, Castries, France, L 160000A Spectrum Two, 2.5 μm to 25 μm, $LiTaO_3$ detector), Raman spectroscopy (Thunder Optics Gurzil Raman Microscope TO-RM-S-785 nm, Montpellier, France), Energy-Dispersive X-Ray Spectroscopy (EDX) (QUANTAX Brucker, Berlin, Germany), Scanning Electron Microscopy (SEM) (Coxem Tabletop EM-30N, Daejeon, Republic of Korea, Tungsten filament (W), detector SE and BSE), Transmission Electron Microscopy (TEM/HRTEM) and Selected Area Electron Diffraction (SAED) were performed using the same instrument (Tecnai G^2 20, FEI Company, Hillsboro, OR, USA, LaB_6 filament, 200 kV), The crystal structure was examined by the powder X-ray diffraction (XRD)(Bruker AXS GmbH, Karlsruhe, Germany, CuK radiation ($k_{1/4}$ 1.54056Å) and a graphitic monochromator), and Diffuse reflectance spectroscopy (DRS) (PerkinElmer UV-Vis spectrometer (OPDI.MA, ODM98) Castries, France, 250–1100 nm, $BaSO_4$).

2.5. Analytical Method

The calibration curve of Venla is presented in Figure 1. It exhibits a characteristic absorption band in the UV region, with maximum absorbance around 225–230 nm. Venla was established over a concentration range of 1–30 mg/L, showing a strong linear relationship between absorbance and concentration. The correlation coefficient (R^2) confirmed excellent linearity, ensuring the accuracy and reliability of quantitative analysis.

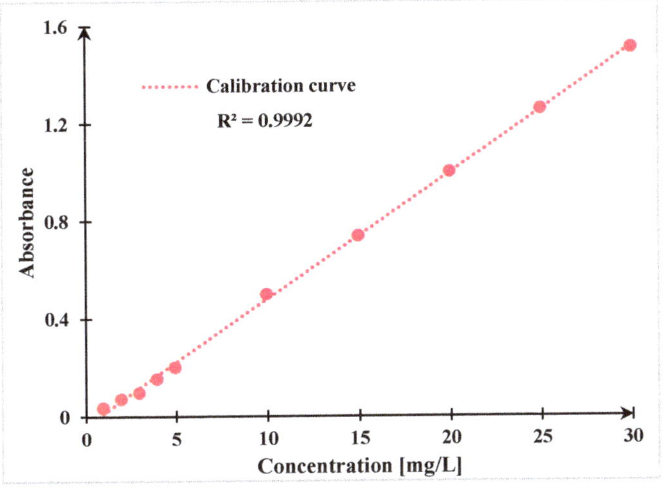

Figure 1. Calibration curve of Venlafaxine.

3. Results and Discussion

3.1. Characterization of Nanocomposites

3.1.1. Fourier Transform Infrared Spectroscopy (FTIR)

FTIR spectroscopy (Figure 2a) was employed to investigate the structural and vibrational features of the synthesized TiO_2/WO_3 nanocomposites with varying WO_3 loadings (5, 10, 20, and 30 wt.%). Spectra of commercial TiO_2 and WO_3 were used as reference materials. The TiO_2 spectrum displays a prominent absorption band between 450 and 700 cm^{-1}, attributed to Ti-O stretching vibrations. This feature was also present in the nanocomposites, though progressively shifted and reduced in intensity with increasing WO_3 content, suggesting modifications to the Ti-O bonding environment. Pure WO_3 showed characteristic bands at 507–610 cm^{-1}, corresponding to W-O bending vibrations, and at 804 cm^{-1} assigned to W-O-W bridging vibrations within WO_6 octahedra. These peaks also appeared in nanocomposites, albeit slightly displaced, which may indicate partial structural interaction or distortion of WO_3 within the TiO_2 matrix. Additional weak bands observed in the 1800–2200 cm^{-1} region were assigned to C=O stretching, most likely arising from adventitious carbonates or atmospheric CO_2 incorporated during synthesis or measurement. A broad absorption band between 3000 and 3600 cm^{-1}, characteristic of O-H stretching, was detected in all samples and is commonly associated with surface hydroxyl groups or absorbed water, particularly in nanocomposites prepared by hydrothermal routes [20].

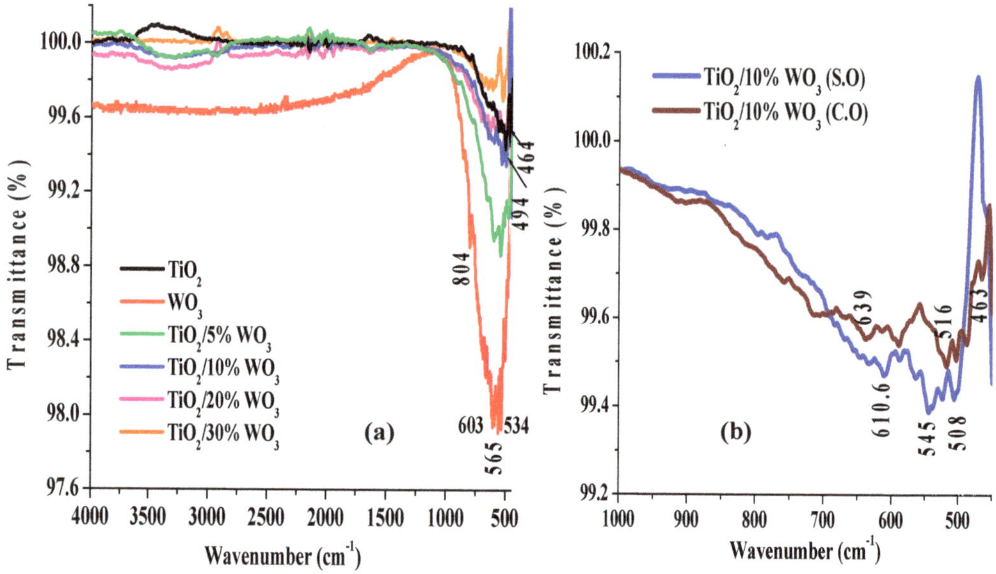

Figure 2. (**a**) FTIR spectra of nanocomposite TiO_2/WO_3 (0, 5, 10, 20, 30, and 100 wt.% WO_3) and (**b**) comparison between TiO_2/10%WO_3 (S.O) and TiO_2/10%WO_3 (C.O).

To assess the influence of the synthesis route, the FTIR spectrum of the TiO_2/10%WO_3 (S.O) nanocomposite was compared with TiO_2/10%WO_3 (C.O) in Figure 2b. The hydrothermal sample exhibited subtle shifts in the Ti-O (450–700 cm^{-1}) and W-O (507–610 cm^{-1}) vibrational bands, together with sharper and more intense absorption features. The spectral changes suggest a stronger interaction between TiO_2 and WO_3, most likely through the formation of Ti-O-W linkages. Such structural modifications are indicative of improved incorporation of WO_3 within the TiO_2 and WO_3 framework and may account for the enhanced photocatalytic performance observed in hydrothermally prepared samples.

3.1.2. X-Ray Diffraction of Synthesized TiO_2/WO_3 Nanocomposites

XRD analyses were conducted to examine the crystalline phases present in the prepared nanocomposites with varying WO_3 loadings. The diffraction pattern of pure TiO_2 confirms predominant crystallization in the anatase phase, with characteristic reflections at 2θ = 26.4°, 38.4°, and 48°, corresponding to the (101), (004), and (200) planes, respectively, in agreement with JCPDS card No. 88–1175. Additional peaks attributable to the rutile phase of TiO_2 (space group $P4_2/mmm$) were also detected, assigned to the (110), (101), (111), (210), and (220) planes (Figure 3). For WO_3, diffraction peaks characteristic for the monoclinic crystalline phase were identified, corresponding to the (001), (020), and (200) planes. The XRD diffractograms of the TiO_2/WO_3 nanocomposites reveal the coexistence of TiO_2 and WO_3 phases, with no evidence of secondary phases or impurities, confirming the high purity of the synthesized nanomaterials. Moreover, the progressive increase in WO_3 content from 5% to 30% leads to a corresponding intensification of the WO_3 diffraction peaks, demonstrating the successful incorporation of WO_3 and its compositional influence on the crystallographic features of the nanocomposites.

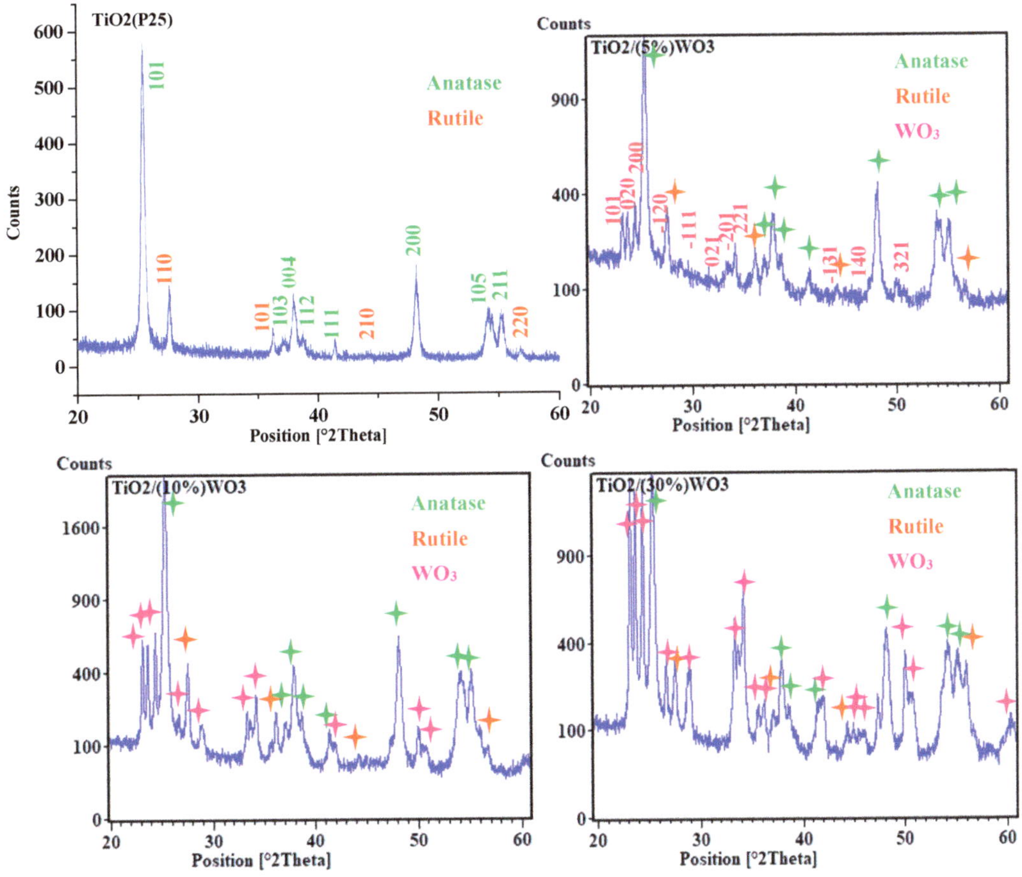

Figure 3. XRD diffractogrammes of TiO_2 and TiO_2/WO_3 nanocomposites. Green stars represent anatase, orange stars represent rutile and purple stars represent WO_3.

3.1.3. Raman Spectroscopy

Raman spectroscopy was used in this study to probe the structural properties and interfacial interactions within $TiO_2/10\%WO_3$ thin film. The Raman-active modes characteristic of anatase are well-defined in the spectra of Figure 4, including a symmetric bending vibration at 399 cm^{-1} (B_1g), and an antisymmetric bending mode around 516 cm^{-1} (A_1g). In addition, a distinct peak at 635 cm^{-1} (Eg), corresponding to the antisymmetric stretching vibration of the rutile phase, was also observed. An intense band confirms the presence of WO_3 at 807 cm^{-1}, attributed to W-O stretching vibrations, together with additional features at 717, 272 cm^{-1}, as well as a lower-frequency band at 325 cm^{-1}, all of which can be assigned to W-O-W bending and stretching modes [21,22]. Notably, a slight shift in the vibrational modes of $TiO_2/10\%WO_3$ film was detected when compared with the reference spectra of TiO_2 and WO_3. This shift is indicative of strong interfacial interactions between the oxides, suggesting possible Ti-O-W bond formation or local structural distortions at the heterojunction.

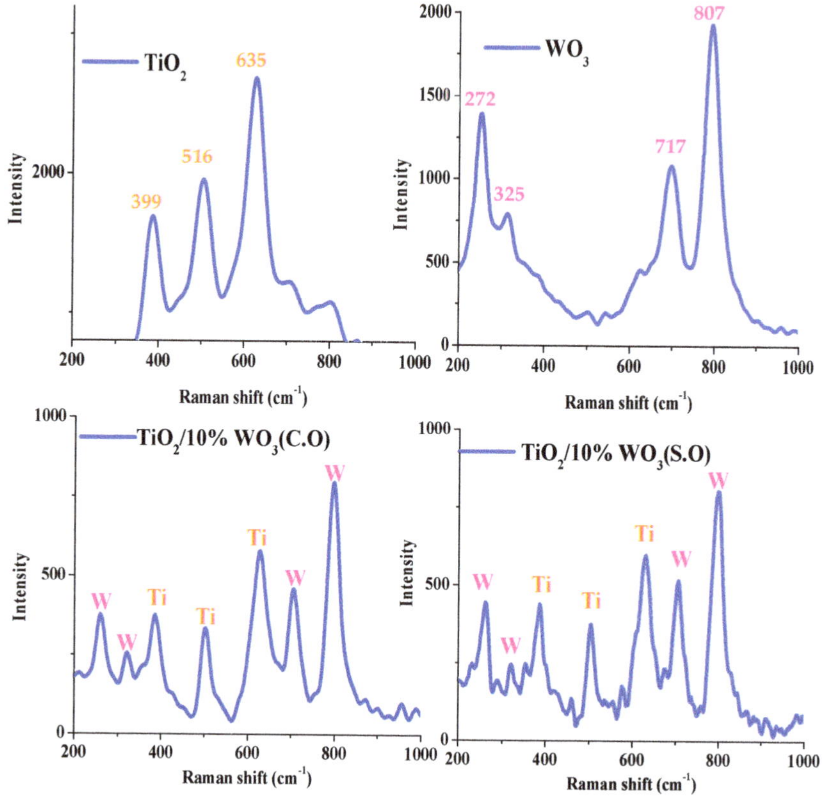

Figure 4. Raman spectra of: TiO_2, WO_3, TiO_2/10%WO_3 (C.O), TiO_2/10%WO_3 (S.O).

3.1.4. Energy Dispersive X-Ray Spectroscopy (EDX)

The EDX spectrum presented in Figure 5 for the pure TiO_2 thin film deposited on a standard glass slide as a substrate indicates titanium (Ti) and oxygen (O) as the principal elements. Distinct Ti peaks were observed at Kα (4.52 keV) and Kβ (4.93 keV). Although the characteristic O Kα peak at 0.52 keV was not detected in the spectrum, oxygen was nevertheless quantified in the elemental composition table, with a normalized mass concentration at 58.99%. This value is consistent with the theoretical stoichiometry of TiO_2, and the apparent absence of the oxygen peak is attributed to the well-recognized limitation of EDX in detecting light elements, which emit low-energy X-rays that are often absorbed before reaching the detector. In contrast, the EDX spectrum of the TiO_2/10%WO_3 nanocomposite thin film revealed, in addition to Ti, distinct peaks associated with tungsten: Mα at 1.75 keV, Lα at 8.4 keV, and a weaker peak of Lβ at 9.65 keV. The corresponding normalized mass concentration for W is 27.24%, confirming the presence of WO_3. In this case, the O Kα peak remained discernible, reflecting contributions from both oxides. Minor signals corresponding to Na, C, and Cl were also detected, most likely arising from trace impurities introduced during the precursor chemistry or film preparation process. Collectively, the EDX results substantiate the successful synthesis of the TiO_2/10%WO_3 composite while preserving the compositional integrity of the thin film.

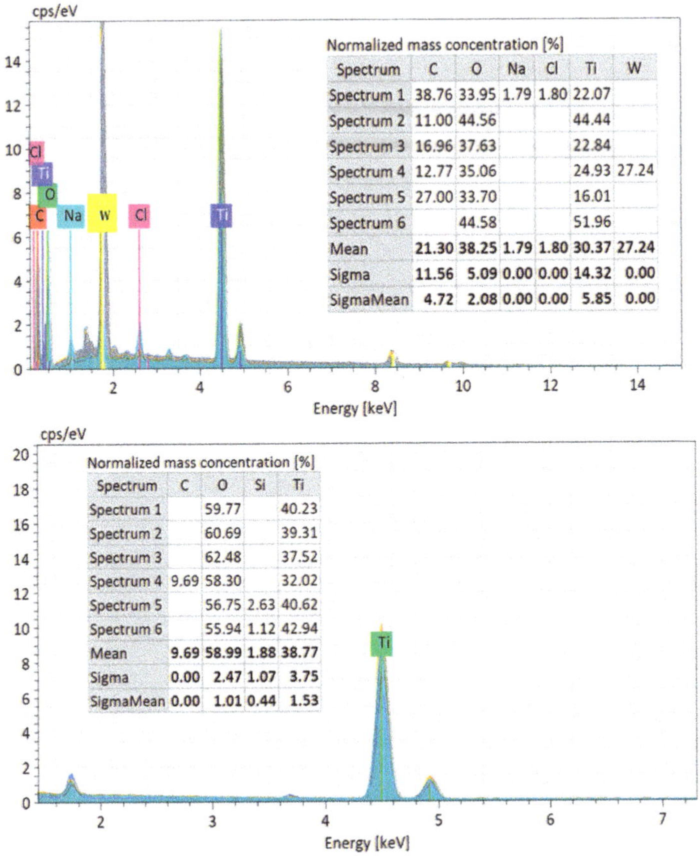

Figure 5. The EDX spectra of hydrothermally synthesized TiO_2/10%WO_3 TiO_2 and pure TiO_2.

3.1.5. Scanning Electron Microscopy (SEM) of TiO_2/WO_3 Composites

The surface morphology of TiO_2/WO_3 thin films with varying WO_3 loadings (0, 5, 10, 20, 30 wt.%) was investigated by SEM imaging to evaluate the homogeneity, particle distribution, and structural integrity of the deposits. As shown in Figure 6, the pure TiO_2 thin film exhibited a relatively uniform coverage but was characterized by the presence of cracks, discernible grain boundaries, and occasional agglomerates. With the incorporation of 5% WO_3, the thin film displayed enhanced particle dispersion and elimination of surface cracks, indicating a positive effect of WO_3 even at a low ratio. The TiO_2/10%WO_3 (S.O) film demonstrated the most compact and homogeneous morphology, consisting of a continuous and smooth layer with uniformly dispersed particles of the smallest dimension, suggesting a strong interaction between the TiO_2 and WO_3 phases and improved interfacial contact [23]. In contrast, the nanocomposite TiO_2/10%WO_3 (C.O) exhibited large agglomerates, poor uniformity, and less homogeneous structures, highlighting the advantage of using synthesized oxides to achieve finer particle distribution and enhanced interfacial interactions. Increasing the WO_3 content to 20% led to some surface irregularities and moderate agglomeration. Whereas at 30% thin film exhibited markedly larger agglomerates and rougher textures. These observations indicate that, although higher WO_3 loading (30%) induces progressively irregular surfaces, larger agglomerates, and rougher textures, the 10 wt.% WO_3 nanocomposite maintains a more uniform morphology with reduced agglomeration, thereby possessing the most favorable structural characteristics and emerging as the most promising candidate for photocatalytic applications.

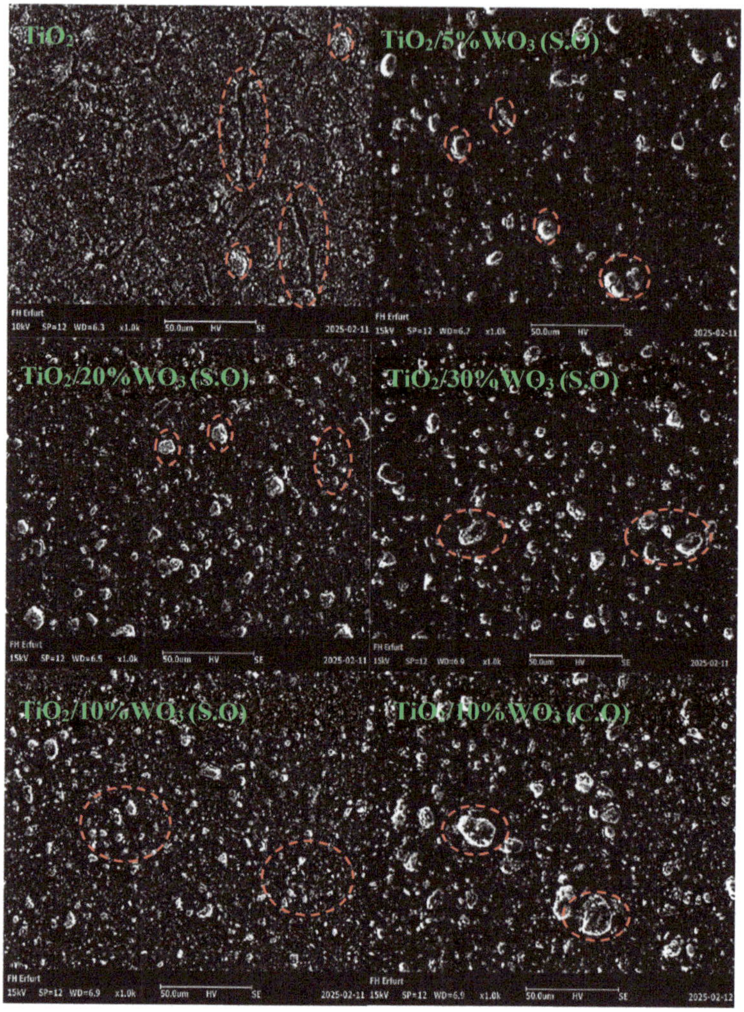

Figure 6. SEM micrographs of TiO_2/WO_3 films, with red circles indicating cracks and agglomerates of different dimensions on the surface.

3.1.6. Transmittance Electronic Microscopy (TEM) of TiO_2/WO_3 Composite

The TEM images of TiO_2/WO_3 nanocomposite are presented in Figure 7a,b. They reveal two distinct morphologies: nanoplatelets and a hexagonal structure. TiO_2 crystallizes predominantly as nanoplatelets, whereas the secondary WO_3 phase adopts a hexagonal form. The average size of the nanoplatelets ranges from 12 to 30 nm, while the hexagonal structures exhibit a characteristic thickness of approximately 7 nm.

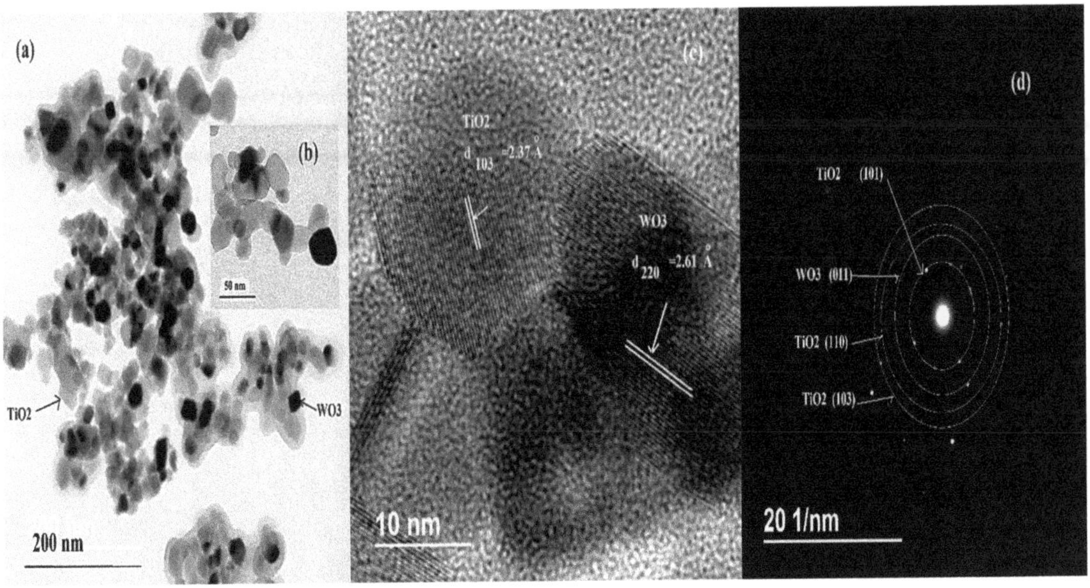

Figure 7. (**a**,**b**) TEM micrograph of TiO_2/WO_3 nanocomposite, (**c**) HRTEM micrograph of TiO_2/WO_3 nanocomposite, and (**d**) SAED of TiO_2/WO_3 nanocomposite.

High-resolution TEM (HRTEM) analysis of the TiO_2/WO_3 nanocomposites (Figure 7c) displays alternating bright and dark fringes. The interplanar spacing of d = 2.38 Å is attributed to the (103) plane of TiO_2, whereas the spacing of d = 2.60 Å corresponds to the (220) plane of WO_3. The selected area electron diffraction (SAED) pattern (Figure 7d) further shows concentric diffraction rings, which can be indexed to the (101), (110), and (103) planes of TiO_2, along with the (011) plane of WO_3, thereby confirming the successful formation of TiO_2/WO_3 nanocomposites.

3.1.7. Optical Band Gap (Eg) Estimation via Kubelka-Munk Analysis

The Kubelka-Munk function, $F(R) = (1 - R)^2/(2R)$, was applied to the reflectance data of TiO_2, WO_3, and TiO_2/WO_3 (S.O) thin films (Figure 8a), and the optical band gap (Eg) was estimated from the Tauc plots $(F(R).\ h\nu)^n$ versus hν. Although TiO_2 and WO_3 are generally reported as indirect band gap semiconductors, the indirect transition model $(F(R).\ h\nu)^{1/2}$ versus hν was adopted in this work. Pure TiO_2 presented a band gap of 3.12 eV, corresponding to an absorption edge at ~400 nm, which confines its photoresponse mainly to the ultraviolet region. While pure WO_3 exhibited a narrower band gap of 2.60 eV, extending absorption well into the visible range and enabling stronger utilization of solar light. The incorporation of WO_3 into TiO_2 progressively narrowed the band gap and red-shifted the absorption edge: 2.99 eV (~415 nm) at 5 wt.% WO_3, 2.91 eV (~426 nm) at 10 wt.% WO_3, 2.83 eV (~438 nm) at 20 wt.% WO_3, 2.75 eV (~451 nm) at 30% wt.% as it was showed in Figure 8b. These shifts, although modest and within the experimental uncertainty (0.01–0.03 eV), clearly demonstrate a systematic trend. These results indicate that low WO_3 contents retain predominantly UV absorption similar to TiO_2, whereas higher WO_3 loading significantly extends absorption further into the visible range. Notably, composites containing 20 and 30 wt.% WO_3 exhibit band gaps of 2.83 eV and 2.75 eV, with absorption edges near 438 nm and 451 nm, respectively, indicating their potential to harvest a larger fraction of sunlight compared to other samples. The observed gradual band gap narrowing is attributed to interfacial electronic interactions between TiO_2 and WO_3. While excessive WO_3 loading may lead to particle agglomeration, as confirmed by SEM analysis, it potentially compromises heterojunction quality. Overall, moderate WO_3 contents (5–10 wt.%) appear optimal for balancing UV and visible absorption, whereas higher loading (20–30 wt.%) may further enhance solar light utilization at the expense of charge separation efficiency [24,25].

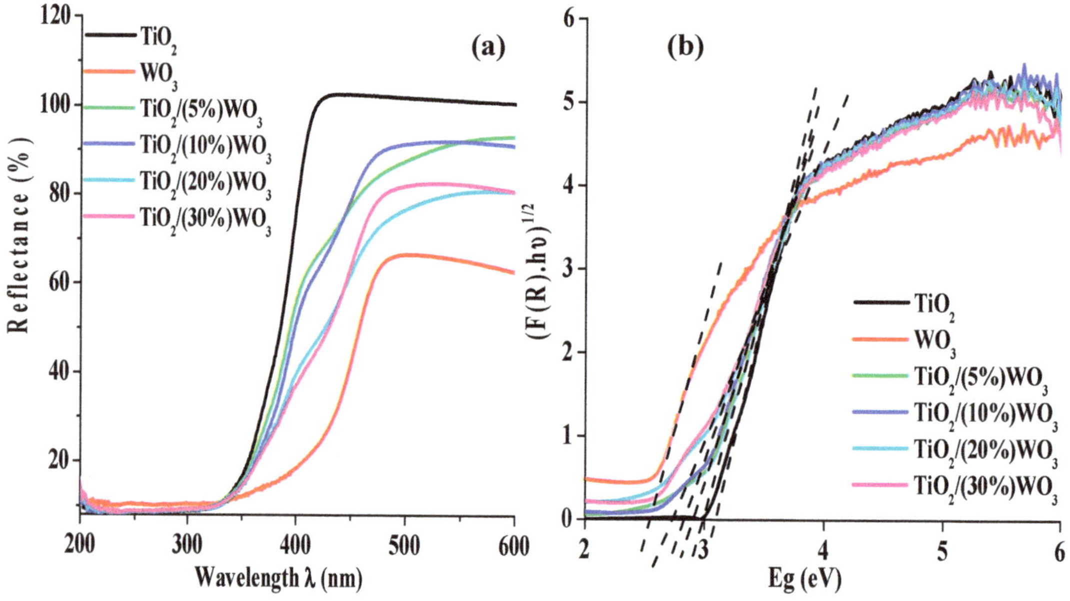

Figure 8. (**a**) UV-Vis diffuse reflectance spectra of TiO_2, WO_3, and TiO_2/WO_3 composites; (**b**) Tauc plots used to estimate the optical band gaps from the Kubelka-Munk-transformed reflectance data, where the dashed lines represent the extrapolation of the linear region to the $(F(R).\ h\nu)^{1/2}$ gives the optical band gap energy.

3.2. *Photocatalytic Activity*

3.2.1. Evaluating Photocatalytic Activity Against Controls

A comparative analysis was carried out to assess photolysis, adsorption, and photocatalysis of the nanocomposite $TiO_2/10\%WO_3$ (S.O) over 30 min. The Venla was selected as a pharmaceutical pollutant in this study, with a concentration of 30 mg/L, and the source light was UV-LEDs. It should be noted that the Venla concentration used in our experiments (30 mg/L) is substantially higher than the levels typically detected in the environment, which generally range from ng/L to low μg/L. This elevated concentration was chosen to ensure accurate and reproducible spectrophotometric measurements, facilitate reliable kinetic analysis, and allow meaningful comparison of photocatalyst performance under controlled laboratory conditions. While the absolute degradation efficiencies reported here may not directly reflect those achievable in natural waters, the observed trends in photocatalytic activity, interfacial charge separation, and material stability are expected to remain relevant at environmentally realistic concentrations. Moreover, the mechanistic insights obtained at higher pollutant loads can inform the design of scaled-up systems, where factors such as longer residence times, enhanced light penetration, or pre-concentration strategies may be employed to achieve effective Venla removal in real-world purification applications. Under identical conditions, as shown in Figure 9a, photolysis (irradiation in the absence of a photocatalyst) and adsorption (dark experiment with a photocatalyst) produced negligible substrate removal, confirming that neither direct irradiation nor physical adsorption plays a significant role. By contrast, the photocatalytic experiment (irradiation in the presence of nanocomposite) led to a pronounced reduction in substrate concentration within 30 min, thereby demonstrating that the degradation originates predominantly from photocatalytic activity rather than from adsorption or photolysis

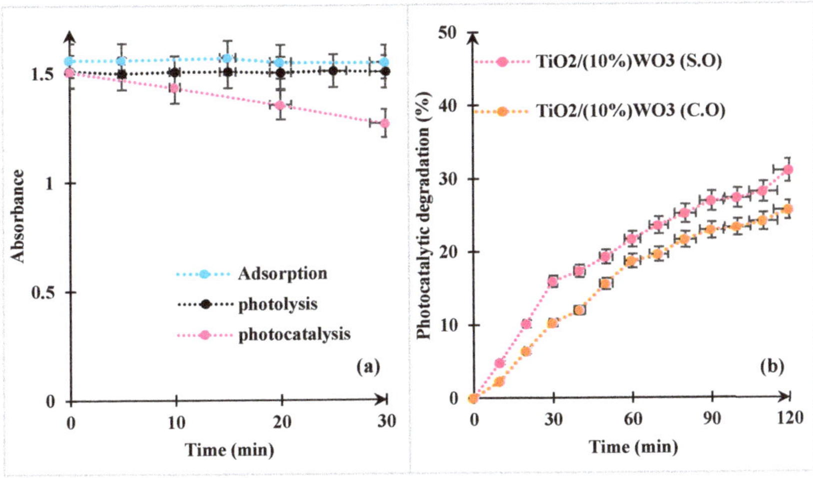

Figure 9. (**a**) Comparison study of adsorption, photolysis, and photocatalysis, and (**b**) The effect of the method of nanocomposite preparation on the photocatalysis activity.

3.2.2. Comparison of the Photocatalytic Performance Between the Synthesized Nanocomposite and That Prepared from Commercial Oxides

An assessment of photocatalytic activity was conducted to evaluate the efficiency of TiO_2/10%WO_3 thin film nanocomposite synthesized Via a controlled method versus a nanocomposite prepared from commercial TiO_2 and WO_3 powders during 120 min. Under identical experimental conditions as the previous test. The thin film TiO_2/10%WO_3 (S.O) achieved a photocatalytic degradation of 31%, as shown in Figure 9b, while the thin film TiO_2/10%WO_3 (C.O) exhibited a photocatalytic efficiency of 25%. The superior performance of the synthesized binary system was attributed to its nanoscale dimensions, markedly smaller than those of commercial oxides, together with its enhanced structural homogeneity, confirmed by SEM and TEM imaging, improved interfacial contact between TiO_2 and WO_3, and more efficient charge separation, which collectively promote higher photocatalytic activity [26,27]. The result highlights the significance of the synthesis method and composite quality in ameliorating the performance of photocatalytic materials.

3.2.3. The Effect of WO_3 Amount on the Photocatalytic Behavior of TiO_2

The photocatalytic degradation of Venla (30 mg/L) under UV-LEDs irradiation was evaluated using pure TiO_2 and TiO_2/WO_3 synthesized nanocomposites with varying WO_3 content (5, 10, 20, 30 wt.%). The results showed in Figure 10a a clear dependence of the photocatalytic efficiency on the WO_3 loading. Pure TiO_2 reached 18% of degradation during 120 min. A slight enhancement to 20% was observed with 5% WO_3, suggesting improved charge separation due to heterojunction formation between TiO_2 and WO_3. The highest photocatalytic activity, 31% was achieved with TiO_2/10%WO_3, indicating that this photocatalyst provides an optimal balance between charge separation and UV-light absorption. However, further increases in WO_3 amount led to a decline in efficiency, 24%

for 20 wt.% WO_3 and 15% for 30% wt.% WO_3. This decrease is likely due to the excessive presence of WO_3, which can act as a recombination center and diminish the effective surface area available for light penetration, thereby limiting the potential for enhanced visible-light absorption, as discussed in the band gap analysis. These findings highlight that while the addition of WO_3 can enhance TiO_2 performance under UV light, optimal loading is crucial to avoid detrimental effects on photocatalytic activity [28].

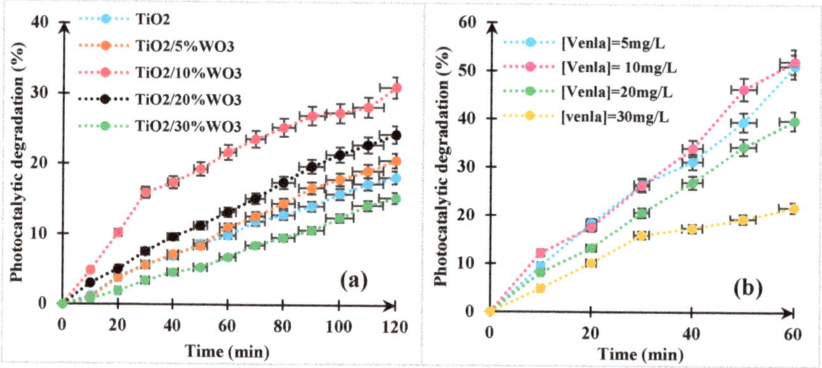

Figure 10. (**a**) The effect of the WO_3 amount on the photocatalytic activity of TiO_2, and (**b**) the effect of the Venlafaxine initial concentration.

3.2.4. Effect of Initial Concentration of the Pollutant

The effect of the initial concentration of Venla on its photocatalytic degradation was investigated using $TiO_2/10\%WO_3$, identified as the most efficient photocatalyst in this study, under UV LEDs irradiation for 60 min. The results in Figure 10b showed a clear inverse relationship between the initial concentration of Venla and the degradation efficiency. At lower concentrations, 5 and 10 mg/L, the photocatalyst achieved a high degradation efficiency of 52%, indicating that a greater proportion of the pollutant molecules could interact with the available active sites on the catalyst surface, and the generation of reactive oxygen species was sufficient for effective degradation. However, as the initial concentration increased to 20 mg/L, the degradation efficiency decreased to 39%, and further dropped to 21% at 30 mg/L. This decline can be attributed to several factors. The higher concentration of the pollutant increases the number of molecules competing for the same active sites, potentially leading to their saturation. Higher concentrations can limit light penetration due to increased solution turbidity, reducing photon absorption by the photocatalyst. Moreover, the fixed amount of generated reactive species may become insufficient to degrade a larger number of pollutant molecules [29,30]. These observations confirm that lower concentrations of pollutants favor more efficient photocatalytic degradation and prove the importance of optimizing operational parameters for real-world wastewater treatment applications.

3.2.5. Kinetic Study

The kinetics of Venla degradation by $TiO_2/10\%WO_3$ under UV-LEDs irradiation were examined at different concentrations (5, 10, 20 mg/L) over one hour. The applicability of pseudo-first-order kinetics was validated by the linear relationship obtained from $\ln(C_0/C)$ versus time plots, with good correlation coefficients. The apparent rate constant (k_{app}) and half-life time ($t_{1/2}$) were calculated for each concentration as presented in Figure 11a. At 5 mg/L, the reaction presented a k_{app} of 0.0114 min^{-1} with a correlation coefficient R^2 of 0.9675, and a $t_{1/2}$ of 57.76 min. For 10 mg/L, k_{app} increased slightly to 0.012 min^{-1}, with a higher R^2 of 0.9818, and $t_{1/2}$ was 60.8 min, indicating a good fit to the pseudo-first-order model and efficient degradation at lower concentrations. However, when the initial concentration was increased to 20 mg/L, k_{app} decreased to 0.0085 min^{-1}, despite a higher R^2 of 0.9955, and the $t_{1/2}$ extended to 81.54 min. This decline in reaction rate at higher concentrations may be attributed to the saturation of active sites on the catalyst surface, reduced availability of hydroxyl radicals per pollutant molecule, and increased light attenuation. These results confirm that the degradation rate is strongly influenced by the initial concentration of Venla, with lower concentrations favoring faster and more efficient photocatalytic reactions.

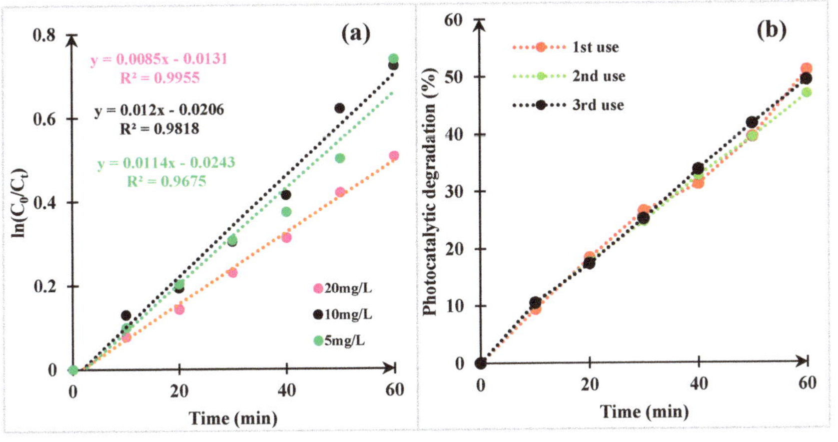

Figure 11. (**a**) Kinetic degradation of Venlafaxine by $TiO_2/10\%WO_3$, (**b**) testing of $TiO_2/10\%WO_3$ stability.

3.2.6. Stability of Photocatalyst

The stability and reusability of the $TiO_2/10\%WO_3$ thin film photocatalyst were assessed over 3 consecutive cycles of Venla degradation (5 mg/L) under UV LEDs irradiation (Figure 11b). The results demonstrated good photocatalytic durability. In the first cycle, the degradation efficiency reached 52%, confirming the high initial activity of the photocatalyst. In the second cycle, a slight decrease was observed, with 46% degradation, likely due to partial surface fouling, minor loss of active sites, or accumulation of intermediate byproducts. However, in the third cycle, the efficiency slightly improved to 49%, suggesting that the photocatalyst retained most of its activity and that any deactivation observed was not progressive or irreversible. This recovery may also indicate the partial desorption or degradation of surface-bound intermediates during the cycle. Overall, the $TiO_2/10\%WO_3$ thin film exhibits good stability and reusability, maintaining consistent performance over multiple uses, which is essential for practical and sustainable wastewater treatment applications.

3.2.7. Photocatalytic Charge Transfer Pathways in $TiO_2/10\%WO_3$ Nanocomposite

The band alignment of the $TiO_2/10\%WO_3$ heterostructure plays a crucial role in its enhanced photocatalytic performance. For a semiconductor, the CB and VB potentials vs. normal hydrogen electrode (NHE) can be estimated using these equations:

$$E_{CB} = \chi - E_e - 0.5E_g \quad (3)$$

$$E_{VB} = E_{CB} + E_g \quad (4)$$

where E_{CB} is the conduction band edge potential (V vs. NHE), E_{VB} is the valence band edge potential (V vs. NHE), χ is the absolute electronegativity of the semiconductor (eV), E_g is the band gap energy (eV), and E_e is the energy of free electrons on the hydrogen scale (4.5 eV).

TiO_2 exhibits a conduction band (CB) potential of −0.25 V and a valence band (VB) potential of +2.87 V (vs. NHE). In comparison, WO_3 shows a CB at +0.79 V and a VB at +3.39 V. Upon illumination, as presented in Figure 12, electrons excited into TiO_2 CB readily transfer to the lower-energy CB of WO_3, leading to electron accumulation in WO_3. In contrast, the corresponding holes remain in the VB of TiO_2. This spatial separation effectively suppresses electron–hole recombination. Owing to its positive potential, the TiO_2 VB (+2.87 V) possesses sufficient power to generate hydroxyl radicals (HO·) from water or hydroxide ions, and can also directly oxidize adsorbed Venla molecules. In contrast, electrons accumulated in the WO_3 CB (+0.79 V) are not thermodynamically capable of reducing O_2 to superoxide radicals ($O_2\cdot^-$), rendering this pathway negligible. Consequently, the degradation of Venla is primarily driven by direct hole oxidation and HO· attack, with the TiO_2/WO_3 heterojunction ensuring more efficient charge separation and sustained photocatalytic activity.

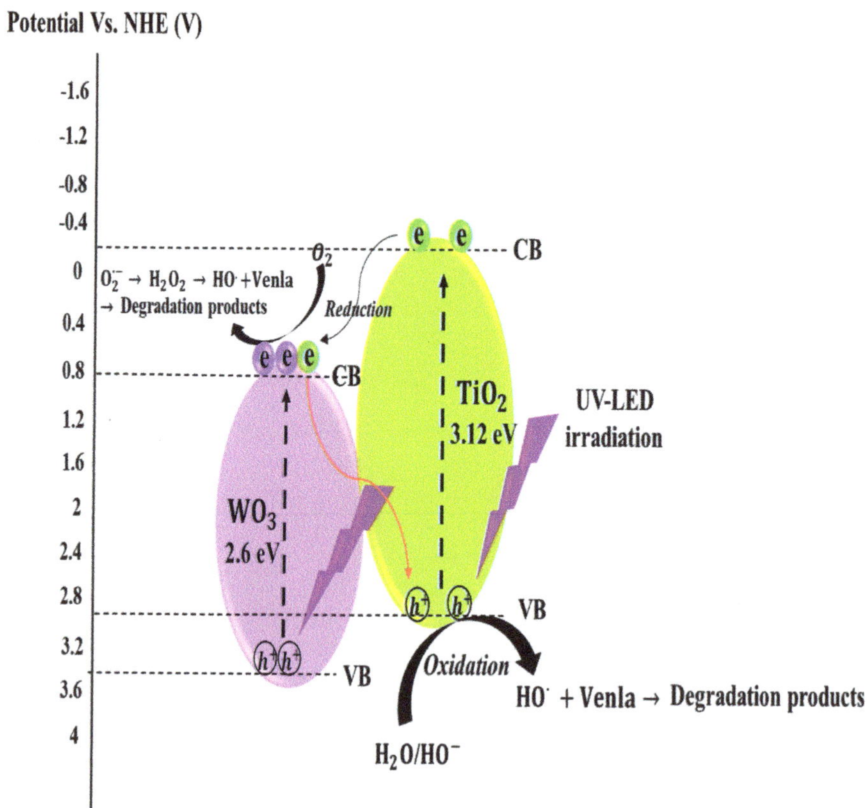

Figure 12. Band alignment and charge transfer in $TiO_2/10\%WO_3$ heterojunction.

4. Conclusions

This study demonstrated that the addition of WO_3 to TiO_2 markedly enhances the photocatalytic degradation of Venlafaxine under UV-LEDs irradiation as a safe, energy-efficient, and compact alternative for water purification. Among the synthesized materials, the hydrothermally prepared $TiO_2/10\%WO_3$ nanocomposite achieved the highest degradation efficiency, outperforming both the pure TiO_2 and the nanocomposite obtained from commercial oxides. The superior activity is attributed to the formation of a heterojunction, which promotes efficient charge separation and suppresses electron–hole recombination. In contrast, higher WO_3 loadings (20-30 wt.%) led to reduced performance, likely due to surface coverage effects and the introduction of recombination sites. Characterization analyses reinforced these findings: Raman and FTIR spectra confirmed the successful formation of TiO_2/WO_3 heterostructure, XRD analysis confirmed the coexistence of TiO_2 phases together with monoclinic WO_3, with no evidence of secondary phases, thereby verifying the structural purity of the synthesized nanocomposites. SEM and TEM micrographs revealed homogeneous thin films with favorable morphology at the nanoscale, as confirmed by TEM, thereby indicating enhanced UV-light absorption. Photocatalytic efficiency decreased with increasing initial pollutant concentration, particularly at 30 mg/L Venla, due to limited active sites and reduced light penetration. Importantly, the $TiO_2/10\%WO_3$ thin film also demonstrated excellent reusability over three consecutive cycles with the same efficiency (~50%/1 h), highlighting practical advantages such as immobilized thin-film configuration, reusability, and avoidance of slurry separation. which underlines its potential as a promising photocatalyst for groundwater purification.

Author Contributions: Investigation, M.G.; Writing original draft, M.G.; Data Curation, M.G.; Conceptualisation, T.S. and S.K.; writing-review and editing, C.S., S.K., S.M. and A.W.; Data Curation, F.T. All authors have read and agreed to the published version of the manuscript.

Funding: This research received no external funding.

Data Availability Statement: The raw data supporting the conclusions of this article will be made available by the authors on request.

Conflicts of Interest: The authors declare no conflict of interest.

References

1. Bergmann, T.A.D.; Witte, F.-A.; Buser, A.; Heberer, S.; Eberlein, I.; Hildebrandt, A.; Küster, A. Pharmaceuticals in the Environment-Global Occurrences and Perspectives. *Environ. Toxicol. Chem.* **2016**, *35*, 823–835. [CrossRef]
2. Corcoran, J.; Jacobs, M.; Winter, M.J.; Tyler, C.R. Pharmaceuticals in the Aquatic Environment: A Critical Review of the Evidence for Health Effects in Fish. *Criti. Rev. Toxicol.* **2010**, *40*, 287–304. [CrossRef] [PubMed]
3. Schlich, K.; Hund-Rinke, K. Influence of Soil Properties on the Effect of Silver Nanomaterials on Microbial Activity in Five Soils. *Environ. Pollut.* **2015**, *196*, 321–330. [CrossRef]
4. Chen, Y.; Wang, J.; Xu, P.; Xiang, J.; Xu, D.; Cheng, P.; Wang, X.; Wu, L.; Zhang, N.; Chen, Z. Antidepressants as Emerging Contaminants: Occurrence in Wastewater Treatment Plants and Surface Waters in Hangzhou, China. *Front. Public Health* **2022**, *10*, 963257. [CrossRef] [PubMed]
5. Ferreiro, C.; Gómez-Motos, I.; Lombraña, J.I.; de Luis, A.; Villota, N.; Ros, O.; Etxebarria, N. Contaminants of Emerging Concern Removal in an Effluent of Wastewater Treatment Plant under Biological and Continuous Mode Ultrafiltration Treatment. *Sustainability* **2020**, *12*, 725. [CrossRef]
6. Hernández-Zamora, M.; Cruz-Castillo, L.M.; Martínez-Jerónimo, L.; Martínez-Jerónimo, F. Diclofenac Produces Diverse Toxic Effects on Aquatic Organisms of Different Trophic Levels, Including Microalgae, Cladocerans, and Fish. *Water* **2025**, *17*, 1489. [CrossRef]
7. Fu, L.A.; McCarthy, R.A.; Hawkins, I.J.; West, C.; Koenig, W.; Smith, H.I.; Martin, A.E.; Adams, D.S. Assessing Pharmaceutical Removal and Reduction in Toxicity Provided by Advanced Wastewater Treatment Systems. *Environ. Sci.: Water Res. Technol.* **2020**, *6*, 62–77. [CrossRef]
8. Elskens, C.J.; Van de Meulebroeck, T.V.; Lübberding, T.; Van den Meersche, P.F.; Kraak, M.H.S. Removal of Psychopharmaceuticals from WWTP Effluent by an Algae–Mussel Trophic Cascade: A Potential Nature-Based Solution? *Environ. Sci. Water Res. Technol.* **2025**, *11*, 1643–1656. [CrossRef]

9. Balasubramanian, S.; Gnanasekar, D.; Anandan, S.; Arulraj, R.; Sankaran, A.K.S. Advanced Photocatalytic Materials-Based Degradation of Micropollutants and Their Use in Hydrogen Production—A Review. *RSC Adv.* **2024**, *14*, 14392–14424. [CrossRef]
10. Antonopoulou, M. Homogeneous and Heterogeneous Photocatalysis for the Treatment of Pharmaceutical Industry Wastewaters: A Review. *Toxics* **2022**, *10*, 539. [CrossRef]
11. Friedmann, D. A General Overview of Heterogeneous Photocatalysis as a Remediation Technology for Wastewaters Containing Pharmaceutical Compounds. *Water* **2022**, *14*, 3588. [CrossRef]
12. Basavarajappa, P.S.; Patil, S.B.; Ganganagappa, N.; Reddy, K.R.; Raghu, A.V.; Reddy, C.V. Recent Progress in Metal-Doped TiO_2, Non-Metal Doped/Co-Doped TiO_2 and TiO_2 Nanostructured Hybrids for Enhanced Photocatalysis. *Int. J. Hydrogen Energy* **2020**, *45*, 7764–7778. [CrossRef]
13. Huang Kong, E.D.; Lai, C.W.; Juan, J.C.; Pang, Y.L.; Khe, C.S.; Badruddin, I.A.; Gapsari, F.; Anam, K. Recent Advances in Titanium Dioxide Bio-Derived Carbon Photocatalysts for Organic Pollutant Degradation in Wastewater. *iScience* **2025**, *28*, 112368. [CrossRef]
14. Tanos, F.; Razzouk, A.; Lesage, G.; Cretin, M.; Bechelany, M. A Comprehensive Review on Modification of Titanium Dioxide-Based Catalysts in Advanced Oxidation Processes for Water Treatment. *Chem. Sus. Chem.* **2024**, *17*, e202301139. [CrossRef]
15. He, X.; Gong, Y.; Niu, L.; Li, C. Development of Defect-Rich WO_{3-x}/TiO_2 Heterojunction Toward Dual-Functional Enhancement: Boosting SERS and Photocatalytic Performance. *Nanomaterials* **2025**, *15*, 521. [CrossRef]
16. Levinas, R.; Podlaha, E.; Tsyntsaru, N.; Cesiulis, H. Composites Based on Electrodeposited WO_3 and TiO_2 Nanoparticles for Photoelectrochemical Water Splitting. *Materials* **2024**, *17*, 4914. [CrossRef]
17. Palma, F.; Baldelli, G.; Schiavano, G.F.; Amagliani, G.; Aliano, M.P.; Brandi, G. Use of Eco-Friendly UV-C LEDs for Indoor Environment Sanitization: A Narrative Review. *Atmosphere* **2022**, *13*, 1411. [CrossRef]
18. Mach, V.; Dvorak, L.; Dvorakova, H.; Mikulikova, L.; Kolarova, K.; Černy, L. Role of Lamp Type in Conventional Batch and Micro-Photoreactor for Photocatalytic Hydrogen Production. *Front. Chem. Sec. Photocatal. Photochem.* **2023**, *11*, 1271410. [CrossRef]
19. Shephered, M.T.; Newman, J.P.; Kubheka, O. Effect of TiO_2 Phase on the Photocatalytic Degradation of Methylene Blue Dye. *Phys. Chem. Earth Parts A/B/C* **2020**, *118–119*, 102900. [CrossRef]
20. Dhanalekshmi, K.I.; Umapathy, M.J.; Magesan, P.; Zhang, X. Biomaterial (Garlic and Chitosan)-Doped WO_3-TiO_2 Hybrid Nanocomposites: Their Solar Light Photocatalytic and Antibacterial Activities. *ACS Omega* **2020**, *5*, 31673–31683. [CrossRef]
21. Odhiambo, V.O.; Ongarbayeva, A.; Kéri, O.; Simon, L.; Szilágyi, I.M. Synthesis of TiO_2/WO_3 Composite Nanofibers by a Water-Based Electrospinning Process and Their Application in Photocatalysis. *Nanomaterials* **2020**, *10*, 882. [CrossRef]
22. Dávidné, N.; Tamás, F.; Eszter, D.; Ágnes, S.; Krisztina, L.; Imre, M.S. Photocatalytic WO_3/TiO_2 Nanowires: WO_3 Polymorphs Influencing the Atomic Layer Deposition of TiO_2. *RSC Adv.* **2016**, *6*, 95369–95377. [CrossRef]
23. Suvarna, R.; Bathe, P.S.P. Electrochemical Behavior of TiO_2 Nanoparticle Doped WO_3 Thin Films. *Materials* **2014**, *5*, 642069. [CrossRef]
24. Yang, Y.; Ma, G.; Hu, X.; Wang, W.; Du, Z.; Wang, Y.; Gong, X.Z.; Tan, H.; Guo, F.; Tang, J. Hollow Flower-Like WO_3@TiO_2 Heterojunction Microspheres for the Photocatalytic Degradation of Rhodamine B and Tetracycline. *RSC Adv.* **2025**, *15*, 12629–12644. [CrossRef] [PubMed]
25. Boga, B.; Székely, I.; Pap, Z.; Baia, L.; Baia, M. Detailed Spectroscopic and Structural Analysis of TiO_2/WO_3 Composite Semiconductors. *J. Spectrosc.* **2018**, *1–7*, 6260458. [CrossRef]
26. Tsay, C.-Y.; Hsu, T.-Y.; Lee, G.-J.; Chen, C.-Y.; Chang, Y.-C.; Chen, J.-H.; Wu, J.J. Hydrothermal Synthesis of Nanocomposites Combining Tungsten Trioxide and Zinc Oxide Nanosheet Arrays for Improved Photocatalytic Degradation of Organic Dye. *Nanomaterials* **2025**, *15*, 772. [CrossRef] [PubMed]
27. Li, W.; Chen, C.; Yang, R.; Cheng, S.; Sang, X.; Zhang, M.; Zhang, J.; Wang, Z.; Li, Z. Efficient and Stable Degradation of Triazophos Pesticide by TiO_2/WO_3 Nanocomposites with S-Scheme Heterojunctions and Oxygen Defects. *Catalysts* **2023**, *13*, 1136. [CrossRef]
28. Lee, W.H.; Lai, C.W.; Hamid, S.B.A. One-Step Formation of WO_3-Loaded TiO_2 Nanotubes Composite Film for High Photocatalytic Performance. *Materials* **2015**, *8*, 2139–2153. [CrossRef]
29. Paiu, M.; Lutic, D.; Favier, L.; Gavrilescu, M. Heterogeneous Photocatalysis for Advanced Water Treatment: Materials, Mechanisms, Reactor Configurations, and Emerging Applications. *Appl. Sci.* **2025**, *15*, 5681. [CrossRef]
30. Sharma, M.; Mishra, M.K.; Patel, S.; Kumar, R.; Dubey, K. Visible-Light-Driven Photocatalytic Degradation of Tetracycline Using Heterostructured Cu_2O–TiO_2 Nanotubes, Kinetics, and Toxicity Evaluation of Degraded Products on Cell Lines. *ACS Omega* **2022**, *7*, 33572–33586. [CrossRef]

environments

MDPI

Article

Spatial and Multivariate Analysis of Groundwater Hydrochemistry in the Solana Aquifer, SE Spain

Víctor Sala-Sala [1,*], José Miguel Andreu [2], Ana Pérez-Gimeno [1], Manuel M. Jordán [1], Jose Navarro-Pedreño [1] and María Belén Almendro-Candel [1,*]

1 Department of Agrochemistry and Environment, University Miguel Hernández of Elche, 03202 Elche, Spain; aperez@umh.es (A.P.-G.); manuel.jordan@umh.es (M.M.J.); jonavar@umh.es (J.N.-P.)
2 Department of Environment and Earth, University of Alicante, 03690 San Vicente del Raspeig, Spain; andreu.rodes@ua.es
* Correspondence: v.salas@umh.es (V.S.-S.); mb.almendro@umh.es (M.B.A.-C.)

Abstract

The Solana aquifer is located in the South-East of the Iberian Peninsula and forms part of the Villena-Benejama groundwater body. It is a limestone and dolomite aquifer that has historically been considered overexploited due to intensive agriculture and urban use. Despite this, the quality of the water has remained stable over time. This study analyses the spatial and temporal variability within the aquifer and identifies the controlling processes. Chemical analyses were conducted on samples taken from 26 wells in July 2024 and February 2025. The results reveal a predominant calcium carbonate facies with minimal seasonal variation. However, sulphate-chloride water was found in the South-Western sector, which is associated with the dissolution of evaporitic materials from the Triassic Keuper. Principal Component Analysis (PCA) and Hierarchical Cluster Analysis (HCA) identified two processes: a salinity gradient linked to lithology, and a second process related to bicarbonates and nitrates, indicating potential nitrate inputs in the eastern half of the aquifer. HCA differentiates four clusters: one highly mineralised group located in the south-western sector near Triassic outcrops, two intermediate groups with slight differences in composition and distribution, and a fourth group with the lowest mineralisation located on the Southern flank of the Solana range.

Keywords: geochemistry; Solana aquifer; spatial variability; Triassic evaporites; water quality

1. Introduction

The quality of groundwater is not just an environmental concern, it is also directly tied to farming, drinking water and the survival of ecosystems that depend on it. Almost all the planet's liquid freshwater, around 99%, is stored in soils and underground. The use of groundwater has increased due to technological and scientific improvements, which makes it possible to identify areas with greater resources and ways to access and extract them. Around half of the world's population relies on groundwater on a daily basis. In agriculture, between a third and a quarter of irrigation comes straight from this source. In fact, this accounts for around 70% of all groundwater extracted globally. This fact alone demonstrates its importance for global water security. UNESCO, in its World Water Development Report (2024), notes that poor handling of groundwater could have serious knock-on effects [1]. Efforts to reduce poverty could falter, food production could be at risk, and many communities would find it harder to cope with the effects of climate change [1–3].

One of the main factors behind this increase in groundwater use is its good quality, especially in carbonate aquifers, which allows it to be suitable for multiple uses. However, being located underground, where it is often perceived as stable and protected, groundwater is still exposed to processes that can reduce its quantity and quality. The most significant threats include overexploitation, marine intrusion in coastal areas and diffuse or point-source pollution resulting from agricultural, urban or industrial activities [4–6]. Excessive pumping may also facilitate the infiltration of pollutants accumulated in soils, while in certain contexts, natural contamination can also occur through the dissolution of highly soluble geological formations. These factors often act together, progressively deteriorating the quality of groundwater to the point where it can no longer be used [7,8], and affecting the health of the ecosystems that depend on it in some cases. Furthermore, extreme weather events such as prolonged droughts and heavy rainfall are altering the recharge patterns of many aquifers, thereby reducing the available groundwater resources.

In semi-arid regions such as South-Eastern Spain, reliance on groundwater is particularly pronounced due to the limited availability of surface water resources. This reliance has often resulted in intensive exploitation and subsequent problems of over-abstraction and chemical deterioration. Several studies have reported a steady intensification of extraction, which has been associated with marked declines in piezometric levels. This has led to the abandonment of wells, either because the water table has dropped to impractical depths or because the quality of the water has deteriorated [9–14]. Cases in nearby aquifers, including those in Crevillente, Cid and Quibas, illustrate how prolonged overexploitation can trigger higher salinity levels by dissolving evaporitic materials. This process has reduced the availability of water suitable for irrigation and domestic use to some extent [15,16].

In this context, this study focuses on the Solana aquifer, also known as the Villena–Benejama aquifer, situated in the North of the province of Alicante, Spain. This Cretaceous aquifer [17] plays a key socio-economic role in supplying water to urban areas and agriculture. This aquifer has historically been considered overexploited due to major agricultural transformations in the area during the 20th century, which saw a shift towards high-yield crops dependent on groundwater usage [18]. In addition to this, a significant proportion of the extracted water is used for urban supply within and outside the basin, including transfers to the coastal area of the province of Alicante. Over the years, this exploitation has led to problems such as spring depletion, ecosystem deterioration and well abandonment. In the past decades, groundwater levels have fallen by almost 180 m, from the elevation of the former springs (=505 m a.s.l.) down to the historical minimum reached in 2018 (=326 m a.s.l.). Since then, a partial recovery of more than 30 m has been observed, with current levels close to 360 m a.s.l. (Figure S1).

Although the decline in the piezometric level has been evident for decades, it was not until 2020 that the Jucar Hydrographic Confederation officially declared that the groundwater body did not meet the criteria for good quantitative status [19]. Following the official declaration, plans have been established for the gradual reduction in extraction. According to the agreed guidelines, the exploitation of wells intended for agricultural use must be reduced each year so that groundwater is reserved exclusively for human consumption. Meanwhile, water for irrigation will come from external sources (Jucar-Vinalopó water transfer). This change is mainly focused on improving the quantitative status of the water body and it is also relevant from a hydrochemical point of view, given that water intended for human consumption is subject to stricter quality criteria and requires more rigorous control.

Nevertheless, recent hydrochemical studies of the Solana aquifer are scarce and were conducted decades ago [20], failing to address the spatial and temporal variability of its chemical composition. Although routine analyses are carried out by irrigation communities and mixed companies responsible for supplying water to humans, these results remain fragmented and are not publicly available, which prevents a comprehensive understanding of the hydrochemical conditions of the aquifer.

For these reasons, the present study aims to provide a general hydrochemical assessment of the Solana aquifer, offering an overview of its spatial and temporal quality. To achieve this, the groundwater of the Solana aquifer was studied during two periods with contrasting climatic characteristics. The main parameters were analysed using descriptive and multivariate statistical techniques, and the gradients were represented through interpolation. Interpreting the results in relation to geology, land use and possible anthropogenic effects enables us to recognise the main processes controlling water composition. Consequently, this work not only contributes to increasing hydrogeological knowledge of this complex system located in South-Eastern Spain, but also provides a chemical basis for the management of a strategic resource in other Mediterranean regions.

2. Materials and Methods

2.1. Study Area Overview

The Solana aquifer is located in the South-East of the Iberian Peninsula, most of it in the province of Alicante, though it also extends into the provinces of Valencia and Albacete. Forming part of the Villena-Benejama groundwater body, it covers an area of approximately 280 km^2. Altitudes range from 500 to 1050 m a.s.l., with a slope running from the higher-altitude Eastern town of Banyeres de Mariola to the lower-altitude Western town of Villena.

2.1.1. Climate and Recharge Processes

The aquifer is located in a typically Mediterranean area with an average annual rainfall of around 500 mm, mostly concentrated in autumn and spring. The average temperature is around 15 °C, with hot summers and often cold winters due to continental influences [21].

The Solana aquifer's main recharge occurs through the infiltration of rainwater into permeable materials (approximately 18 hm^3/year). According to the official data from the Jucar Hydrographic Confederation, average annual extraction is 26.3 hm^3/year and available resources are estimated at 15 hm^3/year [19]. There are no known hidden inflows or outflows, and the springs associated with this aquifer dried up at the beginning of the 20th century.

An important hydrographic feature of the region is the presence of the Vinalopó river, whose course runs on the study area. This river plays a significant role in the recharge of the aquifer [22]. However, the river flow within the aquifer boundaries is limited as most of its water is abstracted and diverted for crop irrigation in the upper basin. The remaining flow continues along the river basin and eventually infiltrates the Quaternary soils and the Cretaceous aquifer. The river only carries flow after significant rainfall, which is quite frequent in autumn.

2.1.2. Geology

From a geological point of view, the aquifer is characterised by a mountainous landscape with significant reliefs (such as the Solana and La Villa ranges), which are aligned with the NE-SW orientation of the Betic structural range (NE-SW). These mountains are primarily composed of limestones, dolomites and marly dolomites, with a total thickness exceeding 400 m, while the valleys below, where the area's agriculture thrives, are filled with detrital material and Miocene marls [17] (Figure 1). The basal impermeable layer is made up of the clayey marls and sands of the Utrillas Formation. The Northern and Southern boundaries are defined by thrusts with NW vergence, which bring the aquifer into contact with impermeable materials of the Lower Cretaceous and Miocene (mainly marls). The North-Eastern boundary is considered open, but it is marked by a piezometric threshold that separates the Solana aquifer from the Volcadores aquifer [17]. Finally, the South-Western boundary is open to flow, and there may be flows from the Quaternary aquifer, made up of gravels, sands and silts. However, the precise relationship between these two aquifers is currently unknown, due to the reduced number of water points in the Quaternary aquifer, resulting from the abandonment of these wells because of sand entrainment, low yields, and increased salinity [23].

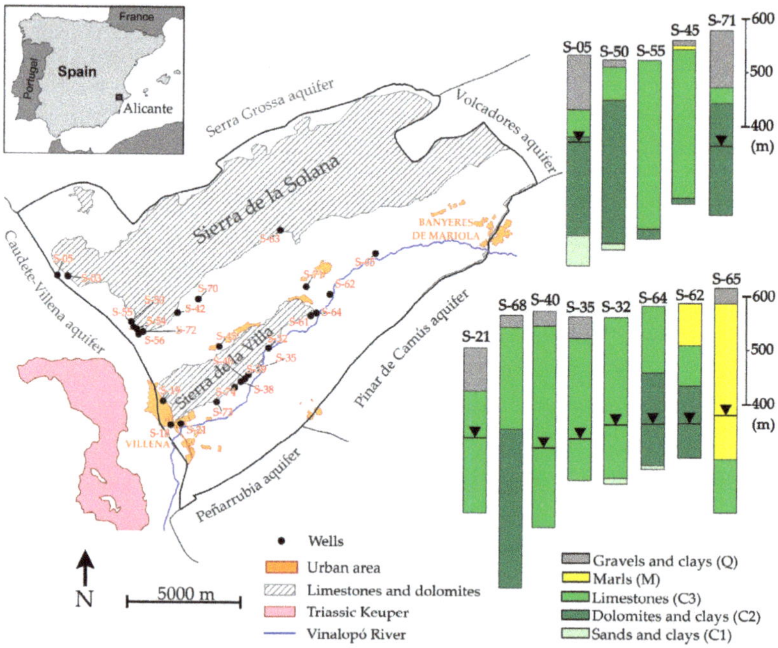

Figure 1. Geological and geographical situation of the Solana aquifer. Lithologic logs of selected wells (black lines indicate the piezometric level in 2022). C1: Sands and clays of the Lower Cretaceous (Utrillas Formation), C2: Dolomites and marly dolomites of the Upper Cretaceous, C3: Limestones of the Upper Cretaceous, M: Marls of the Miocene (*Tap* facies), Q: Undifferentiated Quaternary.

2.2. Sample Collection and Analysis

The hydrochemical data used in this document were collected in July 2024 (summer) and February 2025 (winter) in the Solana aquifer exploitation wells. Wells chosen where that which captured water from the permeable Cretaceous unit and, the selection process was implemented based on two key criteria: the well and the availability of data regarding its structural and lithological column. Additionally, wells were chosen to provide the best possible spatial distribution, allowing an accurate determination of the hydrochemical characteristics of each aquifer sector. Access to the wells was obtained with consent from the well owners, primarily town councils, the Alto Vinalopó General Users' Community, irrigation communities, and drinking water supply companies.

Sampling was carried out early in the morning because of the electricity tariffs contracted by the owners of the wells. This methodology ensured that water representative of the aquifer was obtained and avoided collecting stagnant water in the well column after prolonged periods of time without pumping [24,25].

Samples were collected in polyethylene containers and refrigerated at 6 ± 2 °C (temperatures were recorded using a mini T datalogger to ensure the traceability of the readings). They were transported to the analytical laboratories within a maximum period of 48 h. Analysis began within 24 h of their reception at both the Agrochemistry and Environment laboratory at the University Miguel Hernández of Elche (UMH) and the laboratory of the Hydrogeology Centre at the University of Málaga (CEHIUMA). To meet the preservation requirements outlined in section 1060 C of the Standard Methods [24], a portion of each sample was acidified with nitric acid. This preserved subsample was subsequently sent to a laboratory in Málaga. The pH of the samples was determined according to electrometric method 4500-H^+ B of the Standard Methods [24] with a pH-meter CRISON GLP 21. The analysis of bicarbonates, carbonates and alkalinity was determined immediately according to potentiometric method 2320 [24] in the laboratories of the UMH. In the laboratory of CEHIUMA, a complete analysis was carried out using ion chromategraphy (Metrohm 881 Compact IC Pro for anions and Metrohm 930 Compact IC for cations, both with detection limits <1 μg/L and analytical precision of ±2%). Temperature and electrical conductivity (EC) were measured in situ using a sensION+ EC5 (precision: ±0.2 °C for temperature and ≤0.5% for EC). All laboratory analyses were performed following APHA standard procedures and the analytical method for ion chromatography was based on the EN ISO 10304-1 [24,26]. All samples had an absolute ionic balance error ≤5%, within the acceptable range for hydrochemical studies [27].

2.3. Data Analysis

The hydrochemical data were subjected to statistical analysis in order to characterise the water of the Solana aquifer and to understand the variability. Descriptive statistics were applied (mean, minimum, maximum and standard deviation) to provide an overview of the main parameters and their dispersion.

The results were plotted on a Piper diagram [28], using the software *Diagrammes* [29], which is used to classify the samples according to their hydrochemical facies. The diagram allows the simultaneous interpretation of the relative concentrations of the main cations (Na^+, K^+, Mg^{2+}, Ca^{2+}) and anions (Cl^-, $NO_3{}^-$, $SO_4{}^{2-}$, $HCO_3{}^-$), and the detection of possible trends. The results were also plotted on a Wilcox diagram [30,31], which makes it possible to assess the suitability of water for irrigation on the basis of the percentage of sodium and salinity. It has been demonstrated that high sodium concentrations can lead to losses in permeability and soil structure, and high electrical conductivity can hinder water uptake by plants. In addition, the Irrigation Water Quality Index (IWQI) was calculated, which integrates nine parameters traditionally used: electrical conductivity (EC), sodium adsorption ratio (SAR), residual sodium carbonate (RSC), Kelly's ratio (KR), soluble sodium percentage (SSP), magnesium adsorption ratio (MAR), permeability index (PI), total hardness (TH), and fluoride (F^-). Each parameter was classified into five quality categories (from excellent to very poor), scored from 5 to 1, and added together to obtain the IWQI (range from 9 to 45). This provides a simple complementary assessment to Wilcox's [32]. Likewise, the Water Quality Index (WQI) for human consumption was calculated following World Health Organization (WHO) guidelines.

For optimal results of statistical methods, univariate and multivariate normality were first assessed using the Shapiro–Wilk and Royston tests. Non-normal variables were transformed using the Yeo–Johnson method, and all variables were standardised to z-scores. With the data already transformed, a two-tailed Pearson correlation matrix was calculated (significance threshold $p < 0.05$) to observe significant associations ($|r| \geq 0.8$), allowing the study of linear correlations between the parameters.

To visualise spatial variability, the spatial distribution of the analysed parameters and PCA scores was performed using the Inverse Distance Weighted (IDW) interpolation method within the boundaries of the Solana aquifer for both dates. For the interpolation, a grid resolution of 500 × 500 m was used, with a power parameter of $p = 2$ and k = 5 nearest neighbours. The reliability of the interpolation was assessed using Leave-One-Out Cross-Validation (LOOCV). The resulting relative RMSE values show that the interpolation is adequate and falls within the 'good' category [33].

The suitability of the transformed dataset for multivariate analysis was confirmed through the Kaiser–Meyer–Olkin (KMO) index (>0.7) [34] and Bartlett's sphericity test ($p < 0.001$) [35]. Analyses were performed in Python v3.13 using scikit-learn for PCA and HCA, and SciPy/pingouin for statistical tests. The PCA serves to understand the patterns of variability between the points analysed and thus reduce the complexity of the system [36,37]. This statistical technique transforms the analysed parameters (Na^+, K^+, Mg^{2+}, Ca^{2+}, Cl^-, NO_3^-, SO_4^{2-}, HCO_3^-, EC, pH and temperature) into a coordinate system, formed by principal components (PCs). Therefore, each principal component is the linear combination of the original variables. The PCs are ordered according to the variance they explain [37], so that the first principal component (PC1) concentrates most of the variability among the sampled points, while the following components (PC2 and later) contribute smaller variances.

To complete the analysis, a Hierarchical Cluster Analysis (HCA) was performed with the same parameters used in the PCA, to identify different groups of waters. The analysis was carried out using the Ward method [38], which minimises the total variance within clusters, and the Euclidean distance as a measure of similarity between wells [39,40]. The number of clusters was determined using the elbow method [41], which identifies the maximum number at which adding new groups would provide only a marginal improvement in the explained variance.

3. Results

3.1. Hydrochemical Characteristics of Groundwater

In July 2024, groundwater extracted from 26 wells was analysed (Table 1). The average electrical conductivity (EC) was 659.5 ± 190.8 μS/cm (≈429 mg/L Total Dissolved Solids, TDS), classifying the water as good according to WHO [42], with pH values ranging between 7.7 and 8.1. The main anion was bicarbonate (HCO_3^-), with an average concentration of 270.4 mg/L, a maximum of 324.7 mg/L and a minimum of 239.4 mg/L. Chlorides (Cl^-) had a mean value of 43.7 mg/L, although with high variability, with maximum values reaching 223.7 mg/L and minimum values of no more than 9.0 mg/L. Sulphate (SO_4^{2-}) concentrations ranged between 7.7 and 79.7 mg/L, with a mean value of 26.0 mg/L in July. Regarding the main cations, calcium (Ca^{2+}) and magnesium (Mg^{2+}), stood out with average values of 66.5 and 27.5 mg/L, respectively. The mean value for sodium (Na^+) was 21.7 mg/L, with maximum and minimum values of 100.6 and 5.6 mg/L. Regarding nitrogen species, only nitrate (NO_3^-) has been detected, but without exceeding 50.0 mg/L at any point.

With respect to 2025 winter campaign, 22 wells were analysed on this occasion (Table 1) due to the lack of demand for irrigation during this period of the year. The average EC was 648.2 ± 187.6 μS/cm (≈421 mg/L TDS), classifying the water as good according to WHO [42], and pH values ranged between 7.6 and 8.0. In general, the analyses show again that the majority anion was HCO_3^- with an average value of 226.2 mg/L, slightly lower than that reported in summer. Chlorides continued to show high variability, with a maximum value of 248.2 mg/L, a minimum of 10.0 mg/L and an average of 48.8 mg/L. Sulphate concentrations were slightly higher, with a mean value of 29.1 mg/L. The dominant cation was Ca^{2+} (mean 70.8 mg/L), followed by Mg^{2+} (mean 28.9 mg/L) and Na^{+} (mean 22.1 mg/L). The mean NO_3^- value was 19.8 mg/L, with a minimum value of 3.5 mg/L and a maximum of 53.4 mg/L, exceeding the WHO guideline for drinking water (50 mg/L) [42].

Table 1. Raw analytical results—July 2024 and February 2025. All values are in mg/L except EC (μS/cm at 25 °C), pH, and T (°C).

	July 2024				February 2025			
	Mean	**Min**	**Max**	**SD**	**Mean**	**Min**	**Max**	**SD**
Na^+	21.7	5.6	100.6	20.1	22.1	5.6	109.6	20.6
K^+	1.1	0.6	2.7	0.5	1.1	0.6	2.7	0.5
Mg^{2+}	27.5	15.5	47.0	7.5	28.9	15.4	47.7	7.7
Ca^{2+}	66.5	44.6	95.0	9.9	70.8	47.4	96.6	9.5
Cl^-	43.7	9.0	223.7	46.6	48.8	10.0	248.2	48.7
NO_3^-	17.0	0.0	48.9	12.4	19.8	3.5	53.4	14.4
SO_4^{2-}	26.0	7.7	79.7	17.4	29.1	8.2	88.5	18.0
HCO_3^-	270.4	239.4	324.7	20.3	226.2	198.9	274.7	17.9
EC	659.5	467.0	1363.0	190.8	648.2	458.0	1392.0	187.6
pH	7.8	7.7	8.1	0.1	7.8	7.6	8.0	0.1
T	19.0	16.0	21.5	1.7	16.7	13.2	20.3	2.1

3.2. Hydrochemical Classification and Water Quality Assessment

Hydrochemical classification and water quality assessment was performed mainly considering human consumption and irrigation. The Piper diagram is a hydrochemical tool used to determine the chemical composition of waters [27,28,43,44]. It consists of two triangles showing the relative concentrations of the major cations and anions (calcium, magnesium, sodium plus potassium, sulphates, chlorides, carbonate plus hydrogen carbonate). The result displayed in the central diamond is the projection of the two preceding triangles.

The hydrochemical facies of the Solana aquifer are predominantly calcium bicarbonate, typical of carbonate aquifers. Approximately 95% of the samples fall into this facies group and no major seasonal differences are observed, and the remaining percentage is classified as mixed facies (Figure 2). This pattern suggests high hydrochemical stability and indicates that most samples are derived from the dissolution of carbonate rocks. Only a few wells, specifically those located at the south-western boundary of the aquifer, exhibit a shift towards mixed or calcium-sulphate facies, which could indicate an interaction with Triassic Keuper outcrops [16,45].

The suitability of groundwater for agriculture was assessed using two complementary methodologies that produced consistent results. The Wilcox diagram [30,31], where samples from both the July and February campaigns are plotted, classifies irrigation water quality according to salinity (EC) and sodium hazard. Salinity classes are defined as C1 (<250 μS/cm), C2 (250–750 μS/cm), C3 (750–2250 μS/cm), and C4 (>2250 μS/cm), while sodium hazard classes are defined as S1 (SAR < 10), S2 (10–18), S3 (18–26), and S4 (>26). Most samples are plotted within the C2-S1 and C3-S1 fields, corresponding to good to moderate water quality, whereas wells identified as S-18 and S-21 fall in C3-S2, indicating a slightly higher sodium influence, possibly related to the dissolution of Triassic Keuper evaporites.

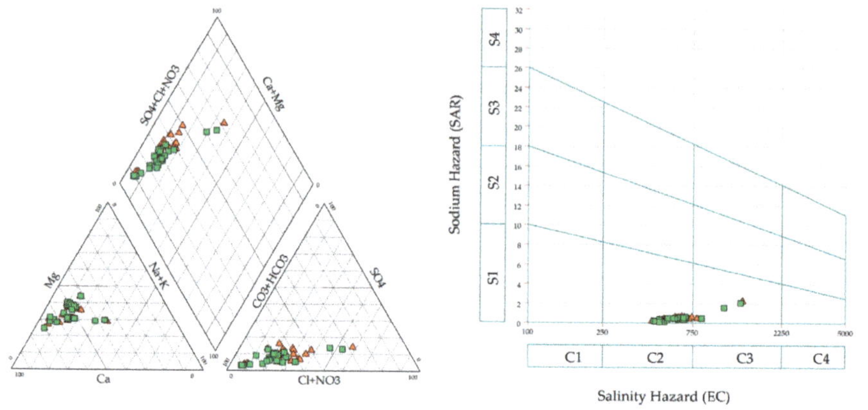

Figure 2. Piper Diagram (**left**) and Wilcox diagram (**right**) of Groundwater Samples—July 2024 (green) and February 2025 (orange).

The irrigation water quality index (IWQI) [32] was applied as a complementary assessment tool. This index integrates nine of the classical parameters used to assess irrigation water quality, including electrical conductivity (EC) and the sodium adsorption ratio (SAR). Each parameter is rated on a scale from very poor (1) to excellent (5), giving a total score between 9 and 45. According to this methodology, IWQI values are grouped into three categories: very poor (0–15), poor (16–30), and good (31–45). Based on this classification, all samples fall within the "good" category in both campaigns, with values ranging from 35 to 42.

To assess the suitability of groundwater for human consumption, the Water Quality Index (WQI) has been calculated according to WHO guidelines [42]. Most samples are classified as "excellent" in both campaigns (96% in July 2024 and 95% in February 2025), with only one well classified as "good" (S-21). The best quality is found in well S-42, located next to the Cretaceous outcrops of the Solana range, and the worst in well S-21, located near the evaporite outcrops. However, although the S-71 well is classified as "excellent" in both campaigns, the nitrate concentrations measured in this well exceed the established maximum (50 mg/L) in February 2025 (53.4 mg/L), making it unsuitable for this use without prior treatment.

3.3. Spatial Distribution of the Hydrochemical Parameters

As a preliminary step to the spatial analysis of the parameters, a correlation matrix was generated for the results of the two field campaigns (Figure 3). The aim was to identify significant statistical relationships and to detect possible hydrogeochemical processes. In both campaigns, strong positive correlations were observed between sodium (Na^+), potassium (K^+) chlorides (Cl^-), sulphates ($SO_4{}^{2-}$) and electrical conductivity (EC), suggesting similar spatial behaviour [46], which is consistent with the dissolution of Triassic evaporites, mainly gypsum ($CaSO_4.2H_2O$) and halite (NaCl). In contrast, the correlations of nitrates ($NO_3{}^-$) and bicarbonates ($HCO_3{}^-$) with other parameters were weak, and pH showed moderate negative correlations ($r = -0.30$ to -0.50) with most of the ions. The Jennrich test indicated statistically significant differences between the July and the February correlation matrices. However, these variations are minor (mean $\Delta r = -0.03$; mean $|\Delta r| = 0.07$; range -0.32 to $+0.18$), and no relevant hydrogeological differences are observed.

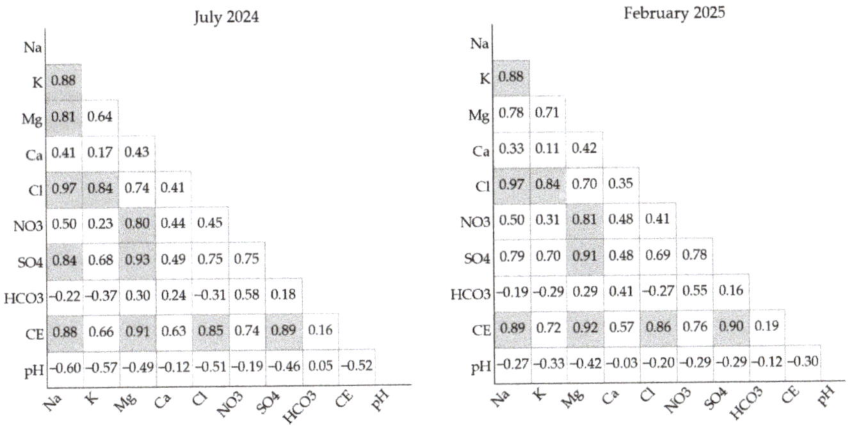

Figure 3. Pearson Correlation Matrices of Hydrochemical Parameters—July 2024 and February 2025.

Figures 4 and 5 were produced using the IDW interpolation method, with the aim of visualising the spatial distribution of the main hydrogeological parameters measured in groundwater during the field campaigns of July 2024 and February 2025.

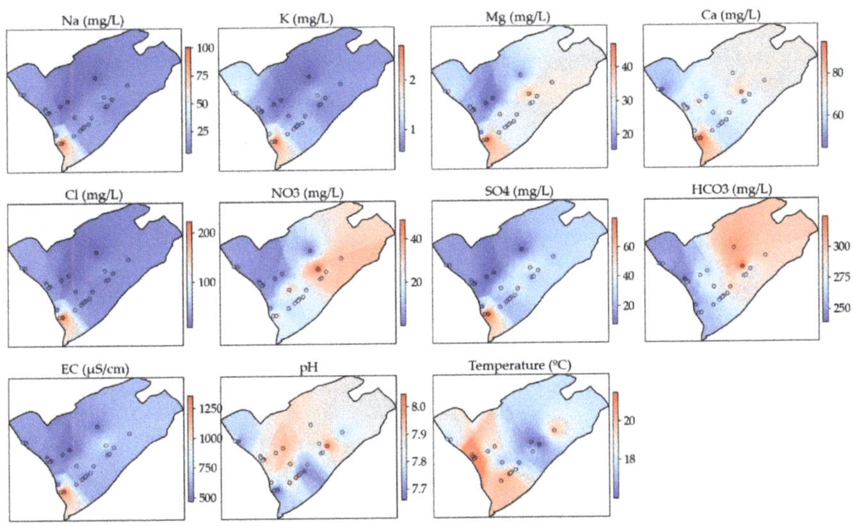

Figure 4. Spatial Distribution of Groundwater Hydrochemical Parameters—July 2024.

As can be seen, the distribution of SO_4^{2-}, Cl^-, Na^+ and K^+ shows a similar pattern in both campaigns. The highest concentrations are located in the south-western sector of the Solana aquifer, where the analysed parameters also show the greatest spatial variability (CV based on original concentration data = 43–106%).

The distribution of calcium and magnesium exhibited greater homogeneity, with CV of 15% and 27% in July, and 14% and 27% in February, respectively. The highest concentrations were located in the Western half, and no major seasonal differences are observed. On the other hand, higher concentrations of bicarbonates were observed towards the North-East of the aquifer in both seasons, showing the lowest spatial variability among all parameters (CV = 7–8%).

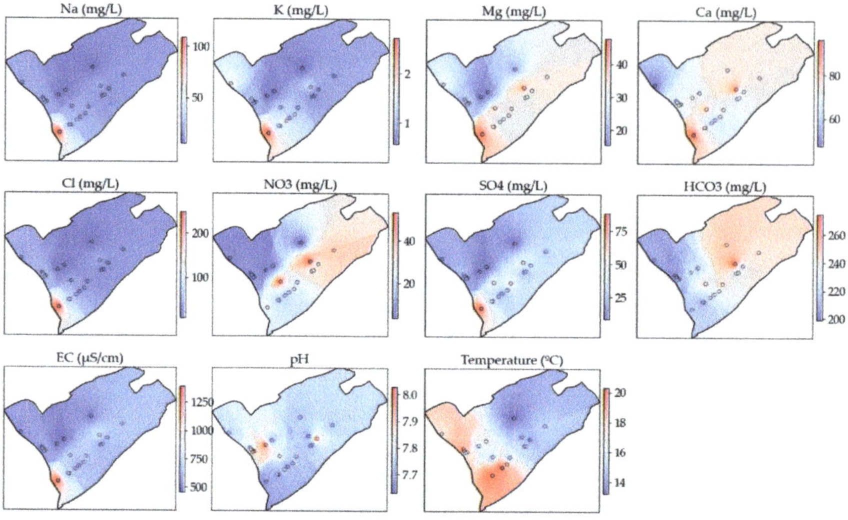

Figure 5. Spatial Distribution of Groundwater Hydrochemical Parameters—February 2025.

Nitrate concentrations were similar in both field campaigns (CV = 73%). While there were no significant changes between the sampling dates, some wells showed higher concentrations at specific times. One well's concentration slightly exceeded the 50 mg/L limit (53.4 mg/L, S-71).

With respect to physicochemical parameters, the electrical conductivity (EC) exhibits the same pattern as the major ions, particularly chlorides and sodium, with a CV of 29%. The pH is slightly more alkaline in July but remains within the typical range for bicarbonate-type groundwater (CV < 1.5%). Temperature reflects the expected seasonal variability, with higher values in July, especially at the western edge (CV = 9–13%).

3.4. Hydrochemical Variability Analysis Using PCA and HCA

In order to identify the main sources of variability in the hydrochemical composition, a Principal Component Analysis (PCA) was carried out for the different field campaigns. This analysis reduces the dimensionality of the hydrochemical information and facilitates interpretation by grouping correlated variables [36,47,48].

In both cases, the principal components (PC1 and PC2) explained 81.4% of the variance in July and 86.1% in February (Figure 6). According to Kaiser's criterion, which re-commends retaining components with eigenvalues greater than 1 [49,50], two principal components are sufficient to explain the variability in both cases. The third component (7.5–8.2% of variance) had an eigenvalue <1 and did not reveal a clear hydrogeochemical process, so it was not interpreted.

For both campaigns, the main component (PC1) appears to be associated with a water mineralisation gradient, with high positive loadings on most ions and EC [51,52]. The second principal axis (PC2) is represented by bicarbonates and nitrates, which could be associated with recharge processes and potential external nitrogen inputs [53,54].

The scores of the first two principal components (PC1 and PC2) were interpolated using the IDW method ($p = 2$, k = 5, grid 500 × 500 m), to visualise their spatial distribution and to identify the areas most influenced by the hydrochemical processes represented by PC1 and PC2 (Figure 7).

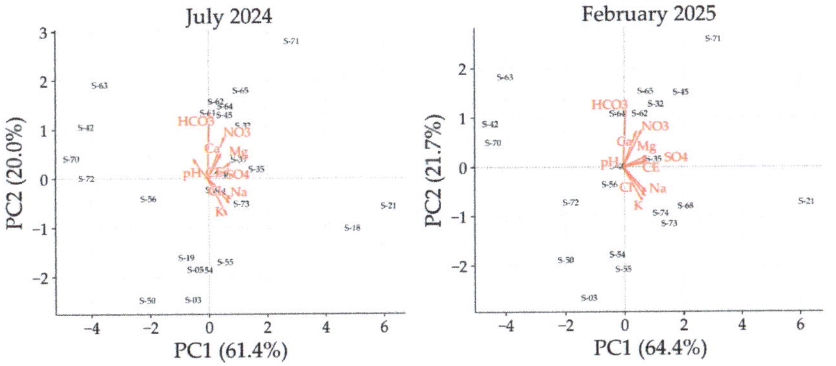

Figure 6. Principal Component Analysis (PCA) of Groundwater Hydrochemistry—July 2024 and February 2025.

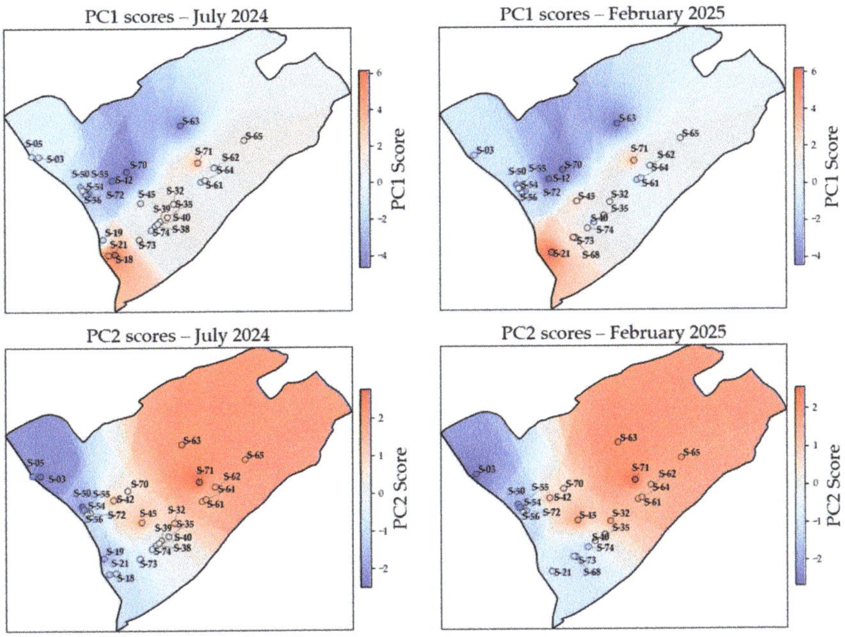

Figure 7. Spatial Distribution of PCA Scores for Groundwater Samples—July 2024 and February 2025.

The values of the first principal component (PC1) are clearly concentrated in the south-western sector of the aquifer, near the wells identified as S-18 and S-21. The area had previously been identified as the most saline due to the high concentrations of major ions such as Na^+, K^+, SO_4^{2-} and Cl^- (Figures 4 and 5). Towards the North and North-east, lower PC1 scores are observed.

The results of the second principal component (PC2), which is interpreted as an indicator of possible recharge zones and anthropogenic inputs, show an inverse spatial distribution to that of PC1. The highest values are located in the eastern half of the aquifer, especially around the wells named as S-63, S-65 and S-71. This pattern is consistent with recharge occurring predominantly in the eastern half of the Solana range. In contrast, the lowest values are found in the western part, reinforcing the idea that this sector is less affected by recent recharge.

To complete the Principal Component Analysis, an HCA (Figure 8) was performed using Ward's method and Euclidean distance to group the wells according to their hydrochemical similarity [40,55,56]. The number of clusters was set at four (red, blue, green and purple), determined using the elbow method [41], which identifies the point after which adding further clusters no longer substantially improves clustering quality (Figure S2).

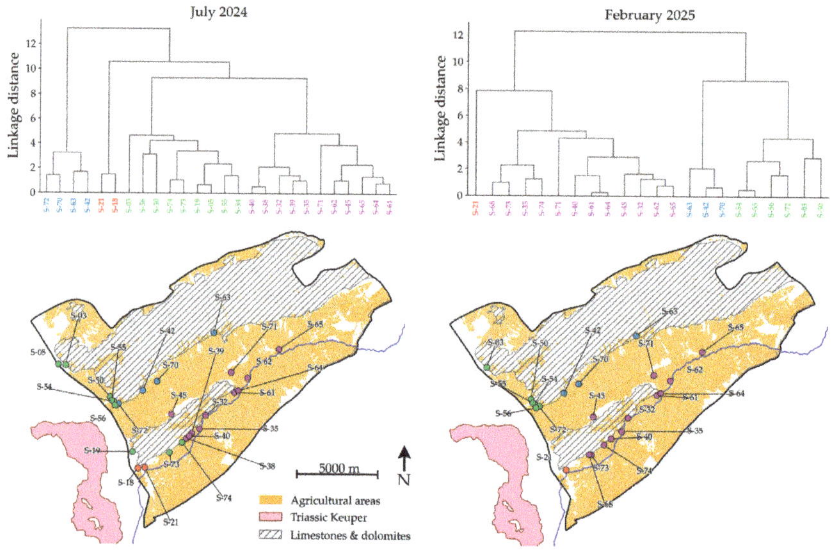

Figure 8. Hierarchical Cluster Analysis (HCA) of Groundwater Samples—July 2024 and February 2025.

In July 2024, the first cluster (blue) included wells with lower mineralisation (Na^+ = 11.0 mg/L, Cl^- = 18.6 mg/L, SO_4^{2-} = 13.2 mg/L, EC = 522.3 μS/cm), corresponding to calcium bicarbonate waters. These wells were located along the southern flank of the Solana range. The second cluster (red) grouped the wells: S-18 and S-21, characterised by the highest concentrations (Na^+ = 85.8 mg/L, Cl^- = 191.8 mg/L, SO_4^{2-} = 74.6 mg/L, EC = 1234.0 μS/cm). The third cluster (green) showed intermediate values (Na^+ = 15.2 mg/L, Cl^- = 32.7 mg/L, SO_4^{2-} = 20.6 mg/L, EC = 611.3 μS/cm). Finally, the fourth cluster (purple) had a similar composition to the green cluster but with higher nitrate (NO_3^- = 18.2 mg/L) and sulphate (SO_4^{2-} = 26.1 mg/L) concentrations. The wells in this last group were located in cultivated valleys.

In February 2025, the blue cluster again contained the least mineralised waters (Na^+ = 13.4 mg/L, Cl^- = 27.2 mg/L, SO_4^{2-} = 14.7 mg/L, EC = 529.7 μS/cm). The green cluster maintained intermediate compositions (Na^+ = 20.0 mg/L, CL^- = 44.7 mg/L, SO_4^{2-} = 28.3 mg/L, EC = 629.5 μS/cm), with wells grouped in the North-Western sector. The red cluster continued to include the most mineralised waters, with higher values than in July (Na^+ = 109.6 mg/L, Cl^- = 248.2 mg/L, SO_4^{2-} = 88.5 mg/L, EC = 1392.0 μS/cm). The purple cluster was characterised by moderate mineralisation with elevated nitrate values (NO_3^- = 21.5 mg/L, EC = 615.4 μS/cm).

When comparing both campaigns, the four clusters remained relatively stable. In February, the compositional differences between clusters were reduced (Table 2), and some wells shifted from the blue cluster to the green cluster, and others from the green cluster to the purple cluster.

Table 2. Mean values of major hydrochemical parameters, pH, and electrical conductivity by cluster–July 2024 and February 2025. All values are in mg/L except EC (μS/cm at 25 °C) and pH.

	July 2024				February 2025			
	C1	C2	C3	C4	C1	C2	C3	C4
Na^+	11.0	85.8	15.2	18.5	109.6	20.0	13.4	17.3
K^+	0.7	2.3	0.8	1.0	2.7	1.2	0.8	0.9
Mg^{2+}	20.7	42.4	25.8	28.2	47.7	28.4	22.0	28.1
Ca^{2+}	59.5	91.8	66.4	65.0	96.6	72.0	67.2	70.9
Cl^-	18.6	191.8	32.7	31.5	248.2	44.7	27.2	38.4
NO_3^-	9.7	24.4	16.6	18.2	28.2	21.1	10.8	21.5
SO_4^{2-}	13.2	74.6	20.6	26.1	88.5	28.3	14.7	25.6
HCO_3^-	268.2	257.7	269.3	276.0	217.0	226.0	221.2	229.6
EC	522.3	1234.0	611.3	634.0	1392.0	629.5	529.7	615.4
pH	7.9	7.7	7.8	7.8	7.7	7.8	7.8	7.8

Note: Colours indicate HCA groups; the same scheme as in Figure 8.

4. Discussion

The chemical characteristics and quality of groundwater depend on a combination of natural factors, such as water-rock interaction and residence time [57,58], and anthropogenic influences, like agricultural practices, urbanization, and groundwater abstraction [8,27,59]. Moreover, in the case of karst aquifers such as the Solana aquifer, other factors are inherent to this type of environment, such as lithological heterogeneity [60–64].

According to the analytical results of the Solana aquifer, the waters are mainly calcium bicarbonate, consistent with the nature of the permeable geological formations (limestone and dolomite) that make up the aquifer. Therefore, the predominant process in the aquifer seems to be the dissolution of these carbonates, a pattern that is consistent with the hydrochemical facies already described more than a decade ago [20], indicating long-term stability of groundwater composition.

However, in the South-Western sector, waters of mixed SO_4^{2-} and Cl^- facies were found in several wells (S-18, S-21), where concentrations of both ions are markedly higher than in the rest of the aquifer (SO_4^{2-} > 70 mg/L and Cl^- > 160 mg/L) together with EC values of approximately 1300 μS/cm. This situation could be explained by the dissolution of gypsum and halite from the Triassic Keuper, whose outcrops are located a short distance from this area (about 2500 m) a process commonly associated with chloride and sulphate rich facies in wells and reported both within and outside the Vinalopó basin [16,45,65].

However, contrasting patterns are observed in adjacent well (S-19), close to those catalogued as more mineralised (S-18, S-21), which shows purely calcium bicarbonate facies and an EC of around 550 μS/cm. These differences could be due to the structural complexity of the Triassic Keuper, whose geometry and extent beneath the Quaternary deposits are not precisely known and which, in many cases, lacks detailed studies due to its chaotic nature. In addition to the structural complexity, these variations can also be attributed to the inherent heterogeneity of Karst aquifers [66–68], to differences in well construction (e.g., screen depth), or even the exploitation regimes, since some wells may capture water that has been in contact with the Triassic, while others receive better quality inputs from within the aquifer.

There are no significant differences between the concentrations observed in both campaigns (paired *t*-tests on transformed data, mean difference = 0.06 SD across parameters), despite >200 mm of accumulated rainfall between campaigns and a 1 m rise in the piezometric level. At the basin scale, carbonate equilibrium buffers the major-ion chemistry: field pH, temperature, alkalinity and Ca yield stability indices indicating waters close to calcite saturation (median Ryznar 6.82, IEB 0.16). This near-equilibrium state explains the limited temporal variability of Ca, Mg and bicarbonate under seasonal recharge and pumping. The slight winter decrease in bicarbonate likely reflects modest dilution by recent recharge, with the aquifer's large storage damping short-term fluctuations, consistent with longer-term records for this system [20].

Regarding the correlation analysis between parameters, very strong relationships ($|r| > 0.8$) have been observed between SO_4^{2-}, Cl^-, K^+, Na^+ and EC, which supports the interpretation of the common origin of all these ions, related to the dissolution of the evaporite materials of the Keuper Triassic (gypsum and halite).

According to the PCA, the first two principal components account for more than 80% of the total variance in both campaigns, indicating that only a few processes control the chemical characteristics of the Solana aquifer [36]. On the one hand, PC1 shows strong positive loadings for most ions and electrical conductivity, except for bicarbonates and nitrates. This suggests that PC1 may represent the salinity gradient within the aquifer. The spatial distribution of PC1 scores aligns with the spatial patterns of these parameters, showing higher values near the Triassic evaporite outcrops. On the other hand, PC2 is positively correlated with bicarbonates and nitrates, with its highest scores found in the eastern half of the aquifer. The maximum bicarbonate concentrations, located in the NE sector, may reflect recharge processes through carbonate outcrops, particularly where Upper Cretaceous limestones and dolomites crop out and the highest rainfall values of the area are recorded (470 mm/year), conditions that enhance carbonate dissolution. Similarly, the occurrence of high nitrate concentrations is consistent with anthropogenic contamination from agriculture, livestock farming, and wastewater. This is especially relevant in the associated valley, the widest in the region, which is extensively occupied by cultivated land (Figures 4, 5 and 8), thereby increasing the likelihood of anthropogenic inputs [69]. Slightly higher nitrate concentrations at certain points in February may reflect localized recharge events that mobilize nitrates from agricultural soils, or alternatively vertical leakages through poorly sealed wells, which allow the entry of nitrate-rich shallow groundwater into the aquifer [70]. Nevertheless, the relationship between PC2, recharge zones, and nitrogen inputs should be confirmed through specific studies based on isotopes and land use.

These nitrate concentrations are surprising given the conceptual model of the aquifer. In this sector, the greatest thicknesses of Miocene marls are found (over 300 m in the centre of the valley), which would theoretically make it less vulnerable. One possible hypothesis is that the nitrates come from agriculture and that the recharge takes place along the edges of the alluvial fans and in a more concentrated form in the areas where the detrital deposits rest directly on the Cretaceous limestones. In order to evaluate this interpretation and determine its chemical composition, water samples from the Quaternary aquifer and the Vinalopó River were collected during both field campaigns in the vicinity of well S-65. The Quaternary showed EC = 676 μS/cm (≈439 mg/L TDS), HCO_3^- = 242 mg/L, Cl^- = 43 mg/L, SO_4^{2-} = 34 mg/L, Ca^{2+} = 71 mg/L and NO_3^- = 35 mg/L, while the river sample presented EC = 832 μS/cm (≈540 mg/L TDS), HCO_3^- = 270 mg/L, Cl^- = 69 mg/L, SO_4^{2-} = 54 mg/L, Ca^{2+} = 86 mg/L and NO_3^- = 8 mg/L. These results show that the quality of these waters is similar to the average of the wells, even showing lower nitrate concentration values than those observed in the wells of the eastern sector. This suggests that this contribution from higher levels does not seem to be the only cause of this nitrate enrichment in this sector.

Another possible source of nitrates could be the infiltration of urban wastewater from nearby urban sites. This hypothesis has the same hydrogeological limitation as the previous one, as the impermeable fill would prevent infiltration from the surface. However, it should be noted that the municipalities in the area, due to their size, do not have their own water treatment plants, so their wastewater is channelled to the Villena wastewater treatment plant, located in the western sector. Although in theory the sewage networks are closed systems, there may be leaks that could reach the Cretaceous limestone.

In any case, the concentration of nitrates is relatively low, and the 50 mg/L limit established by the Water Framework Directive (WFD) and WHO for water intended for human consumption is not reached [42,71], except in February in well S-71, where a concentration of 53.4 mg/L was measured, thus exceeding this threshold, which renders the water unsuitable for human consumption without prior treatment. The lower concentration in the Western part of the aquifer does not seem to be due to natural denitrification processes but rather to higher recharge rates and better water quality in this part of the aquifer compared to the Eastern part.

As for the dendrograms (Figure 8), four groups of water were identified in the aquifer, differentiated mainly by their mineralisation and nitrate concentration. The least mineralised waters (blue and green groups) are located in the Northern and North-Western sectors of the aquifer, near the carbonate outcrops of the Solana range. It is possible that these wells capture younger waters, although this should be verified through specific studies. In contrast, the red group is clearly associated with the Triassic Keuper, reinforcing the strong lithological control on groundwater mineralisation, which becomes even more evident in February. Finally, the purple group is consistently marked by agricultural influence, with an enrichment in nitrates and sulphates that indicates the persistence of anthropogenic inputs in cultivated valleys. Overall, no major differences are observed between campaigns, although some wells with intermediate mineralisation (green and purple clusters) are the most likely to shift from one group to another due to small variations in chemistry possibly linked to recharge processes and nitrate mobilisation.

5. Conclusions

This study provides an integrated diagnosis of the hydrochemical behaviour of the Solana aquifer in two contrasting seasons. Using descriptive and multivariate techniques together with spatial interpolation maps, the relative weights of lithological control and anthropogenic pressures on groundwater quality have been identified.

The PCA reveals two main axes, which explain more than 80% of the variability in both campaigns, suggesting that few hydrochemical processes largely explain the chemistry of the system. The first axis, associated with salinity (PC1), concentrates most of the variance and aligns with a NE–SW gradient and with the influence of Triassic Keuper evaporite outcrops in the south-western sector of the aquifer. The second axis (PC2), related to recharge and anthropogenic influence, reflects both recharge in the Solana range and agricultural influence in the valley floor.

The HCA distinguishes four recurrent hydrochemical groups in both seasons. The least mineralised waters (blue cluster) occur on the southern edge of the Solana range, reflecting recent recharge, while the most mineralised group (red cluster) appears in the south-west near evaporite outcrops, evidencing lithological control. The remaining two clusters (green and purple) show intermediate compositions, differing mainly in nitrate concentration, higher in cultivated valleys. There are no major changes between clusters across seasons, and their spatial distribution remains stable.

The use of PCA and HCA has been particularly useful to identify the main factors controlling groundwater quality in the Solana aquifer. Its chemistry is controlled partly by the lithology, with the dissolution of evaporites raising the concentrations of EC, Na^+, Cl^- and SO_4^{2-} in the south-west, and partly by land use, with nitrate enrichment in the wells in the centre of the valley. The spatial patterns are consistent between the methods used (PCA, HCA, interpolated parameters) and across seasons.

The results also suggest that this is an aquifer with hydrochemical resilience, although with minor seasonal adjustments. Despite differences in exploitation between the summer and winter months, and recharge occurring between dates, the aquifer has a stable hydrochemical signature. This stability is reinforced by the buffering capacity of $CaCO_3$ equilibrium, which maintains waters close to calcite saturation and dampens fluctuations. Statistical tests indicate differences between the correlation matrices, but from a hydrogeological point of view these differences are minor. In general, winter convergence between groups is observed, together with a clearer delimitation of the valley group characterised by nitrates. Despite this, both the dominant processes and the spatial position of the groups remain unchanged.

These contributions also offer practical insights in the context of groundwater-use reorganisation. Firstly, wells supplying water for human consumption should be located within the blue or green zones, avoiding the red zone, where suitability may be compromised by high mineralisation. Wells in the centre of the valley require more thorough nitrate control. In this area, management should focus on identifying contamination sources and pathways into the Cretaceous aquifer. Secondly, this study makes it possible to establish a monitoring network aligned NE–SW, which will enable the hydrochemical evolution of the system to be effectively tracked.

However, the study also has limitations. The analysis is based on two field campaigns separated by one season. Studies with longer data series would be necessary to identify trends and atypical years. Similarly, specific studies are also needed to corroborate the attribution of nitrates to agricultural sources through isotopic analyses ($\delta^{15}N–NO_3$, $\delta^{18}O–NO_3$). All of this reinforces the need for data integration among well owners, so that routine and isolated analyses can be transformed into open and useful products for groundwater management.

Supplementary Materials: The following supporting information can be downloaded at: https://www.mdpi.com/article/10.3390/environments12090323/s1, Figure S1: Evolution of piezometric levels, Figure S2: Elbow method results.

Author Contributions: Conceptualization, V.S.-S., J.M.A. and J.N.-P.; methodology, V.S.-S.; software, V.S.-S.; validation, J.M.A. and J.N.-P.; formal analysis, V.S.-S.; investigation, V.S.-S. and J.M.A.; resources, A.P.-G., M.B.A.-C. and J.M.A.; data curation, V.S.-S.; writing—original draft preparation, V.S.-S.; writing—review and editing, J.M.A., J.N.-P., A.P.-G., M.B.A.-C. and M.M.J.; visualization, V.S.-S.; supervision, J.M.A.; project administration, M.B.A.-C. All authors have read and agreed to the published version of the manuscript.

Funding: This work was carried out within the framework of the REVOKER project (PID2023-151910OB-I00), funded by the Ministry of Science and Innovation.

Data Availability Statement: The original contributions presented in this study are included in the article/Supplementary Materials. Further inquiries can be directed to the corresponding authors.

Acknowledgments: We would like to thank the water supply companies AMAEM and Hidraqua; the General Community of Users of the Alto Vinalopó and the irrigation communities of Huerta y Partidas and Villena; as well as the municipalities of Benejama and Campo de Mirra for the support provided in the completion of this work.

Conflicts of Interest: The authors declare no conflict of interest.

References

1. UNESCO. *The United Nations World Water Development Report 2024: Water for Prosperity and Peace*; UNESCO: Paris, France, 2024; ISBN 978-92-3-100657-9. Available online: https://unesdoc.unesco.org/ark:/48223/pf0000388948 (accessed on 5 August 2025).
2. UNESCO. *The United Nations World Water Development Report 2022: Groundwater: Making the Invisible Visible*; UNESCO: Paris, France, 2022; ISBN 978-92-3-100507-7. Available online: https://unesdoc.unesco.org/ark:/48223/pf0000380721 (accessed on 5 August 2025).
3. Food and Agriculture Organization of the United Nations (FAO). *The State of Food and Agriculture 2023—Revealing the True Cost of Food to Transform Agrifood Systems*; FAO: Rome, Italy, 2023. [CrossRef]
4. Mastrocicco, M.; Colombani, N. The issue of groundwater salinization in coastal areas of the Mediterranean region: A review. *Water* **2021**, *13*, 90. [CrossRef]
5. Campanale, C.; Losacco, D.; Triozzi, M.; Massarelli, C.; Uricchio, V.F. An overall perspective for the study of emerging contaminants in karst aquifers. *Resources* **2022**, *11*, 105. [CrossRef]
6. Alexakis, D.E.; Kiskira, K.; Gamvroula, D.; Emmanouil, C.; Psomopoulos, C.S. Evaluating toxic element contamination sources in groundwater bodies of two Mediterranean sites. *Environ. Sci. Pollut. Res.* **2021**, *28*, 34400–34409. [CrossRef] [PubMed]
7. Custodio, E. Groundwater pollution in Spain: General aspects. *J. Inst. Water Environ. Manag.* **1992**, *6*, 452–458. [CrossRef]
8. Al Haj, R.; Merheb, M.; Halwani, J.; Ouddane, B. Hydrogeochemical characteristics of groundwater in the Mediterranean region: A meta-analysis. *Phys. Chem. Earth.* **2023**, *129*, 103351. [CrossRef]
9. Pulido-Bosch, A.; Vallejos, A.; Martín-Rosales, W.; Molina, L.; Andreu Rodes, J.M.; Calaforra, J.M. La surexploitation dans certains aquifères du sud-est espagnol. In Proceedings of the International Conference on World Water Resources at the Beginning of the 21st Century, Paris, France, 3–6 June 1998.
10. Custodio, E. Aquifer overexploitation: What does it mean? *Hydrogeol. J.* **2002**, *10*, 254–277. [CrossRef]
11. Custodio, E.; Andreu-Rodes, J.M.; Aragón, R.; Estrela, T.; Ferrer, J.; García-Aróstegui, J.L.; Manzano, M.; Rodríguez-Hernández, L.; Sahuquillo, A.; del Villar, A. Groundwater intensive use and mining in south-eastern peninsular Spain: Hydrogeological, economic and social aspects. *Sci. Total Environ.* **2016**, *559*, 302–316. [CrossRef]
12. Sahuquillo, A. La explotación intensa de los acuíferos en la cuenca baja del Segura y en la cuenca del Vinalopó. *Ing. Agua* **2016**, *20*, 13–26. [CrossRef]
13. Mammadova, L.; Negri, S. Understanding the impacts of overexploitation on the Salento aquifer: A comprehensive review through well data analysis. *Sustain. Futures* **2024**, *1*, 100188. [CrossRef]
14. Taheri, K.; Taheri, M.; Parise, M. Impact of intensive groundwater exploitation on an unprotected covered karst aquifer: A case study in Kermanshah Province, western Iran. *Environ. Earth Sci.* **2016**, *75*, 1221. [CrossRef]
15. Andreu Rodes, J.M. Contribución de la Sobreexplotación al Conocimiento de los Acuíferos Kársticos de Crevillente, Cid y Cabeçó d'Or (Provincia de Alicante). Doctoral Thesis, Universidad de Alicante, Alicante, Spain, 6 June 1997. Available online: https://rua.ua.es/dspace/handle/10045/3196 (accessed on 3 August 2025).
16. Andreu, J.M.; Pulido-Bosch, A.; Llamas, M.R.; Bru, C.; Martínez-Santos, P.; García-Sánchez, E.; Villacampa, L. Overexploitation and water quality in the Crevillente aquifer (Alicante, SE Spain). *WIT Trans. Ecol. Environ.* **2008**, *111*, 67–77. [CrossRef]
17. Instituto Geológico y Minero de España; Diputación Provincial de Alicante. *Desarrollos Metodológicos en Geología del Subsuelo para la Caracterización de Recursos Hidrogeológicos Profundos de la Provincia de ALICANTE (HIDROPROAL): Modelo Geológico 3D del Acuífero de Solana–Onteniente–Volcadores y Evaluación de las Reservas Totales de Agua Subterránea*; Instituto Geológico y Minero de España: Madrid, Spain, 2013; ISBN 978-84-96979-40-6.
18. Diputación Provincial de Alicante. *Mapa del Agua. Provincia de Alicante*, 2nd ed.; Diputación Provincial de Alicante: Alicante, Spain, 2007.
19. Ministerio para la Transición Ecológica y el Reto Demográfico. Anuncio de la Confederación Hidrográfica del Júcar sobre declaración en riesgo de no alcanzar el buen estado cuantitativo de la masa de agua subterránea 080.160 Villena-Benejama. *Boletín Oficial Del Estado* **2020**, *266*, 45583–45586. Available online: https://www.boe.es/boe/dias/2020/10/08/pdfs/BOE-B-2020-34158.pdf (accessed on 1 August 2025).
20. Pérez Bielsa, C.; Lambán Jiménez, L.J. Caracterización hidrogeoquímica e isotópica de las aguas subterráneas en el acuífero carbonatado Solana (Alicante). *Bol. Geol. Min.* **2006**, *117*, 589–592.
21. Olcina Cantos, J.; Moltó Mantero, E. *Climas y Tiempos del País Valenciano*; Publicacions Universitat d'Alacant: Alicante, Spain, 2019; ISBN 978-84-9717-659-0.
22. Instituto Geológico y Minero de España; Diputación Provincial de Alicante—Ciclo Hídrico. *Atlas Hidrogeológico de la Provincia de Alicante*; IGME–DPA: Madrid/Alicante, Spain, 2015; ISBN 978-84-7840-959-4.
23. Cantos Conejero, J.M. Estudio hidrogeológico del acuífero detrítico del entorno de Caudete-Villena. *Rev. Estud. Albacetenses* **2001**, *2*, 5–43.
24. American Public Health Association. *Standard Methods for the Examination of Water and Wastewater*, 23rd ed.; American Public Health Association: Washington, DC, USA, 2017; ISBN 978-0-87553-047-5.

25. Chapman, D. *Water Quality Assessments: A Guide to the Use of Biota, Sediments and Water in Environmental Monitoring*, 2nd ed.; Chapman, D., Ed.; CRC Press (Taylor & Francis Group): London, UK, 1996. [CrossRef]
26. *ISO 10304-1*; Water Quality—Determination of Dissolved Anions by Liquid Chromatography of Ions—Part 1: Determination of Bromide, Chloride, Fluoride, Nitrate, Nitrite, Phosphate and Sulfate. International Organization for Standardization: Geneva, Switzerland, 2007.
27. Freeze, R.A.; Cherry, J.A. *Groundwater*; Prentice Hall: Englewood Cliffs, NJ, USA, 1979; ISBN 978-0-13-365312-0.
28. Piper, A.M. A graphic procedure in the geochemical interpretation of water-analyses. *Trans. AGU* **1944**, *25*, 914–928. [CrossRef]
29. Simler, R. *DIAGRAMMES: Logiciel D'hydrochimie Multilangage [software]*, Version 9.3; Université d'Avignon: Avignon, France, 2025.
30. Wilcox, L.V. *Classification and Use of Irrigation Waters*; U.S. Geological Survey: Washington, DC, USA, 1955; Circular 969.
31. U.S. Salinity Laboratory Staff. *Diagnosis and Improvement of Saline and Alkali Soils*; United States Department of Agriculture: Washington, DC, USA, 1954. Available online: https://www.ars.usda.gov/ARSUserFiles/20360500/hb60_pdf/hb60complete.pdf (accessed on 17 August 2025).
32. Thapa, R.; Gupta, S.; Kaur, H. Introducing an irrigation water quality index (IWQI) based on the case study of the Dwarka River basin, Birbhum, West Bengal, India. *Sustain. Water Resour. Manag.* **2020**, *6*, 86. [CrossRef]
33. Jamieson, P.D.; Porter, J.R.; Wilson, D.R. A test of the computer simulation model ARCWHEAT1 on wheat crops grown in New Zealand. *Field Crops Res.* **1991**, *27*, 337–350. [CrossRef]
34. Kaiser, H.F. An index of factorial simplicity. *Psychometrika* **1974**, *39*, 31–36. [CrossRef]
35. Bartlett, M.S. A note on the multiplying factors for various χ^2 approximations. *J. R. Stat. Soc. Ser. B* **1954**, *16*, 296–298. [CrossRef]
36. Jolliffe, I.T. *Principal Component Analysis*, 2nd ed.; Springer: New York, NY, USA, 2002; ISBN 978-0387954424.
37. Abdi, H.; Williams, L.J. Principal component analysis. *Wiley Interdiscip. Rev. Comput. Stat.* **2001**, *2*, 433–459. [CrossRef]
38. Ward, J.H. Hierarchical grouping to optimize an objective function. *J. Am. Stat. Assoc.* **1963**, *58*, 236–244. [CrossRef]
39. Johnson, R.A.; Wichern, D.W. *Applied Multivariate Statistical Analysis*, 6th ed.; Pearson Prentice Hall: Upper Saddle River, NJ, USA, 2007; ISBN 978-0131877153.
40. Otto, M. *Chemometrics: Statistics and Computer Application in Analytical Chemistry*, 3rd ed.; Wiley-VCH: Weinheim, Germany, 2017; ISBN 978-3-527-34097-2.
41. Thorndike, R.L. Who belongs in the family? *Psychometrika* **1953**, *18*, 267–276. [CrossRef]
42. World Health Organization. *Guidelines for Drinking-Water Quality*, 4th ed.; World Health Organization: Geneva, Switzerland, 2017; ISBN 978-92-4-154995-0.
43. Helsel, D.R.; Hirsch, R.M. Graphical data analysis. In *Statistical Methods in Water Resources*; United States Geological Survey: Reston, VA, USA, 2020. [CrossRef]
44. Sánchez, D.; Barberá, J.A.; Mudarra, M.; Andreo, B. Hydrogeochemical tools applied to the study of carbonate aquifers: Examples from some karst systems of Southern Spain. *Environ. Earth Sci.* **2015**, *74*, 199–215. [CrossRef]
45. Gil-Márquez, J.M.; Barberá, J.A.; Andreo, B.; Mudarra, M. Geochemical evolution of groundwater in an evaporite karst system: Brujuelo area (Jaén, S Spain). *Procedia Earth Planet. Sci.* **2017**, *17*, 336–339. [CrossRef]
46. Hounslow, A.W. *Water Quality Data: Analysis and Interpretation*; CRC Press: Boca Raton, FL, USA, 1995; ISBN 0-8493-9090-2.
47. Güler, C.; Thyne, G.D.; Mccray, J.E.; Turner, K.A. Evaluation of graphical and multivariate statistical methods for classifcation of water chemistry data. *Hydrogeol. J.* **2002**, *10*, 455–474. [CrossRef]
48. Huang, F.; Vasic, L.; Wu, X.; Cao, J.; Milanovic, S. Hydrochemical features and their controlling factors in the Kucaj-Beljanica Massif, Serbia. *Environ. Earth Sci.* **2019**, *78*, 498. [CrossRef]
49. Kaiser, H.F. The application of electronic computers to factor analysis. *Educ. Psychol. Measur.* **1960**, *20*, 141–151. [CrossRef]
50. Davis, J.C.; Sampson, R.J. *Statistics and Data analysis in Geology*, 3rd ed.; Wiley: New York, NY, USA, 2002; ISBN 978-0471172758.
51. Helena, B.; Pardo, R.; Vega, M.; Barrado, E.; Fernández, J.M.; Fernández, L. Temporal evolution of groundwater composition in an alluvial aquifer (Pisuerga River, Spain) by principal component analysis. *Water Res.* **2000**, *34*, 807–816. [CrossRef]
52. Aydoğan, D.; Şen, Z.; Öztürk, T. Assessment of Groundwater Quality through Hydrochemistry Using Principal Component Analysis in the Kızılırmak Delta (Turkey). *Water* **2024**, *16*, 1570. [CrossRef]
53. Domenico, P.A.; Schwartz, F.W. *Physical and Chemical Hydrogeology*, 2nd ed.; Wiley: New York, NY, USA, 1998; ISBN 978-0471597629.
54. Appelo, C.A.J.; Postma, D. *Geochemistry, Groundwater and Pollution*, 2nd ed.; CRC Press/Balkema: Leiden, The Netherlands, 2005. [CrossRef]
55. Adams, M.J. *Chemometrics in Analytical Spectroscopy*; Royal Society of Chemistry: Cambridge, UK, 1998; ISBN 978-0-85404-555-6.
56. Everitt, B.S.; Landau, S.; Leese, M.; Stahl, D. *Cluster Analysis*, 5th ed.; Wiley: Chichester, UK, 2011; ISBN 978-0-470-74991-3.
57. Elango, L.; Kannan, R. Rock–water interaction and its control on chemical composition of groundwater. In *Developments in Environmental Science*; Sarkar, D., Datta, R., Hannigan, R., Eds.; Elsevier: Amsterdam, The Netherlands, 2007; Volume 5, pp. 229–243. [CrossRef]

58. Salvadori, M.; Pennisi, M.; Masciale, R.; Fidelibus, M.D.; Frollini, E.; Ghergo, S.; Parrone, D.; Preziosi, E.; Passarella, G. Isotopic study for evaluating complex groundwater circulation patterns, hydrogeological zoning, and water-rock interaction in a Mediterranean coastal karst environment. *Sci. Total Environ.* **2024**, *955*, 176850. [CrossRef]
59. Leduc, C.; Pulido-Bosch, A.; Remini, B. Anthropization of groundwater resources in the Mediterranean region: Processes and challenges. *Hydrogeol. J.* **2017**, *25*, 1529–1547. [CrossRef]
60. Bakalowicz, M. Karst groundwater: A challenge for new resources. *Hydrogeol. J.* **2005**, *13*, 148–160. [CrossRef]
61. Mudry, J. Apport du Traçage Physico–Chimique Naturel à la Connaissance Hydrocinématique des Aquifères Carbonatés. Doctoral Thesis, Université de Franche-Comté, Besançon, France, May 1987. Available online: https://theses.hal.science/tel-00654518 (accessed on 20 August 2025).
62. Ford, D.C.; Williams, P.W. *Karst Hydrogeology and Geomorphology*, 2nd ed.; John Wiley & Sons Ltd.: Chichester, UK, 2007. [CrossRef]
63. Goldscheider, N.; Drew, D.P. (Eds.) *Methods in Karst Hydrogeology*; CRC Press: London, UK, 2007. [CrossRef]
64. Narany, T.S.; Bittner, D.; Disse, M.; Chiogna, G. Spatial and temporal variability in hydrochemistry of a small-scale dolomite karst environment. *Environ. Earth Sci.* **2019**, *78*, 273. [CrossRef]
65. Andreu Rodes, J.M.; García Sánchez, E.; Pulido Bosch, A.; Jorreto, S.; Francés, I. Influence of Triassic deposits on water quality of some karstic aquifers to the south of Alicante (Spain). *Estud. Geol.* **2010**, *66*, 131–138. [CrossRef]
66. Jiménez-Gavilán, P. Caracterización Hidrogeológica de Acuíferos Carbonáticos del sur de España a Partir de sus Respuestas Naturales. Ph.D. Thesis, Universidad de Granada, Granada, España, 12 July 2010. Available online: https://digibug.ugr.es/handle/10481/5641 (accessed on 7 August 2025).
67. Pardo-Igúzquiza, E.; Durán, J.J.; Robledo-Ardila, P.A. A Three-Dimensional Karst Aquifer Model: The Sierra de Las Nieves Case (Málaga, Spain). In *Hydrogeological and Environmental Investigations in Karst Systems. Environmental Earth Sciences*; Andreo, B., Carrasco, F., Durán, J., Jiménez, P., LaMoreaux, J., Eds.; Springer: Berlin/Heidelberg, Germany, 2015; pp. 285–291. [CrossRef]
68. Pulido-Bosch, A. Investigación y exploración de acuíferos kársticos. *Bol. Geol. Min.* **2001**, *112*, 25–42.
69. Foster, S.S.D.; Chilton, P.J. Groundwater: The processes and global significance of aquifer pollution. *Philos. Trans. R. Soc. Lond. B. Biol. Sci.* **2003**, *358*, 1957–1972. [CrossRef] [PubMed]
70. Koh, D.C.; Kim, Y.-C.; Lee, J.-Y.; Ko, K.-S.; Lee, S.-G.; Kim, K.-M.; Kim, T.-Y.; Lee, H.-J.; Kim, C.-W.; Kim, D.-Y.; et al. Impact of leaky wells on nitrate cross-contamination in a layered aquifer system: Methodology for and demonstration of quantitative assessment and prediction. *J. Hydrol.* **2016**, *540*, 1068–1082. [CrossRef]
71. European Commission. Directive 2000/60/EC of the European Parliament and of the Council establishing a framework for Community action in the field of water policy. *Off. J. Eur. Communities* **2000**, *L327*, 1–73. Available online: https://eur-lex.europa.eu/legal-content/EN/TXT/?uri=CELEX%3A32000L0060 (accessed on 4 August 2025).

environments

MDPI

Article

Improving Groundwater Quality Through Biosphere Reserve Management: Insights from the Anaga Reserve, Tenerife

Joselin S. Rodríguez-Alcántara [1], Noelia Cruz-Pérez [1], Jesica Rodríguez-Martín [2], Alejandro García-Gil [3], Jelena Koritnik [4] and Juan C. Santamarta [1,*]

1 Department of Agricultural and Environmental Engineering, Universidad de La Laguna (ULL), 38200 Tenerife, Spain; jrodralc@ull.edu.es (J.S.R.-A.); ncruzper@ull.edu.es (N.C.-P.)
2 Department of Engineering and Architectural Techniques and Projects, Universidad de La Laguna (ULL), 38200 Tenerife, Spain; jrodrima@ull.edu.es
3 Geological Survey of Spain (IGME), Spanish National Research Council (CSIC), 28003 Madrid, Spain; a.garcia@igme.es
4 Department of Geology, Faculty of Science, University of Zagreb, Horvatovac 102a, 10000 Zagreb, Croatia; alu0101616728@ull.edu.es
* Correspondence: jcsanta@ull.es

Abstract: The Canary Islands, an outermost Spanish territory in the Atlantic Ocean, are renowned for their subtropical climate and significant tourism. However, substantial areas are designated for environmental protection, notably the Anaga Rural Park in Tenerife, a UNESCO Biosphere Reserve, which is the focus of this study. This research investigates the influence of Biosphere Reserve designation on groundwater quality, a crucial resource for Tenerife's population. We analysed the physicochemical properties of groundwater within the Anaga region over a decade (2007–2016). Our findings demonstrate that groundwater quality consistently meets regulatory standards, exhibiting no evidence of pollution. This high quality is attributed to several factors, including the low population density, limited tourism impact within the reserve, and crucially, the effective soil protection measures implemented within the Biosphere Reserve. The compact geology of the region further limits infiltration and potential pollution. The sustained high quality of groundwater, even in the absence of detectable pollution, highlights the importance of ongoing monitoring to maintain this valuable resource and support local biodiversity. This case study provides a valuable model for sustainable groundwater management and soil protection strategies in other areas of Tenerife and beyond.

Keywords: groundwater quality; biosphere reserve; soil conservation; island vulnerability; sustainable groundwater management

1. Introduction

Groundwater is a vital natural resource for humanity, serving as a critical component in the provision of potable water, agricultural production, and industrial processes. The quality of groundwater is of paramount importance for the preservation of public health and the environment [1]. However, the mounting pressure exerted by human activities and natural phenomena has given rise to considerable challenges in terms of their preservation and management, particularly in geographically vulnerable contexts such as islands. It is therefore becoming increasingly necessary to plan the use, management and protection of this resource in a given territory [2].

Historically, the management of water resources in Spain has been characterised by a prioritisation of the increase in water supply [3], with the objective of meeting the demand

https://doi.org/10.3390/environments12020053

derived from agricultural expansion and massive urbanisation, particularly in coastal areas with scarce water resources [4,5]. The conventional strategy of intensive freshwater production, encompassing desalination and the utilisation of abandoned wells, has resulted in a number of challenges, including marine intrusion [6] and a decline in water quality, in addition to public health concerns pertaining to human and animal pollution [7,8]. This approach has been subject to revision over time, with a paradigm shift towards demand-driven water management, environmental protection and cost recovery.

The management of groundwater on islands presents a number of unique challenges, largely due to the limited hydrogeological and geological characteristics of these environments and their exposure to extreme weather events [9–11]. In such environments, groundwater is of vital importance, and its protection from pollution is of the utmost importance to ensure the well-being of the population and the sustainability of economic activities [12]. This context emphasises the necessity for the implementation of efficacious measures to ensure the maintenance of water quality and the protection of island ecosystems [13,14].

A principal element of this protection is soil conservation, which represents a fundamental measure in the sustainable management of groundwater resources. The protection of soil not only preserves soil fertility and structure but also acts as a barrier to groundwater pollution [15]. Several studies conducted globally have demonstrated that the implementation of effective soil management strategies can effectively reduce the infiltration of pollutants, particularly in regions characterised by low population density and low levels of tourism and industrial activity [16–22]. In Europe, this integrated approach has become particularly important in protected areas, where the regulation of land use and the promotion of sustainable practices have been instrumental in maintaining the quality of aquifers and protecting biodiversity [23,24].

This context provides the backdrop for an investigation of the Anaga Rural Park in Tenerife, a UNESCO-designated Biosphere Reserve. The protected area has implemented land management strategies that limit pollution and promote the maintenance of groundwater quality, due to low population density and controlled tourism. The compact geology of the region, characterised by areas of low permeability interspersed with basaltic fissures and fractured volcanic rocks, impedes widespread infiltration and reinforces the aquifer's natural protection against pollution. However, the aquifer is recharged mainly by the infiltration of atmospheric precipitation into zones of higher permeability. This case study presents an analysis of the impact of specific factors, including Biosphere Reserve status, low population density and low levels of agriculture, on water quality. The approach taken is aligned with the framework of the second and third cycles of the Tenerife Hydrological Plan.

Study Area

The Canary Islands are an archipelago located in the Atlantic Ocean, close to the Western Sahara. They are part of the Kingdom of Spain and are divided into two provinces: Tenerife and Las Palmas de Gran Canaria. In the province of Tenerife, from largest to smallest territories are Tenerife, La Palma, La Gomera and El Hierro. In the province of Las Palmas de Gran Canaria, from greater to lesser extent of territory are Fuerteventura, Gran Canaria, Lanzarote and La Graciosa [25]. The population of the region exceeds 2 million inhabitants. It is a region where the tourist industry predominates; therefore, the service sector but also the primary sector has some influence, with important agriculture exports, although they are in decline [26].

The average rainfall in the Canary Islands can be considered 400 mm per year, with great differences between islands and even within each island. Different areas must be considered, such as the south-facing area, with lower rainfall, compared to the north-facing area, with higher rainfall, influenced by the trade winds [27]. The humidity carried by the trade winds is used by the forest formations through a phenomenon known colloquially in the islands as the "sea of clouds", which can manifest as horizontal precipitation or mist.

The forestry sector has the potential to become a vital contributor to the global effort to mitigate climate change. This is because it can play a role in carbon sequestration by increasing stocks in forests, provided that the forest in question is undergoing expansion or increasing organic mass. Furthermore, it can serve to mitigate carbon emissions by substituting materials that result in higher CO_2 emissions with more environmentally friendly alternatives. It is, however, important to note that a forest in an ecological steady state is carbon neutral [28–30]. Its role in regulating the hydrological cycle is also key in the Canary Islands, for example, by improving the recharge of aquifers, preventing the silting up of reservoirs, reducing erosion and damage to infrastructure, and controlling flooding due to torrential rainfall. The positive effect of the mountains in regulating the horizontal precipitation or fog that occurs in the Canary Islands' forests is also worth noting [31,32].

The Anaga Biosphere Reserve (Figure 1) is located in the northwest of the island of Tenerife; it forms a well-differentiated area within the island due to its abrupt orography and because it is one of the oldest areas of the island, in geological terms [33]. The ancient edifices of Anaga, between 5 and 6 million years old (Middle Miocene-Lower Pliocene) [34] are large shield volcanic edifices, with deep ravines and cliffy coastlines. They are made up of different superimposed volcanostratigraphic sequences, mostly basaltic in composition. Likewise, the constituent materials are highly altered in those areas where later edifices have been superimposed and, in these areas, there may be intense tectonic fracturing/deformation induced by more recent volcanic activity [35].

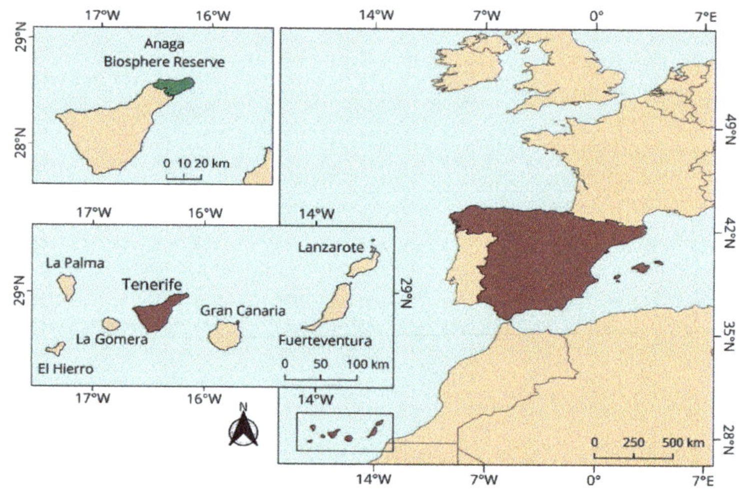

Figure 1. Location of the study area.

The climate of Anaga, as with the remainder of the Canary Islands, is influenced by its location off the northwest coast of Africa, its proximity to the tropics and the cold ocean currents. The islands are subject to the influence of the East Atlantic high-pressure area, which remains stable for the majority of the year. The trade winds, which blow primarily from the north to northeast, bring cool, moist air to the islands. This layer of moist air is situated above a drier layer, forming a phenomenon known as a 'sea of clouds'. This creates a stable atmosphere and limits convective movements.

Within the central portion of the Anaga mountain range, a wooded ridge extends over the elevated peaks of the massif, while on the northern slopes of the Anaga Massif, within the lower midlands, the terrain is characterised by a rugged valley that is exposed to the humid north winds. This environment is conducive to the development of a thermophilic forest of scattered junipers at intermediate altitudes and a dense scrubland at the upper edge, where orographic cloud formation is prevalent. The very humid atmosphere on the higher slopes and ridges of the massif, due to the orographic effect, manifests itself in the form of mist and fog on many days of the year.

This meteorological phenomenon, which includes dew and fog precipitation, plays a significant role in the annual water balance of these areas. Water occasionally flows through the ravines during autumn or winter, following heavy rainfall caused by the arrival of Atlantic depressions. Additionally, stratiform orographic clouds form in the high midlands and peaks of the massif, contributing to increased humidity in the region.

The small population residing within the Anaga Massif Biosphere Reserve is primarily concentrated in the transition zones. These zones consist of areas designated for moderate, special, and general use, as outlined in the Master Plan for the Use and Management of the Anaga Rural Park [33,36]. Anaga, the district with the largest area in the municipality of Santa Cruz de Tenerife, spans 119.55 km^2 and has a population of 11,851 inhabitants, resulting in a population density of 99.13 inhabitants/km^2. Situated within the Anaga Massif, the district's mountainous and rural character has led to a dispersed settlement in small hamlets, with a coastal strip featuring urban centres and an industrial estate. The district is predominantly encompassed by the Anaga Massif Biosphere Reserve, a protected natural area of significant environmental value. In addition, the natural beauty of the park attracts tourists from all over the world. The local population heavily relies on the tourist industry for their livelihood due to the various services required, including restaurants, accommodation, and nature-related activities provided by the industry [26,36]. This dynamic has contributed to the growth of the local economy but has also posed challenges, such as environmental protection, land conservation, and sustainable management. The objective is to attain the proper equilibrium that endorses the responsible development of tourism whilst preserving the forests and the vulnerable natural environment of Anaga.

The geological material forming Anaga is highly impermeable due to its age (in comparison with other parts of the island), alteration, and degree of compaction. The absence of recent volcanic eruptions has contributed to the lack of fissures that facilitate water recharge of the aquifers [33,35]. The permeability resulting from micro-fissures is not very significant, as they have been sealed by materials eroded from adjacent rocks, preventing the free circulation of water. In general, this type of geology does not permit significant seepage, and its internal structure inhibits the formation of independent aquifers above the basement [37,38]. Occasionally, however, hanging aquifers form at the junctions with dykes and emerge to the exterior via springs. These properties generate a terrain that promotes surface runoff and gullies, resulting in most of the conventional, horizontal rainfall finding its way to the sea.

The objective of this study is to examine the relationship between the primary attributes of Anaga—such as its geology, soils, ecosystems, and land use patterns—and the quality of its water resources. Anaga is distinguished by its designation as a Biosphere Reserve, its unique soil features, its relatively low population density, and its agricultural practices that are less intensive than those observed in other regions of the island. The park's status serves to restrict activities that could have a deleterious effect on the soil and aquifers, thereby reducing the risk of pollution. The following sections present a comprehensive analysis of the groundwater quality data for the Anaga Rural Park. In this context, the challenges, health risks and opportunities for sustainable water management in the area are discussed in detail.

2. Materials and Methods

The proposed methodology is aimed at establishing a relationship between the characteristics of the Anaga region and the quality of the water in the area. As previously stated, the area in question is situated in the northern region of the island of Tenerife and has been designated a Biosphere Reserve. This protected zone exemplifies the significance of land management in the preservation of water quality. Additionally, this region is distinguished both by a relatively low population density and a lack of industrial activity, while agricultural activity has become a supplementary activity. It is anticipated that these conditions will be demonstrated to be beneficial in terms of promoting good water quality.

The water samples analysed in the present study were exclusively sourced from groundwater, as the Anaga basin is entirely reliant on this resource for its water supply, with no alternative sources available. The region's water supply is entirely dependent on groundwater, which is sufficient to meet the needs of the local population. Groundwater extraction in Anaga is facilitated by traditional horizontal galleries, a distinctive and unique feature in the Canary Islands. These galleries, which are horizontal tunnels dug into the mountainous terrain, intercept aquifers formed by basaltic fissures and fractures, thereby allowing the natural flow of groundwater without the need for active pumping. In addition to the galleries, small-scale boreholes are sometimes employed to supplement water collection from the galleries. The galleries are engineered to collect and convey water to storage tanks or directly into the distribution system. This method of groundwater management, adapted to the region's geology and topography, is particularly efficient and helps prevent issues such as saline intrusion.

Samples were taken from two categories of reservoir: headwater and distribution, as well as from the water supply network following treatment. The six sampling points within the area of interest (Anaga) encompassed Almaciga; Ijuana I, II, III; Igueste de San Andrés; Pico Inglés; Roque Negro I–II and Taganana (approximate locations shown in Figure 2).

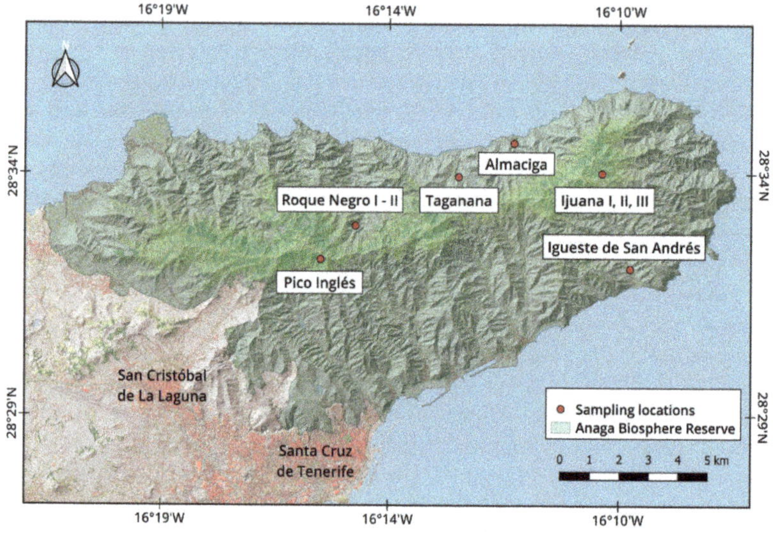

Figure 2. Sampling locations for water quality analysis, Anaga Biosphere Reserve (Tenerife, Canary Islands).

Empresa Mixta de Aguas de Santa Cruz de Tenerife SA (EMMASA) is responsible for the water supply in Santa Cruz de Tenerife, which includes the designated study area.

The company is responsible for the disinfection and distribution of the water, which is subjected to rigorous quality controls.

In addition, EMMASA's laboratory possesses the international accreditation UNE-EN ISO/IEC 17025, bestowed by the National Accreditation Entity (ENAC), which acknowledges the laboratory's technical expertise in performing environmental tests. The laboratory has the necessary infrastructure, technology and personnel to analyse water throughout the entire cycle, facilitating efficient management.

The company made data from these monitoring analyses available for this research project for the 10-year period from 2007 to 2016.

The sampling points were selected in accordance with the guidelines laid down in Royal Decree 3/2023 of Ministerio de la Presidencia, [39]: (i) At least one sampling point has been installed at the outlet of the treatment plant. If this is not possible, the sampling point shall be located at the outlet of the collecting tank associated with the treatment plant. (ii) An additional sampling point has been located at the outlet of each regulating or distribution reservoir within each supply zone. (iii) At least one sampling point has been installed at each of the existing distribution points in the supply zone. (iv) One sampling point has been established in each distribution network, with an additional point being added for each 20,000 m^3/day increase in network capacity. (v) Within each supply zone, as many distribution networks have been identified as there are regulating or distribution reservoirs. (vi) In cases where a tank supplies more than one population unit or nucleus, the sampling point was located in the unit or nucleus with the largest population.

In order to characterise the groundwater quality of the study area, a total of 406 samples were analysed. This dataset included 347 control analyses with 14 measured parameters: free residual chlorine, in-situ chlorine, coliform bacteria, colony count at 22 °C, *Escherichia coli*, laboratory turbidity, in-situ turbidity, ammonium, nitrate, pH, conductivity, colour, odour and taste. These parameters were selected based on their relevance to monitoring drinking water quality in accordance with the mandatory requirements established by Spanish legislation (Royal Decree 2/2023) and the European Union's Drinking Water Directive. They are critical indicators for assessing water safety, pollution risks, and treatment efficiency. Furthermore, 60 complete analyses with 24 measured parameters, the 14 parameters mentioned above, and 10 additional ones, including residual combined chlorine, chloride, aluminium, iron, manganese, sodium, sulphate, oxidisability, Enterococcus and Clostridium perfringens. Although heavy metals were not included in the control analyses, some, such as aluminium and iron, were assessed as part of the complete analyses to provide a broader characterization of the water quality. This approach ensured that the study focused on key indicators of potability while also incorporating additional parameters to capture a more comprehensive understanding of water quality in specific cases.

The study analysed 347 water control samples collected between 2007 and 2016, except for a bacteriological control sample from Almáciga (5/2011). The analysis of the data revealed that 53% of the samples originated from the head tank (185), while the remaining 47% were drawn from the distribution network and reservoirs (162). For each location, between 28 and 76 data sets were collected over the period, with between 19 and 46 sets per year for all locations combined. The number of measurements per location ranged from 1 to 11 per year, with an average of 6 per year, except at Pico Inglés, where no measurements were taken in 2015.

This information serves to illustrate the temporal and spatial scope of the data, thereby underscoring its relevance as a baseline for future analysis and monitoring.

The distribution of the samples according to year and location is shown in the following table (Table 1).

Table 1. Distribution of number of samples by year and location.

Sampling Point	2007	2008	2009	2010	2011	2012	2013	2014	2015	2016	$\sum$
ALMACIGA	5	4	6	7	10	5	6	8	7	6	64
IJUANA	2	3	5	2	4	2	2	3	3	2	28
PICO INGLES	7	1	8	9	9	7	9	11	-	9	70
ROQUE NEGRO	3	3	3	4	5	2	5	6	4	4	39
SAN ANDRES	4	3	8	9	9	6	7	8	9	7	70
TAGANANA	5	5	6	8	9	7	9	10	9	8	76
$\sum$	26	19	36	39	46	29	38	46	32	36	347

The data were then analysed, considering the maximum permissible values of the applicable regulations for the study area. Specifically, they are: (i) RD 3/2023 [39] and (ii) The Health Monitoring Programme for Water Intended for Human Consumption in the Autonomous Community of the Canary Islands [40].

Furthermore, it has been concluded that it is not necessary to calculate the Water Quality Index, since the limits applicable under Royal Decree 3/2023 are stricter than those established by the methodology used to calculate the WQI, after the preliminary results of the samples.

For the purpose of facilitating the analysis, the data were classified into three groups of parameters: (i) microbial parameters, including coliform bacteria, colony count at 22 °C, *Escherichia coli*, Enterococci and Clostridium perfringens, (ii) indicator parameters, including residual free chlorine, residual combined chlorine, in-situ chlorine, laboratory turbidity, in-situ turbidity, pH, conductivity, colour, odour, taste and oxidisability and (iii) chemical parameters, which include nitrate, ammonium, sulphate, chloride, sodium, aluminium, iron, and manganese.

The study has been constrained by the unavailability of more contemporary data. However, it has utilised all records available for the period 2007–2016, which represent the most robust evidence for the analysis of water quality conditions at that time. These limitations emphasise the necessity for the establishment of more consistent monitoring systems with open access to data but do not detract from the relevance of the findings as a basis for further research.

3. Results and Discussion

3.1. Microbial Parameters

Starting with the microbial parameters, the presence of coliform bacteria is the most commonly used indicator of disease-causing bacteria in water [41]. It is associated with improper maintenance of water distribution networks and indoor installation, as coliforms can survive and multiply within water distribution systems; water is only safe to drink if the test for coliform bacteria is negative. It is important to note that other sources of coliform bacteria, such as *Escherichia coli*, include animals, particularly through fecal pollution from livestock, wildlife, or domestic pets. Similarly, Enterococci species can originate from animal feces, serving as indicators of fecal pollution in water systems. Clostridium perfringens, a spore-forming bacterium, may also have diverse origins, including animals, decaying vegetation, and soil, thereby highlighting its natural presence in the environment and its potential to indicate pollution. Over a period of 10 years in the Anaga area, coliform bacteria were found in only two samples. All tests for *Escherichia coli*, Enterococci and Clostridium perfringens were negative. These parameters normally are indicative of wastewater pollution or contact with faeces in the system, so its presence would have posed a risk to public health [42,43]. Therefore, the absence of these indicators in the analyses suggests that the treatment system that comprises the Anaga Rural Park is adequately designed and dimensioned to prevent this type of pollution, thus guaranteeing water quality and safety for users. The colony count at 22 °C is the total number of culturable bacteria in a given sample and gives a general idea of how polluted the water is. The maximum allowable value is 100 ufc/1 mL and the highest value measured in any sample was 87 ufc/1 mL. In addition, 62% of the samples had a value of 0, 26% of the samples had values between 1 and 4 and 10% of the samples had values between 10 and 50 ufc/1 mL. After 2011, the colony count at 22 °C did not exceed 10 for all sampling sites. In the first part of the analysis, there is no reason to worry about the microbial parameters, as most of the results are positive in terms of the limit of detection values.

3.2. Indicator Parameters

Moving on to the second classification, statistical analysis of indicator parameters of all samples is shown in Table 2.

Table 2. Statistical analysis of potable water quality indicator parameters and its coherence with RD 3/2023 and The Health Monitoring Programme for Water Intended for Human Consumption in the Autonomous Community of the Canary Islands regulations.

PARAMETER (Unit)	Min	Max	Range	Median	Mean	SD	CV%	Limit (RD 3/2023)	% Missing Values
pH	6.87	8.97	2.1	8.31	8.29	0.26	3	6.5–9.5	0.3
Conductivity (µs/cm)	277	963	686	401	421.6	97.89	23	2500	0.3
oxidisability (mg O_2/L)	0.1 *	1.76	1.66	0.4	0.55	0.35	63	5	86
turbidity in-situ (FNU)	0.05 *	6.2	6.15	0.3	0.42	0.49	119	1	0
laboratory turbidity (NTU)	0.05 *	0.89	0.84	0.19	0.23	0.16	69	5	73
color (mg Pt-Co/L)	0.1 *	13.9	13.8	0.8	1.19	1.51	127	15	0.5
odour (in. dil)	0	1	1	-	-	-	-	3	0.3
flavour (in. dil.)	0	1	1	-	-	-	-	3	0.5
chlorine in-situ (mg/L)	0.2	1	0.8	0.8	0.73	0.14	20	0.2–1	3
residual free chlorine (mg/L)	0.05	0.99	0.94	0.68	0.64	0.18	28	0.2–1	70
residual combines chlorine (mg/L)	0	0.58	0.58	0.025	0.053	0.09	172	2	86

* $0.1 \leq 0.2$; $0.05 \leq 0.1$; [CV% of mean = 1 SD/mean × 100]; [% of missing values = number of samples without measured value/406 × 100].

Although pH does not normally have a direct effect on the consumer, it is one of the most important operational parameters of water quality [7]. Attention must be paid to pH control at all stages of water treatment to ensure satisfactory clarification and disinfection of water. It is known that pH can be affected by dissolved minerals and chemicals. All pH values measured were within acceptable limits, as can be seen in the following figure (Figure 3). In addition, it was observed that the Ijuana sampling location tended to have the lowest pH values. The samples with the highest pH values also showed the highest conductivity measurements [44,45]. Conductivity is one of the most sensitive indicators for detecting possible external pollution in the distribution network; comparing the conductivity at different points in the network can determine whether an indoor installation is well maintained. Conductivity measures the ability of water to conduct an electrical current, where dissolved salts and inorganic chemicals contribute to electrical conductivity. All conductivity values measured were within acceptable limits. Samples from the Pico Inglés tended to have the lowest values, while samples from San Andres tended to have the highest values. This trend may be due to the fact that Pico Inglés is at a higher altitude and San Andres is located on the coast. Surface and groundwater from precipitation have more time to dissolve the solids they come into contact with as they pass through terrestrial materials [46].

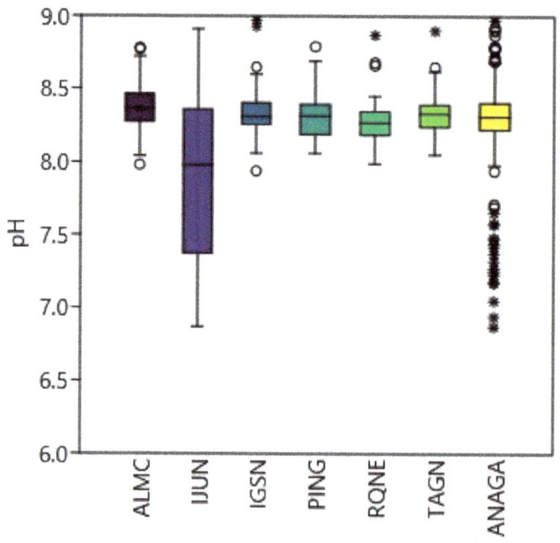

Figure 3. The pH values of the samples for every location and Anaga as a whole during the study period (2007–2016).

Turbidity is caused by the presence of suspended particles in the water. The particles responsible for turbidity vary in size from 1 nm to 1 mm and are mostly due to erosion of the soil surface [47]. High turbidity levels are commonly associated with poor drinking water quality and also interfere with the disinfection process, reducing its effectiveness [48]. Turbidity was measured both in situ and in the laboratory; the measurement can be seen in the next figure (Figure 4). In situ turbidity measurements were conducted using sensors installed at various sampling points, while laboratory measurements followed standardized analytical protocols. It is important to note that turbidity values obtained in the field are expressed in FNU (Formazin Nephelometric Unit), in accordance with the European ISO 7027 standard, whereas laboratory measurements are typically expressed in NTU (Nephelometric Turbidity Unit), as specified by the USEPA Method 180.1 or the Standard Methods for the Examination of Water and Wastewater. There were four samples with in situ turbidity levels exceeding the limit in 2007 and 2009. Of these, two samples were from the Ijuana location, which also tended to have the highest turbidity levels in the laboratory. The observed differences between field and laboratory measurements can be attributed to the distinct methodologies, equipment used and potential sample handling effects during laboratory analysis.

Ideally, drinking water should have no colour. The colour of the water is due to coloured organic matter (humic and fulvic acids) and the presence of iron or manganese. In drinking water, colour can be caused by the dissolution of iron or copper in indoor installations. If every value does not exceed the limitation, water quality can be considered to be good, in terms of this parameter. The variability is shown in the following figure (Figure 5).

The values obtained for odour and flavour were 0 or 1 (in dilution) for all samples. Therefore, no statistical parameters were calculated. An odour value of 1 was detected in 11% of the samples; a flavour value of 1 was also found in 11% of the samples. With the exception of one sample, all samples with a flavour value of 1 also had an odour value of 1. Most of the samples with odour and flavour values of 1 were collected at the Pico Inglés sampling location. The colour measurements with the highest values were mainly from the year 2009 and from the Ijuana sampling location. There are several factors that can cause a perceived change in the odour and/or taste of the water for the consumer. The most common causes are: (i) natural compounds related to the origin of the water; (ii) reagents used in the drinking water treatment process or by-products generated during the process; (iii) materials used in pipes, assemblies and installations; (iv) polluting discharge; and (v) long residence time of the water in the network.

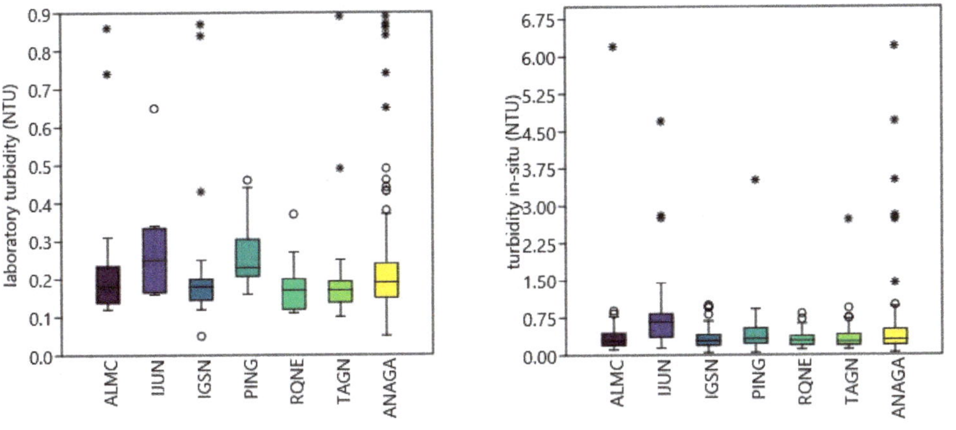

(a) (b)

Figure 4. The laboratory turbidity (**a**) and turbidity in situ and (**b**) values of the samples for every location and Anaga as a whole during the study period (2007–2016).

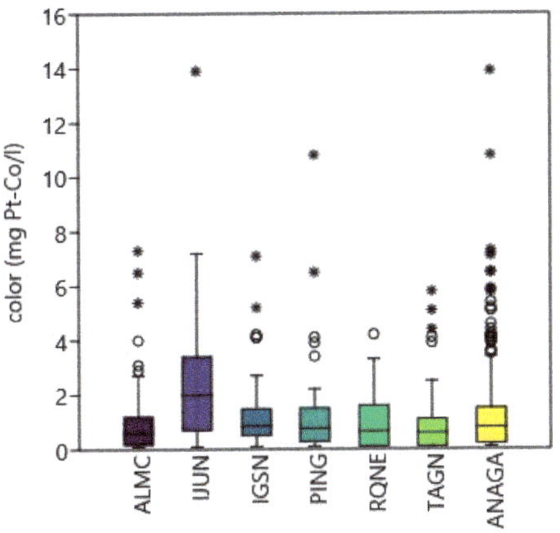

Figure 5. The color values of the samples for every location and Anaga as a whole during the study period (2007–2016).

In a water distribution system that uses chlorination as a disinfection treatment, it is necessary to measure residual chlorine to ensure that the system remains bacteria-free. A decrease in residual chlorine may indicate an increase in pollutants in the system that reacted with the residual chlorine. There were three samples with free residual chlorine values below the limit, all from 2007 (March–June) and located in Almaciga and Taganana. However, all the other parameters measured in these samples were within the permitted limits.

The presence of chloride in drinking water is due to natural causes, industrial effluents and marine intrusion, among others [49–51]. Excessive levels of chloride increase the corrosion of metals in pipes, depending on the alkalinity of the water. The WHO has not proposed any reference values for chloride in drinking water from a health point of view, but high concentrations of chloride can cause a detectable taste. However, the available data are insufficient to draw definitive conclusions regarding residual free chlorine and residual combined chlorine. Specifically, 349 samples were below the detection limit of the method and thus did not show detectable levels. This substantial number of non-detections is consistent with the characteristics of the water supply system in the study area, where chlorine levels dissipate rapidly due to factors such as the small size of the system, low storage volumes and limited distribution distances. These non-detections reflect the dynamics of chlorine behaviour in the system rather than sampling or analytical deficiencies. The values for these parameters can be seen in the following figure (Figure 6), specifically for each location, and a total of all the samples represented in yellow.

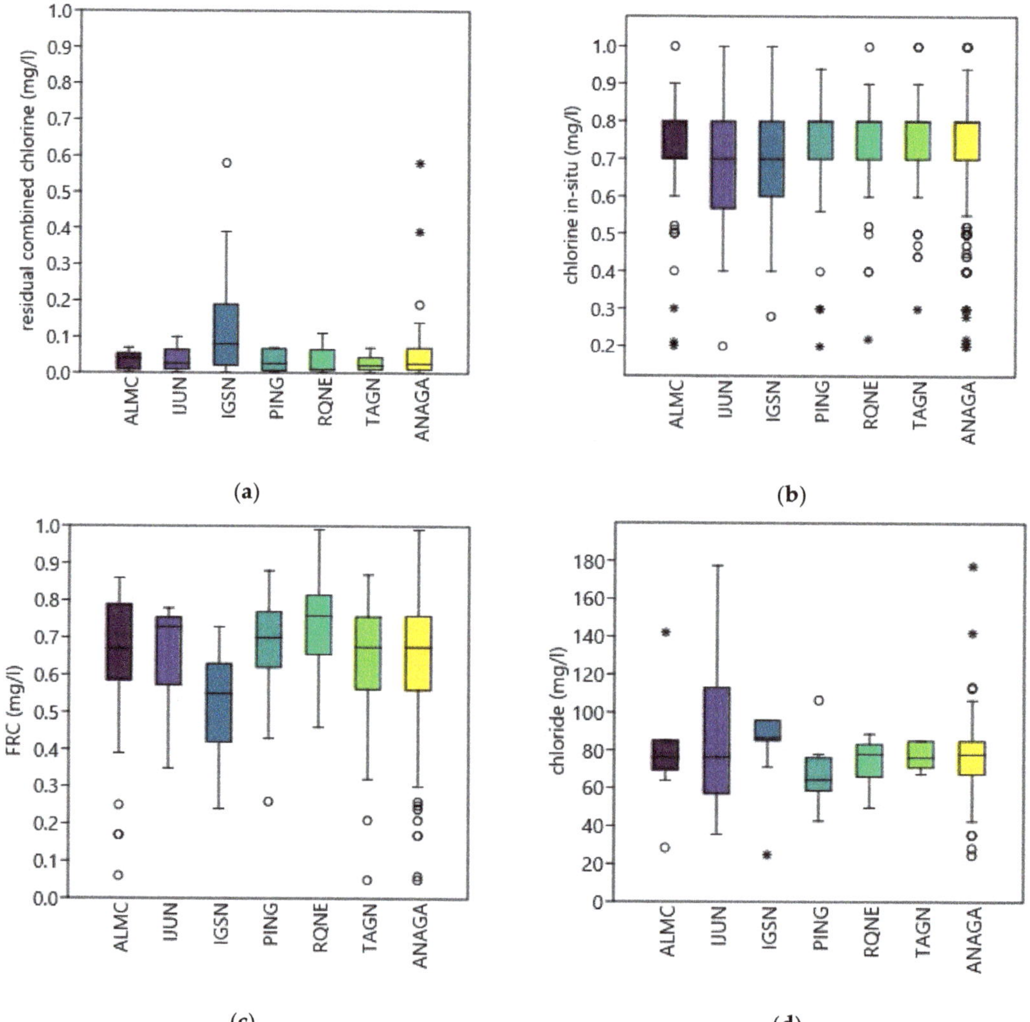

Figure 6. The residual combined chlorine (**a**), chlorine in-situ (**b**), free residual chlorine (**c**) and chloride (**d**) values of the samples for every location and Anaga as a whole during the study period (2007–2016).

3.3. Chemical Parameters

The chemical parameters statistical analysis is shown in Table 3.

Table 3. Statistical analysis of potable water quality chemical parameters and its coherence with RD 3/2023 and The Health Monitoring Programme for Water Intended for Human Consumption in the Autonomous Community of the Canary Islands regulations.

PARAMETER (Unit)	Min	Max	Range	Median	Mean	SD	CV%	Limit (RD 3/2023)	% Missing Values
NO_3^- (mg/L)	0.25 *	25.1	24.85	5.1	5.58	3.72	66.76	50	41
SO_4^{2-} (mg/L)	2.5 *	31	28.5	14.5	15.47	6.08	39.30	250	86
Cl^- (mg/L)	24.8	177.5	152.7	78.1	78.30	23.99	30.63	250	86
NH_4^+ (mg/L)	0.005 *	0.1	0.09	0.005 *	0.02	0.02	119.08	0.5	15
Na (mg/L)	26.25	113.3	87.05	44.02	51.01	20.13	39.46	200	86
Al (µg/L)	10 *	129	119	10	12.05	15.49	128.53	200	86
Fe (µg/L)	10 *	167	157	22.6	32.2	33.1	103	200	86
Mn (µg/L)	2.5 *	12.2	9.7	2.5 *	2.78	1.53	54.90	50	86

* 41 samples had a nitrate value <2 → transformed to 1 for statistics (2 samples had <0.5 → 0.25); 1 sample had sulfate value <5 → 2.5; 216 samples had ammonium value <0.1 → 0.005; 57 samples had aluminum value <20 →10; 28 samples had iron value <20 → 10; 56 samples had manganese value <5 → 2.5.

Ammonium is present in raw water as a result of agriculture, industry and chlorination. The presence of high levels of ammonium can compromise the effectiveness of disinfection or fail to remove manganese in filters, causing taste and odour problems. The presence of ammonium can be an indicator of faecal, agricultural or industrial pollution [52,53]. The analysis of this parameter can be seen in the next figure (Figure 7). It was observed that the ammonium measurements in the samples were generally higher in 2010 (April–August). In July 2009, a sample from Ijuana had the highest ammonium value (0.25 mg/L). However, five days later the ammonium value at this sampling site was below the detection limit (<0.01 mg/L). In these two samples, it can be observed that, as the ammonium value decreases, the in situ chlorine value also decreases and the pH value increases. The sample with the highest ammonium value had one of the lowest sulphate values.

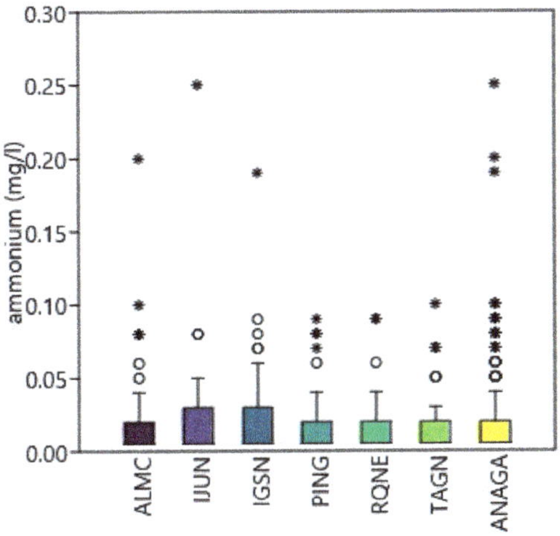

Figure 7. Ammonium values of the samples for every location and Anaga as a whole during the study period (2007–2016).

Sulphates enter the water from industrial waste and atmospheric precipitation, but the highest concentrations are usually found in groundwater and come from natural sources [54,55]. The sulphate measurements in the samples were well below the allowable limit, shown in the following figure (Figure 8). The sample with the lowest sulphate value had one of the highest iron values. It should be noted that, in May 2010, the sulphate values were highest at different sites (Almaciga and Roque Negro), and these samples also had one of the highest nitrate values.

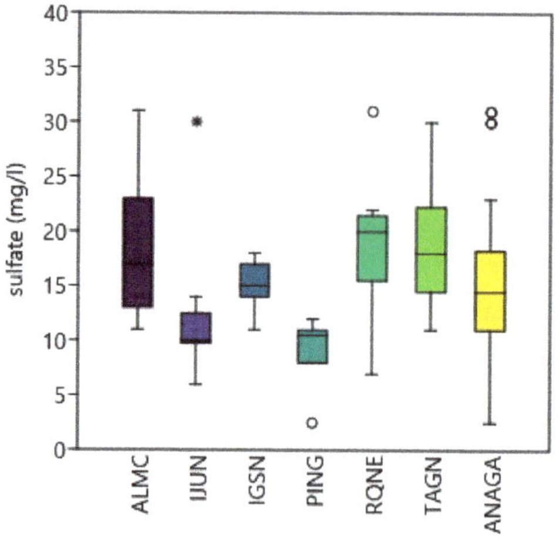

Figure 8. Sulphate values of the samples for every location and Anaga as a whole during the study period (2007–2016).

Nitrate is mainly used in inorganic fertilisers [56–60]. The toxicity of nitrate in humans is attributed to its reduction to nitrite. The main health risk is methaemoglobinaemia in infants, leading to cyanosis and, only at higher concentrations, asphyxia. Other risks from prolonged exposure have been associated with stomach cancer, although there is no evidence of a causal relationship. This is consistent with the conclusion of the IARC, which has classified nitrate and nitrite intakes under conditions leading to endogenous nitrosation in Group 2A (probably carcinogenic to humans) but not nitrate alone. Nitrate measurements in the samples were below the permitted limit and 18% of the samples had values below the detection limit, represented in the next figure (Figure 9). The lowest values were recorded in June 2010 and the highest values in April and May 2010 at all sampling sites. The highest value measured was in March 2012 in a sample from San Andres, which also had one of the highest pH and conductivity values. Samples from other sites (Pico Inglés and Taganana) had significantly higher nitrate values in March 2012 compared to April 2012, which were below the detection limit.

The presence of aluminium in drinking water is mainly due to the use of aluminium salts in drinking water treatment, in the flocculation-coagulation stage. A high residual concentration can give the water an undesirable colour and turbidity. Iron is one of the most abundant metals in the Earth's crust. Iron can also be present in drinking water due to the use of iron coagulants or corrosion of steel or cast-iron pipes during water distribution [61]. The taste and appearance of drinking water may be affected by the presence of iron below 2 mg/L. The WHO has not proposed reference values for iron in drinking water. Manganese is a commonly occurring metal in the Earth's crust, although it is less abundant than elements such as magnesium, calcium and potassium. In highly oxygenated water, deposits of manganese compounds can form, causing colour problems in the water. At levels above 0.1 mg/L, manganese in drinking water can cause an unpleasant taste and stains in laundry. Sodium is present in virtually all foods and this is the main source of exposure. The WHO has not proposed reference values for sodium in drinking water from a health point of view. However, concentrations above 200 mg/L may cause unacceptable taste. No firm conclusions can be drawn from these parameters, as we could not obtain further values to analyse from the samples collected. Nevertheless, none of the measurement values are in excess of the limit value in the regulations. The values for sodium and iron of all the samples are represented in the following figure (Figure 10).

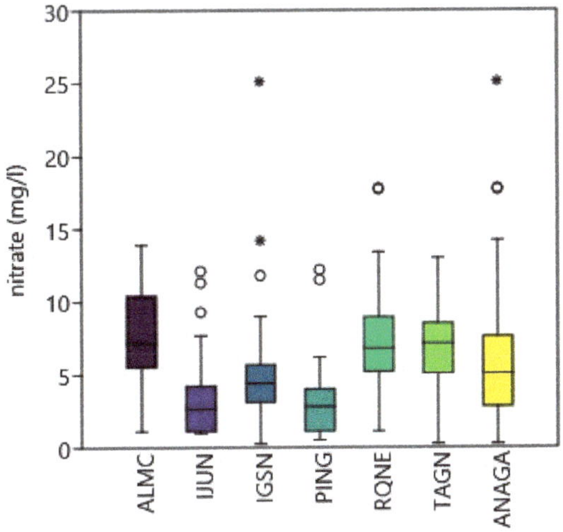

Figure 9. Nitrate values of the samples for every location and Anaga as a whole during the study period (2007–2016).

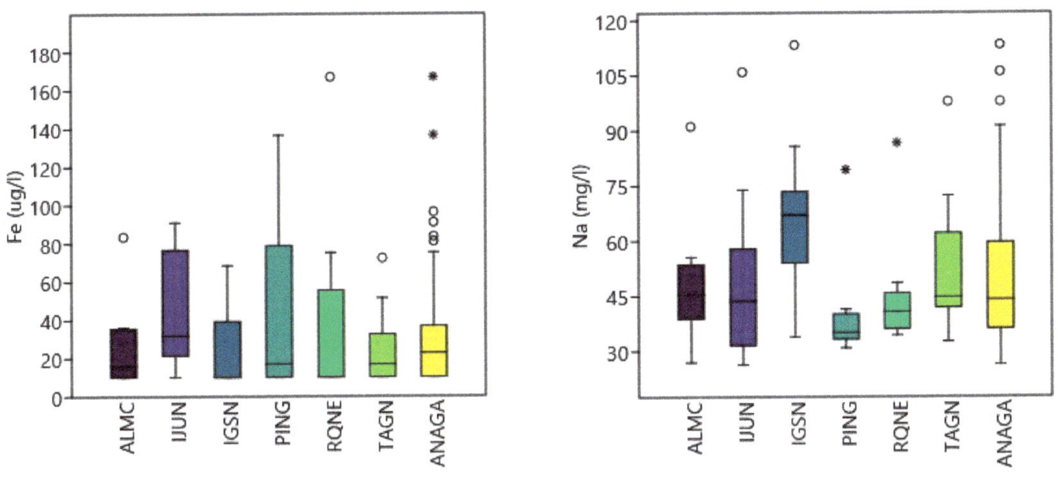

(a) (b)

Figure 10. The Fe (**a**) and Na (**b**) values of the samples for every location and Anaga as a whole during the study period (2007–2016).

Almost all samples analysed over this 10-year period had measured parameter values within the permitted limit, but there are two samples that stand out in terms of water quality, compared to the others.

In December 2007, a sample from Ijuana tested positive for coliform bacteria and had one of the highest colony counts at 22 °C (23 cfu/1 mL). The measurement of aluminium in this sample was 129 µg/L and was the only one not below the detection limit. This sample also had the highest or one of the highest values for colour (13.9 mg Pt-Co/L), in situ chlorine (1 mg/L), sodium (105.9 mg/L), oxidisability (1.68 mg O_2/L), in situ turbidity (4.7 UNF), and manganese (12.2 µg/L). It had one of the lowest pH values (7.25), although the odour and taste values in the dilution were 0.

In November 2009, a sample from Ijuana had the only other manganese measurement not below the detection limit (9.29 µg/L), the highest chloride (177.5 mg/L) and conductivity (809 µs/cm), and one of the highest in situ turbidity values (1.45 UNF), with odour and taste values of 0 in dilution.

As previously mentioned, soil conservation plays a vital role in preserving natural habitats and maintaining sustainable use of resources. This is particularly evident in the study area, where protecting soil has had a positive impact on various environmental and ecological factors that are essential for biodiversity preservation.

Furthermore, the implementation of soil conservation measures in this region has led to a noteworthy reduction in soil erosion, thus preventing the depletion of fertile soil and subsequent degradation of habitats. In addition, the maintenance of healthy soil serves a pivotal role in mitigating climate change by serving as a carbon sink. The deposition of organic carbon in soils helps decrease the concentration of atmospheric carbon dioxide, consequently reducing the impact of global warming.

Additionally, the forest sector and safeguarding soil in the Anaga Massif have favourable impacts on the region's ability to maintain crops and vegetation over an extended period. It is crucial to preserve soil productivity for agricultural output and maintenance of terrestrial ecosystems. Moreover, this safeguard averts unsystematic urbanization, preserving natural scenery and verdant areas in an urban setting. This is crucial for developing ecotourism and fostering appreciation of nature for locals and visitors [12,62].

While the study highlights the positive outcomes of the reserve and its influence on water quality, it is important to acknowledge the significant challenges that the management of these areas is currently facing. Ecotourism, when managed effectively, can contribute to sustainability. However, it can also have adverse effects on ecosystems, such as habitat alteration or increased pollutants from tourism infrastructure. These factors have the potential to undermine the conservation achievements and water quality gains that have been made in the long term [60,63].

Primarily, soils are crucial for water quality, functioning as innate filters as water seeps through them, purifying water by detaching impurities and pollutants. This process is vital for maintaining water quality in streams and aquifers within the Anaga Biosphere Reserve. This, in turn, benefits the aquatic ecosystems and availability of freshwater for local flora, fauna and people. The study results confirm that the water was in a good state during this period. Protection of the soil, the hydrogeology of the area, and the remarkable maintenance of facilities are responsible for these commendable results.

4. Conclusions

In conclusion, this study provides a comprehensive overview of groundwater quality in the Anaga Rural Park and its importance for the sustainable management of this exceptional ecosystem. The designation of the area as a Biosphere Reserve has been pivotal in the protection of water quality, as it has enabled the implementation of rigorous regulations that, although restrictive, have proven to be effective. The distinctive characteristics of the soil, in conjunction with the reserve status and the associated safeguards, have been a significant factor in the conservation of this natural area. This has led to the preservation of biodiversity, an enhancement of water quality, climate change mitigation, the maintenance of agricultural sustainability and the preservation of natural landscapes. These outcomes have been achieved within the context of sustainable development and environmental preservation.

During the course of the research, the overwhelming majority of the samples analysed were found to comply with the rigorous limits and quality standards prescribed by the regulations currently in force in the region. The analysis of microbial parameters indicates that the maintenance practices in the distribution networks are adequate, as no coliform bacteria or signs of sewage pollution were found. This ensures that there is no threat to public health.

The evaluation of water quality using the indicator parameters revealed that the pH values were within the acceptable range in all samples, with a slight tendency towards lower values at the Ijuana site. Furthermore, the conductivity values were within the acceptable range, although the samples from Pico Inglés exhibited lower values and those from San Andrés demonstrated higher values. The presence of suspended particles resulted in turbidity levels exceeding the permitted threshold, with the highest concentrations observed at the Ijuana site. However, the water colour remained within the acceptable range, indicating that the water quality was satisfactory. While the majority of samples exhibited satisfactory levels of odor and taste, some, particularly those from Pico Inglés, displayed aberrant characteristics.

Three samples exhibited a chlorine residual below the established threshold, indicating a potential reduction in water disinfection efficacy. However, no other noteworthy quality issues were identified. The presence of chloride was deemed to be of natural origin and did not indicate concentrations that would pose a public health concern. However, the lack of sufficient data precludes definitive conclusions regarding free and combined chlorine parameters.

The Anaga Rural Park attracts tourists from across the globe and relies on the local population to provide tourist services. Preserving the environment in this context of increasing tourism is a challenge, as the influx can negatively affect regional culture and biodiversity. Therefore, effective sustainable management, especially through constant monitoring, is crucial to safeguard the park's essential natural and cultural heritage.

The present study provides a robust foundation for future research on sustainable groundwater management in protected areas. The results demonstrate the efficacy of conservation measures and geological characteristics in maintaining water quality, while also underscoring the complexities associated with the management of ecotourism areas.

The findings will serve as a reference point for monitoring long-term changes and for developing strategies to balance the benefits of ecotourism with the need to preserve vulnerable ecosystems, thereby ensuring water and environmental sustainability in the region.

Author Contributions: Conceptualization, N.C.-P. and A.G.-G.; Data curation, J.K.; Formal analysis, J.S.R.-A.; Investigation, J.S.R.-A. and J.K.; Methodology, J.S.R.-A. and J.C.S.; Supervision, J.C.S.; Validation, J.R.-M. and A.G.-G.; Writing—original draft, N.C.-P. and J.K.; Writing—review and editing, J.S.R.-A. and J.R.-M. All authors will be updated at each stage of manuscript processing, including submission, revision, and revision reminder, via emails from our system or the assigned Assistant Editor. All authors have read and agreed to the published version of the manuscript.

Funding: This research was supported by the European Union's Horizon 2020 Research and Innovation Program under grant agreement 101037424, project ARSINOE (Climate-resilient regions through systemic solutions and innovations).

Data Availability Statement: Some or all data, models or codes that support the findings of this study are available from the corresponding author upon reasonable request.

Conflicts of Interest: The authors declare no conflicts of interest.

References

1. Unigwe, C.O.; Egbueri, J.C. Drinking water quality assessment based on statistical analysis and three water quality indices (MWQI, IWQI and EWQI): A case study. *Environ. Dev. Sustain.* **2023**, *25*, 686–707. [CrossRef]
2. Stričević, L.; Pavlović, M.; Filipović, I.; Radivojević, A.; Martić Bursać, N.; Gocić, M. Statistical analysis of water quality parameters in the basin of the Nišava River (Serbia) in the period 2009–2018. *Geografie* **2021**, *126*, 55–73. [CrossRef]
3. Llamas, M.R.; Custodio, E.; de la Hera, A.; Fornés, J.M. Groundwater in Spain: Increasing role, evolution, present and future. *Environ. Earth Sci.* **2015**, *73*, 2567–2578. [CrossRef]
4. Teatini, P.; Ferronato, M.; Gambolati, G.; Gonella, M. Groundwater pumping and land subsidence in the Emilia-Romagna coastland, Italy: Modeling the past occurrence and the future trend. *Water Resour. Res.* **2006**, *42*, 1–16. [CrossRef]
5. Hussein, H.A.; Al Baidhani, J.H.; Alshammari, M.H. Evaluation the effects of some parameters on the operational efficiency of the main water pipe in Karbala city. *J. Phys. Conf. Ser.* **2021**, *1973*, 012118. [CrossRef]
6. Parizi, E.; Hosseini, S.M.; Ataie-Ashtiani, B.; Simmons, C.T. Vulnerability mapping of coastal aquifers to seawater intrusion: Review, development and application. *J. Hydrol.* **2019**, *570*, 555–573. [CrossRef]
7. World Health Organization. Guidelines for Drinking-Water Quality: Small Water Supplies. 2024. Available online: https://www.who.int/publications/i/item/9789240088740 (accessed on 10 January 2025).
8. Gibb, J.P. Water Quality and Treatment of Domestic Groundwater Supplies. *Circular 118* **1973**, *State of Illinois, Department of Registration and Education*. Available online: https://www.isws.illinois.edu/pubdoc/C/ISWSC-118.pdf (accessed on 10 January 2025).
9. Santamarta Cerezal, J.C. Hidrología y Recursos Hídricos en Islas y Terrenos Volcánicos: Métodos, Técnicas y Experiencias en las Islas Canarias. *Col. Ing. Montes* **2013**, *1*, 227–245. [CrossRef]
10. García-Gil, A.; Poncela Poncela, R.; Skupien Balon, E.; Morales González-Moro, Á.; Lario-Báscones, R.J.; Marazuela, M.Á.; Cruz-Pérez, N.; Santamarta, J.C. Heterogeneity-Driven Hydrodynamics Conditions the Hydrochemistry of Spring Water in Volcanic Islands. *Groundwater* **2023**, *61*, 375–388. [CrossRef]
11. Marazuela, M.Á.; Baquedano, C.; Cruz-Pérez, N.; Martínez-León, J.; Laspidou, C.; Santamarta, J.C.; García-Gil, A. Dyke-impounded fresh groundwater resources in coastal and island volcanic aquifers: Learning from the Canary Islands (Spain). *Sci. Total Environ.* **2023**, *899*, 165638. [CrossRef]
12. Piñar Álvarez, Á.; Nava Tablada, M.E.; Viñas Oliva, D.K. Migración y ecoturismo en la Reserva de la Biosfera de Los Tuxtlas (México). *PASOS. Rev. De Tur. Y Patrim. Cult.* **2011**, *9*, 383–396. [CrossRef]
13. Brennan, A.; Naidoo, R.; Greenstreet, L.; Mehrabi, Z.; Ramankutty, N.; Kremen, C. Functional connectivity of the world's protected areas. *Science* **2022**, *376*, 1101–1104. [CrossRef]
14. Hernandez, Y.; Guimarães Pereira, Â.; Barbosa, P. Resilient futures of a small island: A participatory approach in Tenerife (Canary Islands) to address climate change. *Environ. Sci. Policy* **2022**, *80*, 28–37. [CrossRef]
15. European Commission. EU Soil Strategy for 2030. 2021. Available online: https://www.eea.europa.eu/data-and-maps/dashboards/land-take-statistics#tab-based-on-data (accessed on 21 February 2024).
16. Cole, S. A political ecology of water equity and tourism: A Case Study From Bali. *Ann. Tour. Res.* **2012**, *39*, 1221–1241. [CrossRef]
17. Hall, C.M.; Gössling, S. *Tourism and Global Environmental Change*; Routledge: London, UK; New York, NY, USA, 2006.
18. Gössling, S.; Peeters, P. Assessing tourism's global environmental impact 1900–2050. *J. Sustain. Tour.* **2015**, *23*, 1–21. [CrossRef]
19. Kairis, O.; Karavitis, C.; Kounalaki, A.; Salvati, L.; Kosmas, C. The effect of land management practices on soil erosion and land desertification in an olive grove. *Soil Use Manag.* **2013**, *29*, 597–606. [CrossRef]
20. Malik, A.A.; Puissant, J.; Buckeridge, K.M.; Goodall, T.; Jehmlich, N.; Chowdhury, S.; Gweon, H.S.; Peyton, J.M.; Mason, K.E.; van Agtmaal, M.; et al. Land use driven change in soil pH affects microbial carbon cycling processes. *Nat. Commun.* **2018**, *9*, 3591. [CrossRef] [PubMed]
21. Santamarta, J.C.; Rodríguez-Martín, J.; Poncela, R.; Fontes, J.C.; Lobo de Pina, A.; Cruz-Pérez, N. Integrated Water Resource Management in the Macaronesia. *Int. Rev. Civ. Eng.* **2022**, *13*, 290. [CrossRef]
22. Tesfahunegn, G.B.; Vlek, P.L.G.; Tamene, L. Management strategies for reducing soil degradation through modeling in a GIS environment in northern Ethiopia catchment. *Nutr. Cycl. Agroecosystems* **2012**, *92*, 255–272. [CrossRef]
23. Kordas, R.L.; Harley, C.D.G.; O'Connor, M.I. Community ecology in a warming world: The influence of temperature on interspecific interactions in marine systems. *J. Exp. Mar. Biol. Ecol.* **2011**, *400*, 218–226. [CrossRef]
24. McLeod, E.; Salm, R.; Green, A.; Almany, J. Designing marine protected area networks to address the impacts of climate change. *Front. Ecol. Environ.* **2009**, *7*, 362–370. [CrossRef]
25. Luis González, M.; Fernández-Pello Martín, L.; Quirantes González, F. La influencia de los factores topoclimáticos en la organización geográfica de los sabinares de Anaga (Tenerife, Islas Canarias). *Investig. Geográficas* **2016**, *66*, 117. [CrossRef]
26. Simancas Cruz, M.R. Los modelos de uso turístico de las áreas protegidas de Canarias: Una propuesta metodológica. *Investig. Geográficas* **2006**, *39*, 25–45. [CrossRef]
27. Aguilera-Klink, F.; Pérez-Moriana, E.; Sánchez-García, J. The social construction of scarcity. The case of water in Tenerife (Canary Islands). *Ecol. Econ.* **2000**, *34*, 233–245. [CrossRef]

28. Barroso Castillo, S.M.; de Martín-Pinillos Castellanos, I.; Cruz-Pérez, N.; Santamarta, J.C. The Most Expensive Agricultural Land Prices in Europe: An Economic Analysis of Tenerife, Canary Islands, Spain. *Isl. Stud. J.* **2023**, *Early access*, 1–17. [CrossRef]
29. Cruz-Pérez, N.; Santamarta, J.C.; Álvarez-Acosta, C. Water footprint of representative agricultural crops on volcanic islands: The case of the Canary Islands. *Renew. Agric. Food Syst.* **2023**, *38*, e36. [CrossRef]
30. Santamarta, J.C.; Perdomo Molina, A.; Suárez Moreno, F.; Rodríguez-Martín, J.; Cruz-Pérez, N. Water Harvesting Strategies for Agriculture in the Canary Islands. *Hum. Ecol. Rev.* **2022**, *27*, 131–144. [CrossRef]
31. Arévalo, J.R.; Fernández-Palacios, J.M. Treefall Gaps and Regeneration Composition in the Laurel Forest of Anaga (Tenerife): A Matter of Size? *Plant Ecol.* **2007**, *188*, 133–143. [CrossRef]
32. Pérez, Z.; Arévalo, J.R. Tree Species Composition and Structure near Road Borders in the Laurel Forest of Anaga (Tenerife—Islas Canarias). Bulletin of University of Agricultural Sciences and Veterinary Medicine Cluj-Napoca. *Agriculture* **2016**, *73*, 145. [CrossRef]
33. Gobierno de Canarias. Memoria y Plan de Acción Reserva de la Biosfera del Macizo de Anaga. 2015. Available online: https://reservabiosfera.tenerife.es/wp-content/uploads/pdf/Memoria_y_plan_de_accion.pdf (accessed on 21 February 2024).
34. Casillas Ruiz, R.; Martín Luis, M.C.; Coello Bravo, J.J.; Balcells Herrera, R.; Colmenero Navarro, J.R. El delta de lava de Igueste de San Andrés (Anaga, Tenerife, Islas Canarias). *Geogaceta* **2019**, *66*, 99–102.
35. SGE-IGME. Canarias y el vulcanismo neógeno peninsular. In *Geología de España*; SGE-IGME: Madrid, Spain, 2004; pp. 635–669.
36. Rodríguez Darias, A.J. Desarrollo, gestión de áreas protegidas y población local. El Parque Rural de Anaga (Tenerife, España). *PASOS Rev. Tur. Patrim. Cult.* **2007**, *5*, 17–29. [CrossRef]
37. Petrella, E.; Aquino, D.; Fiorillo, F.; Celico, F. The effect of low-permeability fault zones on groundwater flow in a compartmentalized system. Experimental evidence from a carbonate aquifer (Southern Italy). *Hydrol. Process.* **2015**, *29*, 1577–1587. [CrossRef]
38. Wang, P.; Li, J.; An, P.; Yan, Z.; Xu, Y.; Pu, S. Enhanced delivery of remedial reagents in low-permeability aquifers through coupling with groundwater circulation well. *J. Hydrol.* **2023**, *618*, 129260. [CrossRef]
39. Ministerio de la Presidencia, R. con las C. y M. D. RD 3/2023. Available online: https://www.boe.es/buscar/pdf/2023/BOE-A-2023-628-consolidado.pdf (accessed on 9 June 2023).
40. Martín Delgado, M.M.; Fernández González, M.C.; Pita Toledo, M.L. Programa de Vigilancia Sanitaria del Agua de Consumo Humano. Comunidad Autónoma de Canarias. 2008. Available online: www.gobiernodecanarias.org/sanidad/scs (accessed on 27 June 2023).
41. Lagomasino, D.; Price, R.M.; Herrera-Silveira, J.; Miralles-Wilhelm, F.; Merediz-Alonso, G.; Gomez-Hernandez, Y. Connecting Groundwater and Surface Water Sources in Groundwater Dependent Coastal Wetlands and Estuaries: Sian Ka'an Biosphere Reserve, Quintana Roo, Mexico. *Estuaries Coasts* **2015**, *38*, 1744–1763. [CrossRef]
42. Phillips, J.; Scott, C.; O'Neil, S. Assessing the Vulnerability of Wastewater Facilities to Sea-Level Rise King County Wastewater Treatment Division. *Mich. J. Sustain.* **2015**, *3*, 127–133. [CrossRef]
43. Rodríguez-Alcántara, J.S.; Cruz-Pérez, N.; Rodríguez-Martín, J.; García-Gil, A.; Santamarta, J.C. Effect of tourist activity on wastewater quality in selected wastewater treatment plants in the Balearic Islands (Spain). *Environ. Sci. Pollut. Res.* **2024**, *31*, 15172–15185. [CrossRef]
44. Houben, G.J.; Stoeckl, L.; Mariner, K.E.; Choudhury, A.S. The influence of heterogeneity on coastal groundwater flow—Physical and numerical modeling of fringing reefs, dykes and structured conductivity fields. *Adv. Water Resour.* **2018**, *113*, 155–166. [CrossRef]
45. Liu, K.-H.; Fang, Y.-T.; Yu, F.-M.; Liu, Q.; Li, F.-R.; Peng, S.-L. Soil Acidification in Response to Acid Deposition in Three Subtropical Forests of Subtropical China. *Pedosphere* **2010**, *20*, 399–408. [CrossRef]
46. Salvia, M.; Ceballos, D.; Grings, F.; Karszenbaum, H.; Kandus, P. Post-Fire Effects in Wetland Environments: Landscape Assessment of Plant Coverage and Soil Recovery in the Paraná River Delta Marshes, Argentina. *Fire Ecol.* **2012**, *8*, 17–37. [CrossRef]
47. Zounemat-Kermani, M.; Alizamir, M.; Fadaee, M.; Sankaran Namboothiri, A.; Shiri, J. Online sequential extreme learning machine in river water quality (turbidity) prediction: A comparative study on different data mining approaches. *Water Environ. J.* **2021**, *35*, 335–348. [CrossRef]
48. Neupane, S.; Vogel, J.R.; Storm, D.E.; Barfield, B.J.; Mittelstet, A.R. Development of a Turbidity Prediction Methodology for Runoff–Erosion Models. *Water Air Soil Pollut.* **2015**, *226*, 415. [CrossRef]
49. Barbeau, B.; Desjardins, R.; Mysore, C.; Prévost, M. Impacts of water quality on chlorine and chlorine dioxide efficacy in natural waters. *Water Res.* **2005**, *39*, 2024–2033. [CrossRef]
50. Fish, K.E.; Reeves-McLaren, N.; Husband, S.; Boxall, J. Uncharted waters: The unintended impacts of residual chlorine on water quality and biofilms. *Npj Biofilms Microbiomes* **2020**, *6*, 34. [CrossRef]
51. Kwio-Tamale, J.C.; Onyutha, C. Influence of physical and water quality parameters on residual chlorine decay in water distribution network. *Heliyon* **2024**, *10*, e30892. [CrossRef]

52. Doussan, C.; Ledoux, E.; Detay, M. River-Groundwater Exchanges, Bank Filtration, and Groundwater Quality: Ammonium Behavior. *J. Environ. Qual.* **1998**, *27*, 1418–1427. [CrossRef]
53. Markogianni, V.; Kalivas, D.; Petropoulos, G.P.; Dimitriou, E. An Appraisal of the Potential of Landsat 8 in Estimating Chlorophyll-a, Ammonium Concentrations and Other Water Quality Indicators. *Remote Sens.* **2018**, *10*, 1018. [CrossRef]
54. Rafter, T. Oxygen isotopic composition of sulphates part 1, A method for the extraction of oxygen and its quantitative conversion to carbon dioxide for the isotope ratio measurements. *N. Z. J. Sci.* **1967**, *10*, 493–510. Available online: https://cir.nii.ac.jp/crid/1573950399220983808 (accessed on 3 April 2024).
55. Sakai, H.; Krouse, H. Elimination of memory effects in 18O/16O determinations in sulphates. *Earth Planet. Sci. Lett.* **1971**, *11*, 369–373. [CrossRef]
56. Bruthans, J.; Kůrková, I.; Kadlecová, R. Factors controlling nitrate concentration in space and time in wells distributed along an aquifer/river interface (Káraný, Czechia). *Hydrogeol. J.* **2019**, *27*, 195–210. [CrossRef]
57. Gavio, B.; Palmer-Cantillo, S.; Mancera, J.E. Historical analysis (2000–2005) of the coastal water quality in San Andrés Island, SeaFlower Biosphere Reserve, Caribbean Colombia. *Mar. Pollut. Bull.* **2010**, *60*, 1018–1030. [CrossRef] [PubMed]
58. Rahmati, O.; Samani, A.N.; Mahmoodi, N.; Mahdavi, M. Assessment of the Contribution of N-Fertilizers to Nitrate Pollution of Groundwater in Western Iran (Case Study: Ghorveh–Dehgelan Aquifer). *Water Qual. Expo. Health* **2015**, *7*, 143–151. [CrossRef]
59. Zhang, H.; Yang, R.; Wang, Y.; Ye, R. The evaluation and prediction of agriculture-related nitrate contamination in groundwater in Chengdu Plain, southwestern China. *Hydrogeol. J.* **2019**, *27*, 785–799. [CrossRef]
60. Liptrot, T.; Hussein, H. Between Regulation and Targeted Expropriation: Rural-to-Urban Groundwater Reallocation in Jordan. *Water Altern.* **2020**, *13*, 864–885. Available online: www.water-alternatives.org (accessed on 10 January 2025).
61. Puig, A.; Olguín Salinas, H.F.; Borús, J.A. Relevance of the Paraná River hydrology on the fluvial water quality of the Delta Biosphere Reserve. *Environ. Sci. Pollut. Res.* **2016**, *23*, 11430–11447. [CrossRef] [PubMed]
62. Chao, Y.-L.; Chao, S.-Y. Resident and visitor perceptions of island tourism: Green sea turtle ecotourism in Penghu Archipelago, Taiwan. *Isl. Stud. J.* **2017**, *12*, 213–228. [CrossRef]
63. Kløve, B.; Ala-aho, P.; Bertrand, G.; Boukalova, Z.; Ertürk, A.; Goldscheider, N.; Ilmonen, J.; Karakaya, N.; Kupfersberger, H.; Kvœrner, J.; et al. Groundwater dependent ecosystems. Part I: Hydroecological status and trends. *Environ. Sci. Policy* **2011**, *14*, 770–781. [CrossRef]

environments

MDPI

Article

Coupling Advanced Geo-Environmental Indices for the Evaluation of Groundwater Quality: A Case Study in NE Peloponnese, Greece

Panagiotis Papazotos [1,2,*], Maria Vlachomitrou [3], Despoina Psarraki [1], Eleni Vasileiou [1] and Maria Perraki [1]

1 School of Mining and Metallurgical Engineering, Division of Geo-Sciences, National Technical University of Athens, 9 Heroon Polytechniou St., 15773 Zografou, Greece

2 School of Engineering, Department of Mineral Resources Engineering, University of Western Macedonia, 50100 Kozani, Greece

3 Faculty of Geology and Geoenvironment, National and Kapodistrian University of Athens, Panepistimiopolis, Zografou, 15784 Athens, Greece

* Correspondence: papazotos@metal.ntua.gr; Tel.: +30-2107722059

Abstract: Water and its management have played a pivotal role in the evolution of organisms and civilizations, fulfilling essential roles in personal use, industry, irrigation, and drinking from ancient times to the present. This study seeks to evaluate groundwater quality for irrigation and drinking in the Northern Peloponnese region, specifically the wells of Loutraki and Schinos areas and the springs of the Gerania Mountains (Mts.), using geo-environmental indices and ionic ratios. For the first time, geo-environmental indices have been applied to a region where groundwater serves multiple purposes, addressing the challenge of understanding their dynamics to optimize their application in environmental science and groundwater pollution research. To achieve this, 68 groundwater samples from the study area were utilized, and a total of 25 geo-environmental indices were calculated to assess water quality. These indices examined: (i) drinking suitability (NPI, RI, PIG, WQI, and WPI), (ii) irrigation suitability (SAR, KR, %Na, PS, MAR, RSC, SSP, TH, PI, IWQI, and TDS), (iii) potentially toxic element (PTE) loadings (Cd, HEI, and HPI), and (iv) major hydrogeochemical processes, expressed as ionic ratios (Ca/Mg, Ca/SO_4, Ca/Na, Cl/NO_3, Cl/HCO_3, and Si/NO_3). Data processing involved descriptive statistics, hydrogeochemical bivariate plots, Spearman correlation coefficients, and multivariate statistical analyses, including factor analysis (FA) and R-mode hierarchical cluster analysis (HCA). Results revealed that all groundwater samples (100%) from the Loutraki area and the Gerania Mts. were of good quality for both drinking and irrigation purposes. In contrast, groundwater from the Schinos area exhibited lower quality, with most samples (93.9%) considered suitable only for irrigation. The deterioration in the coastal aquifer of the Schinos area is attributed to elevated concentrations of Cl^-, Na^+, NO_{3^-}, As, and Cr resulting from salinization and relatively limited anthropogenic influences. The study highlights that relying on individual geo-environmental indices can yield misleading results due to their dependence on factors such as researcher expertise, methodological choices, and the indices' inherent limitations. Consequently, this research emphasizes the necessity of combining indices to enhance the reliability, accuracy, and robustness of groundwater quality assessments and hydrogeochemical evaluations. Last but not least, the findings demonstrate that calculating all available geo-environmental indices is unnecessary. Instead, selecting a subset of indices that either reflect the impact of specific elemental concentrations or can be effectively integrated with others is sufficient. This streamlined approach addresses challenges in optimizing geo-environmental index applications and contributes to improved groundwater resource management.

 https://doi.org/10.3390/environments12010014

Keywords: geo-environmental indices; groundwater quality; drinking suitability; irrigation suitability; potentially toxic elements; environmental monitoring

1. Introduction

Groundwater is a crucial resource for various human activities, playing a vital role in agriculture, industry, and domestic water supply [1,2]. It serves as the primary source of irrigation for crops, especially in regions with limited surface water availability, ensuring food security [3]. Industries rely on groundwater for processes such as cooling, cleaning, and manufacturing, while it also provides drinking water for millions of people worldwide [4,5]. Effective groundwater management is essential to prevent over-extraction, contamination, depletion, scarcity, and stress [6–8]. This includes practices like monitoring water levels, implementing sustainable withdrawal limits, protecting recharge areas, and promoting water conservation and groundwater sustainability to maintain its availability for future generations [2,9].

To date, research on groundwater resource quality has primarily concentrated on identifying the origins of dissolved solutes and other chemical elements, distinguishing between geogenic and anthropogenic sources. Many studies have employed statistical methods and geographic information system (GIS) tools (e.g., [10–13], along with advanced techniques such as machine learning (ML) algorithms [14–16] and spatial autocorrelation [17–19]. These combined methods have been used to analyze key factors influencing water quality. Additionally, several studies have utilized various geo-environmental indices (the term 'geo-environmental indices' is defined in this study as all indices assessing water quality status) to evaluate the chemical suitability of water for specific uses [20–22]. While these geo-environmental indices have been applied in numerous studies worldwide [14,22–24], their application in groundwater quality assessments in Greece remains limited and is mentioned only in a few case studies [25–27]. However, no comprehensive evaluation of these geo-environmental indices has been conducted to highlight their strengths and weaknesses within a specific case study. The lack of such thorough investigations may lead to potentially erroneous or misleading conclusions, indicating that the assessment of water quality through these methodologies is still incomplete. Therefore, it is crucial to explore the optimal application of geo-environmental indices in areas where (i) water quality data are available and (ii) groundwater resources are used for various purposes. Therefore, a study that draws informed conclusions regarding the effective use of these indices in a study area—while clearly identifying their weaknesses, issues, shortcomings, and poor practices—is highly important.

The Loutraki–Schinos–Gerania Mountains (Mts.) region has been extensively studied by various Greek research institutions, including the National Technical University of Athens (NTUA), the National Kapodistrian University of Athens (NKUA), and the Hellenic Survey of Geology and Mineral Exploration (HSGME), as well as by numerous researchers. This attention is due to the region's significant geological, mineralogical, and hydrogeological features, alongside concerns related to groundwater quality. Papadopoulos and Lappas [28] were first to identify elevated levels of chromium (Cr), hexavalent chromium [Cr(VI)], and nitrate (NO_{3^-}) in the coastal alluvial aquifer of the Schinos area. Subsequent research by Pyrgaki et al. [29] confirmed elevated groundwater concentrations of these contaminants in the Loutraki, Schinos, and Gerania Mts. areas. Papazotos et al. [30] further established a statistical relationship between Cr, Cr(VI), and NO_{3^-} concentrations, and also reported increased levels of arsenic (As) in the groundwater. Both Papazotos et al. [30] and Pyrgaki et al. [31] emphasized the geogenic origin of Cr in the Loutraki area, attributing it to the ultramafic rocks of the Gerania Mts. Additionally, Kelepertzis et al. [32] and Pyrgaki et al. [33] conducted isotopic analyses using $\delta^{18}O_{NO3}$ and $\delta^{15}N_{NO3}$ in the Schinos area, suggesting that the likely sources of NO_3- in the groundwater are septic tanks and nitrogen (N)-bearing fertilizers; this fact supports the findings of Papazotos et al. [30], confirming the hydrogeochemical findings about the role of the nitrogen (N) cycle in the mobilization of potentially toxic elements (PTEs) in the groundwater resources. To date, there remains a research gap regarding the suitability of groundwater resources for various uses based on their quality. A water suitability study is therefore necessary for this area; calculating geo-environmental indices could offer valuable insights into optimal water use and management by uncovering hidden hydrogeochemical properties.

This study aims to assess the groundwater quality in Northern Peloponnese, specifically in the broader area of Loutraki–Schinos–Gerania Mts., for various uses including drinking, irrigation, and evaluating PTE loads. The assessment employs geo-environmental indices and ionic ratios, contributing to the fields of environmental geochemistry, hydrogeology, and groundwater resource management. In this research, we calculated geo-environmental indices: (i) to determine the suitability of irrigation water, (ii) to assess the suitability of drinking water, (iii) to evaluate PTE loads, (iv) to compare values across three different subcases, and (v) to conduct a comprehensive evaluation to ensure the holistic, safe, and sustainable use of water. The significance of this paper lies in the application of chemical analyses and the subsequent calculation of geo-environmental indices to evaluate water suitability using mathematical equations. This is particularly important in the Northern Peloponnese, where groundwater resources serve both drinking and irrigation needs. For the first time, geo-environmental indices have been calculated for an area where groundwater is used for multiple purposes. This approach ensures efficient groundwater management, protects human health, and optimizes water demand.

2. Materials and Methods

2.1. Description of the Study Area

The study area is located in the Corinth prefecture of the NE Peloponnese, Greece, between latitudes 37°56′00″ N and 38°04′00″ N and longitudes 22°57′00″ E and 23°08′00″ E; it encompasses Loutraki Town, the Gerania Mts., and Schinos Village. Elevations were recorded up to 90 m in the Schinos area, up to 190 m in the Loutraki area, and up to 1351 m in the Gerania Mts. A simplified lithological map showing the groundwater sampling sites for the three subareas (i.e., wells in Loutraki, Gerania springs, and wells in Schinos) is presented in Figure 1.

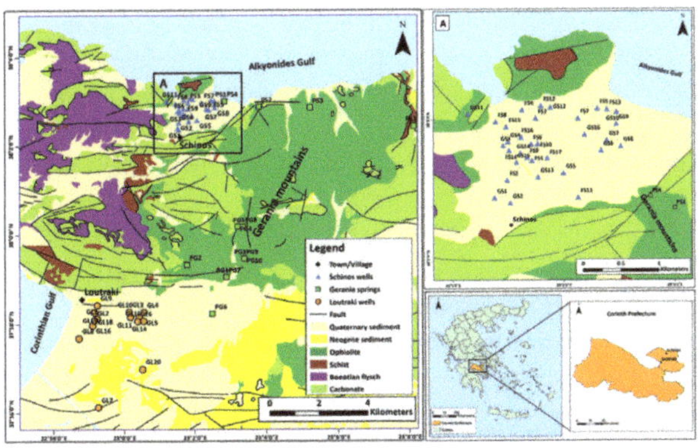

Figure 1. A simplified geological map of the study area [34,35] with the groundwater sampling sites. An enlarged image of the Schinos area is given in (A) [30].

The area's geological and hydrogeological characteristics significantly influence groundwater quality. In Schinos, located in the Northern Corinth prefecture, dominant formations include Triassic–Jurassic carbonate rocks (limestones), Upper Jurassic–Lower Cretaceous Boeotian flysch, Lower Cretaceous ophiolites, and Quaternary sediments [36]. The aquifer systems here consist of a karstic aquifer in the south and west, a fractured ultramafic aquifer in the south and SE, and a granular alluvial unconfined coastal aquifer in the central part. The main focus is on the alluvial aquifer, composed of sands, gravels, and conglomerates, recharged by lateral inflows, karstic aquifer inflows, and direct precipitation infiltration [37]. The vadose zone in the Schinos area reaches up to 15 m, with piezometric elevations ranging from −0.9 to 10.2 m [38]. The local economy is driven by tourism and, to a lesser extent, agriculture and livestock.

In Loutraki, geological formations include Lower Cretaceous ophiolites, Triassic–Jurassic carbonate rocks, Neogene marls, and Quaternary sediments. Aquifer systems include a karstic aquifer in the north and west, a fractured ultramafic aquifer in the northwest, and a major unconfined alluvial aquifer. This aquifer is recharged by precipitation, surface runoff infiltration, and lateral inflows. Its vadose zone thickness varies from 40 m in the east to over 100 m in the west [39]. Hydraulic disconnection is maintained by intercalating marls. Loutraki's thermal springs, linked to the karstic aquifer, are influenced by the South Aegean volcanic arc, with a probable deep geothermal reservoir at 80 °C [40]. Furthermore, it is noted that approximately 10 km from Loutraki Town lies the inactive Sousaki volcano, hosting a geothermal reservoir and characterized by an active hydrothermal activity with low-temperature gas emissions that have extensively altered the surrounding rocks [41].

The Gerania Mts., forming the northern part of the Corinth prefecture, include sedimentary and volcanic rocks, Triassic–Jurassic carbonate rocks, Upper Jurassic radiolarian sediments, Upper Jurassic–Lower Cretaceous Boeotian flysch, and Lower Cretaceous ultramafic rocks (mainly peridotite and serpentinite) [35,42]. The area is tectonically active, characterized by normal faults and steep slopes, which enhance the development of a highly permeable carbonate aquifer and semi-permeable fractured ultramafic spring systems [39]. The degree of weathering in peridotite is more pronounced along the fault zones. The dominant lithology is spinel lherzolite, with local occurrences of dunite and gabbro. It is worth noting that magnesite and other carbonate minerals are commonly found along fractured zones. Furthermore, the surface water network is dendritic in structure, consisting of two primary streams.

The study area is geodynamically situated within the complex neotectonics of the Gulf of Corinth, one of the most tectonically active regions in Greece, marked by significant seismic activity. The geological structure of central mainland Greece, including the Gulf of Corinth, is predominantly controlled by a series of E–W to NW–SE-trending normal faults, which have played a key role in shaping the region's topography [43,44]. Along the southern margin of the Gulf of Corinth, a set of prominent normal faults, ranging in length from 15 to 25 km, exhibit a strike of approximately N100° and dip northward at angles of about 50°. The SE lowland region of the Loutraki area is part of the eastern Corinthian graben, while the mountainous region corresponds to the Gerania tectonic horst [43,44].

Taking into account demographic data, the Municipality of Corinth spans a total area of 611.02 Km^2 and, according to the 2021 census conducted by the Hellenic Statistical Authority [45], has a population of 55,941, resulting in a population density of 91.55 inhabitants per Km^2. The population in the study area exerts pressure on the groundwater resources, which serve multiple purposes. In the Loutraki area, wells primarily supply drinking water, while in the Schinos area, groundwater is mainly used for irrigation. The region is a popular summer destination, attracting significant tourism and driving increased water demand during the warmer months. It experiences a typical Mediterranean climate, classified as Csa (hot-summer Mediterranean) under Köppen's classification, characterized by mild, wet winters and hot, dry summers with sharply reduced precipitation in the summer months. According to the Hellenic National Meteorological Service (HNMS), 33 years of data (1988–2010), from the Velo Corinthia meteorological station [46] indicate that the highest average monthly temperature is observed in July (28.7 °C), while the lowest occurs in January (9.1 °C), with the mean temperature being closer to the average maximum temperature. Precipitation peaks in December at 78.6 mm and drops to its lowest in July, averaging 5 mm. The wet season spans from October to mid-March, while the dry season extends from mid-March to mid-October.

According to the Corine Land Cover (CLC) 2018 data [47], the Prefecture of Corinth comprises four main land-use categories. Forest and semi-natural areas dominate, covering 68.4% of the region, followed by agricultural areas at 21.8%, artificial surfaces at 5.8%, and water bodies at 3.9%. These land uses are classified into four distinct categories: (i) forest and seminatural areas (coniferous forests, natural grasslands, sclerophyllous vegetation, and transitional woodland shrubs), (ii) agricultural areas (non-irrigated arable land, fruit trees and berry plantations, olive groves, pastures, complex cultivation patterns, and land primarily occupied by agriculture), (iii) artificial surfaces (continuous and discontinuous urban fabrics, commercial or industrial units, road and rail networks and associated land, and sport and leisure facilities), and (iv) water bodies (marine waters).

2.2. Sampling and Analysis

A total of 68 groundwater samples were collected from the study area: (i) 33 from irrigation wells in the Schinos area, (ii) 15 from Gerania springs, and (iii) 20 from drinking and irrigation wells in the Loutraki area. Sampling occurred during two campaigns in November 2016 and June 2017. Well samples were collected after purging residual water for at least 15 min, while spring samples were obtained via natural discharge. Water was collected in pre-cleaned 1000 mL polyethylene bottles, which were rinsed with sample water before collection. Samples were divided into three subsamples: 500 mL for major cations (Ca^{2+}, Mg^{2+}, Na^{+}, K^{+}) and anions (Cl^{-}, SO_4^{2-}, HCO_3-, NO_3-, PO_4^{3-}) 100 mL for PTEs and other trace elements (e.g., As, Cd, Cr, Ni, Si, Pb, Zn, etc.); and 25 mL for Cr(VI). Physical parameters such as electrical conductivity (EC), dissolved oxygen (DO), pH, and oxidation–redox potential (ORP/Eh) were measured in situ using a YSI Professional Digital Sampling System (ProDSS). Samples were stored in a cooler and refrigerated at 4 °C prior to analysis. The 500-mL subsamples were filtered and analyzed within 8 h to determine the dissolved forms of major ions. Major ions were determined using atomic absorption spectrometry (AAS), titration, spectrophotometry, or turbidimetry, while PTEs, other trace elements, and Cr(VI) were measured by inductively coupled plasma mass spectrometry (ICP-MS). All analyses followed international quality control protocols—using blanks, standards, and duplicate samples to maintain charge balance errors within ±10%—and were conducted in an ISO/IEC 17025:2005 (UNE-EN ISO/IEC/17025:2005, 2005 [48]. General Requirements for the Competence of Testing and Calibration Laboratories) certified laboratory, ensuring high precision, accuracy, and reliability. The data of this study were previously published in our earlier research [30], which also provides detailed information on the sampling and analysis procedures.

2.3. Data Treatment

This section includes the calculation of geo-environmental indices with equations that utilize the results of chemical determinations and the statistical treatment of various variables.

2.3.1. Evaluation of Water Quality for Various Purposes

The suitability of water is directly related to its intended use. Numerous geo-environmental indices are available to evaluate water quality for different purposes; several of those used in this study are shown in Figure 2. In general, these indices offer valuable information on water suitability (e.g., for drinking or irrigation), water quality assessment (e.g., PTEs and other major/minor elements), and the identification of key hydrogeochemical processes (e.g., contributions from geological formations, anthropogenic activities, and geochemical reactions) (Figure 2).

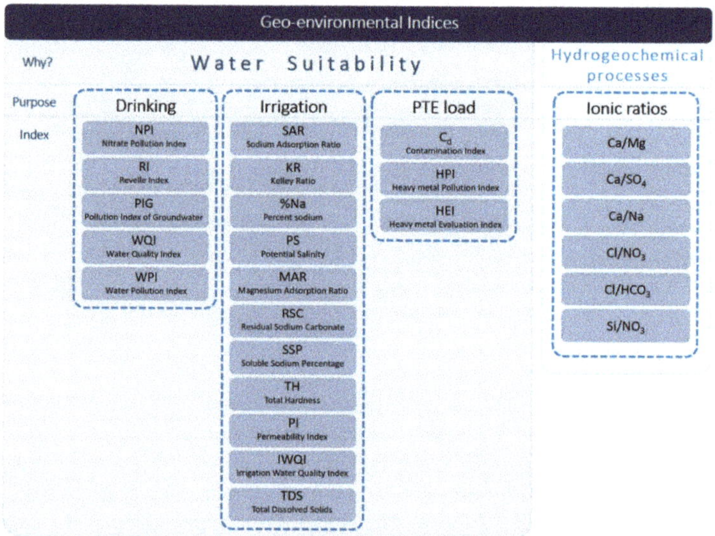

Figure 2. Classification of the 25 calculated geo-environmental indices of this study.

Drinking

This section includes the calculation of geo-environmental indices with equations that utilize the results of chemical determinations and the statistical treatment of various variables.

The evaluation of groundwater quality for drinking purposes was conducted using the following five geo-environmental indices: nitrate pollution index (NPI), Revelle index (RI), pollution index of groundwater (PIG), water quality index (WQI), and water pollution index (WPI). Detailed descriptions and information for each index are provided below:

Nitrate pollution index (NPI): The NPI is a single parameter water quality index that is used to calculate NO_3- contamination. The NPI shows that human activity has contributed to NO_3- pollution in groundwater. To calculate NPI, Equation (1) was used [49]:

$$NPI = \frac{C_s - HAV}{HAV} \quad (1)$$

where C_s represents the NO_3- concentration in each groundwater sample, while HAV (human-affected value) refers to the threshold level of NO_3- in groundwater, set at 10 mg/L according to Panno et al. [50]. Based on NPI values, water quality was classified into five categories: (a) clean water (NPI < 0), (b) lightly polluted water (NPI: 0–1), (c) moderately polluted water (NPI: 1–2), (d) significantly polluted water (NPI: 2–3), and (e) very significantly polluted water (NPI > 3).

Revelle index (RI): The RI was used to assess groundwater salinity in the study area. It was calculated using Equation (2) [51], with all ion concentrations expressed in meq/L:

$$RI = \frac{Cl^-}{HCO_3^- + CO_3^{2-}} \quad (2)$$

Based on RI values, water quality was classified into three categories: (a) unaffected (RI < 0.5), (b) slightly affected (RI = 0.5–6.6), and (c) strongly affected (RI > 6.6) by salinity. In general, RI values below 1 suggest good water quality.

Pollution index of groundwater (PIG): The PIG was calculated following the method proposed by Subba Rao [52], which evaluates the relative influence of individual chemical parameters (i.e., pH, TDS, TH, Ca^{2+}, Mg^{2+}, Na^+, K^+, HCO_3-, SO_4^{2-}, Cl^-, and NO_3-). The computation of the PIG involved a total of five steps:

- Step I: Assigning relative weights (R_w): Each chemical parameter was assigned a relative weight on a scale from 1 to 5, reflecting its impact on human health. Potassium (K^+) received the minimum weight of 1, while TDS, SO_4^{2-}, and NO_3- were assigned the maximum weight of 5 [52].
- Step II: Calculating weight parameters (W_p): The weight parameter for each chemical was calculated to assess its contribution to overall groundwater quality, using Equation (3):

$$W_p = \frac{R_w}{\sum R_w} \quad (3)$$

- Step III: Determining status of concentration (S_c): The concentration status for each parameter was established by dividing the concentration (C) of the parameter in each sample by its corresponding drinking water quality standard limit (D_s) (Equation (4)):

$$S_c = \frac{C}{D_s} \quad (4)$$

- Step IV: Calculating overall groundwater quality (O_w): The overall groundwater quality was calculated by multiplying the weight parameter (W_p) with the concentration status (S_c) (Equation (5)):

$$O_w = W_p \times S_c \quad (5)$$

- Step V: Calculating PIG: Finally, the PIG was obtained by summing all values of O_w (Equation (6)), providing a comprehensive assessment of pollution's impact on the groundwater system.

$$PIG = \sum O_w \quad (6)$$

- Based on the calculated PIG values, the water pollution status was classified into five categories: (a) no significant pollution (PIG < 1.0), (b) low pollution (PIG = 1.0–1.5), (c) moderate pollution (PIG = 1.5–2.0), (d) high pollution (PIG = 2.0–2.5), and (e) very high pollution (PIG > 2.5).

Water quality index (WQI): The WQI, as defined by Horton [53] and Brown et al. [54], utilizes the same relative weights assigned to each chemical parameter in the PIG, as outlined in Equation (3) using a total of 9 parameters (TDS, Ca^{2+}, Mg^{2+}, Na^+, K^+, HCO_3-, SO_4^{2-}, Cl^-, and NO_3-). The quality rating scale (q_i) for each parameter is calculated by dividing its concentration in each water sample by the corresponding standard, then multiplying the result by 100 as shown in the following Equation (7).

$$q_i = S_c \times 100 \quad (7)$$

To compute the final stage of the WQI, the SI for each chemical parameter (SI_i) is determined using Equation (8).

$$SI_i = W_p \times q_i \quad (8)$$

The WQI is then calculated by summing all the SI values, as expressed in Equation (9):

$$WQI = \sum SI_i \quad (9)$$

Based on the calculated WQI values, water quality levels are classified into five categories: (a) excellent water (WQI < 50), (b) good water (WQI = 50–100), (c) poor water (WQI = 100–200), (d) very poor water (WQI = 200–300), and (e) unsuitable for drinking (WQI > 300).

Water pollution index (WPI): The WPI offers a flexible and rational alternative approach to existing indexing systems by consolidating multiple chemical variables into a single value index [55]. This method employs a total of ten water quality parameters (pH, EC, Ca^{2+}, Mg^{2+}, Na^+, K^+, HCO_{3^-}, $SO_4{}^{2-}$, Cl^-, and NO_{3^-}) for calculating WPI. The calculation involves two steps. The first step is to compute the pollution load (PL_i) (Equation (10)):

$$PL_i = 1 + \frac{C_i - S_i}{S_i} \quad (10)$$

where PLi represents the pollution load, C_i is the observed ionic concentration, and S_i is the highest desirable limit prescribed for drinking water quality.

The PL_i depends on pH. If the pH is <7, the PL_i is computed using Equation (11):

$$PL_i = 1 + \frac{C_i - 7}{S_a - 7} \quad (11)$$

where S_a is the lowest desirable limit of pH (6.5) recommended for drinking water quality. If the pH is >7, PL_i is calculated using Equation (12):

$$PL_i = 1 + \frac{C_i - 7}{S_b - 7} \quad (12)$$

where S_b is the maximum desirable limit of pH (8.5) for drinking water quality.
In the second step, WPI is determined by averaging the PLi using Equation (13):

$$WPI = \frac{1}{n}\sum_{i=1}^{n} PL_i \quad (13)$$

where n is the number of chemical variables.
The WPI values can be classified into four categories: (a) excellent water (WPI < 0.5), (b) good water (WPI = 0.5–0.75), (c) moderately polluted water (WPI = 0.75–1.0), and (d) highly polluted water (WPI: >1.0).

Irrigation

The assessment of groundwater quality for drinking purposes was conducted using 11 geo-environmental indices: SAR, KR, %Na, PS, MAR, RSC, SSP, TH, PI, IWQI, and TDS. A detailed explanation of each index, including key information, is provided below:

Sodium adsorption ratio (SAR): The SAR is a significant irrigation quality index that points out Na^+ hazard and evaluates the cations concentrations expressed in meq/L according to Equation (14) [56]:

$$SAR = \frac{Na^+}{\sqrt{\frac{(Ca^{2+}+Mg^{2+})}{2}}} \times 100 \quad (14)$$

Excessive Na^+ concentration in irrigation water can reduce soil permeability by altering its structure. Significant changes in Na^+ concentrations in irrigation water often lead to modifications in soil composition. When water contains high levels of both Na^+ and Ca^{2+}, ion exchange occurs, causing the soil to become saturated with Na^+ and depleted of Ca^{2+}; this imbalance negatively affects soil quality and reduces crop yields [57]. The classification of SAR values includes four categories: (a) excellent (SAR < 10), (b) good (SAR = 10–18), (c) doubtful—unsuitable for most crops due to elevated sodium levels (SAR = 18–26), and (d) unsuitable—unsatisfactory for all crops due to very high Na^+ levels (SAR > 26).

Kelley ratio (KR): The KR is a method used to assess the impact of Na^+ on water quality, particularly for irrigation purposes. It evaluates Na^+ levels in relation to alkaline earth (i.e., Ca^{2+} and Mg^{2+}) concentrations in groundwater, as proposed by Kelley [58,59]. Water is considered unsuitable for irrigation if the Na^+ content exceeds the concentration of alkaline earths in groundwater. Finally, KR is determined using Equation (15) [58,59]:

$$KR = \frac{Na^+}{Ca^{2+} + Mg^{2+}} \quad (15)$$

A KR value greater than 1 indicates excess Na^+ and is not recommended for irrigation due to alkali hazards, while water with a KR value less than 1 is suitable for irrigation [58].

Percent sodium (%Na): %Na concentration is a factor to assess its suitability for irrigation purposes [60]. The Na^+ concentration is vital for classifying irrigation water, as it bonds with soil particles, reducing water movement [61]. Sodium ions (Na^+) form alkaline soils with ${CO_3}^{2-}$ and saline soils with chloride, both inhibiting crop growth [57]. High Na^+ levels in irrigation water cause a base-exchange reaction, replacing Ca^{2+} and Mg^{2+} ions, further limiting water movement. This results in compacted soils that restrict air and water flow, especially in wet conditions [62]. The %Na values are calculated using Equation (16) [60]:

$$\%Na = \frac{Na^+}{Ca^{2+} + Mg^{2+} + Na^+ + K^+} \times 100 \quad (16)$$

The classification of water is based on %Na as excellent (<20%), good (20–40), permissible (40–60), doubtful (60–80), and unsuitable (>80%).

Potential salinity (PS): To assess the salinity of the groundwater, the PS index—another indicator of irrigation quality—was calculated based on Cl^- and ${SO_4}^{2-}$ concentrations. A high $Cl^-/{SO_4}^{2-}$ ratio can lead to scale formation in irrigation systems, which may affect water distribution to crops [63,64]. The PS index, expressed in meq/L, is determined using Equation (17) as follows:

$$PS = Cl^- + \frac{SO_4^{2-}}{2} \quad (17)$$

A PS value greater than 3 indicates unsafe water quality for irrigation, while a value less than 3 signifies safe water quality [65].

Magnesium adsorption ratio (MAR): The MAR index, developed by Szabolcs and Darab [66], is used to assess Mg hazards, as Mg increases soil alkalinity, thereby degrading soil quality and crop yields [23,67]. This index operates on the principle that Ca^{2+} and Mg^{2+}

ions are typically in equilibrium in most groundwater [68,69]. Equation (18) was used to calculate the MAR index:

$$MAR = \frac{Mg^{2+}}{Ca^{2+} + Mg^{2+}} \times 100 \quad (18)$$

Groundwater is considered suitable for irrigation when the MAR value is below 50, but it becomes harmful and unsuitable when the MAR value exceeds 50.

Residual sodium carbonate (RSC): In alkaline water, the high concentration of HCO_3^- increases the likelihood of Ca^{2+} and Mg^{2+} precipitating as carbonates, especially when compared to water with lower HCO_3^- levels. To quantify this effect, ionic concentrations are measured in meq/L, and an index known as RSC is used [70]. Essentially, RSC provides an additional metric for determining whether water is suitable for irrigation, and it is represented by Equation (19) [70]:

$$RSC = (HCO_3^- + CO_3^{2-}) - (Ca^{2+} + Mg^{2+}) \quad (19)$$

RSC values can categorize water quality into three classes: (a) good (RSC < 1.25), (b) moderate (RSC = 1.25–2.50), and (c) unsuitable (RSC > 2.50).

Soluble sodium percentage (SSP): The SSP represents the ratio of Na and K ions to the total cation concentration and is used to assess irrigation water quality, particularly concerning Na^+ hazards. High Na^+ levels, as noted above, can reduce soil permeability and inhibit plant growth. SSP was calculated using Equation (20) from Todd [57], with ionic concentrations expressed in meq/L:

$$SSP = \frac{Na^+ + K^+}{Ca^{2+} + Mg^{2+} + Na^+ + K^+} \times 100 \quad (20)$$

Groundwater is considered suitable for irrigation when the SSP value is below 50, but it is unsuitable when the SSP value exceeds 50.

Total hardness (TH): TH is primarily influenced by the concentrations of Mg^{2+} and Ca^{2+} in water [71], both of which are essential for human health. However, high TH levels in drinking water are associated with various health issues, including arterial calcification, urinary stones, kidney and bladder disorders, and digestive problems [72]. TH was calculated in mg/L by means of Equation (21) [57]:

$$TH = 2.497Ca^{2+} + 4.11Mg^{2+} \quad (21)$$

Based on TH values, water was classified into four categories: (a) soft water (TH < 60 mg/L), (b) moderately hard water (TH = 60–120 mg/L), (c) hard water (TH = 120–180 mg/L), and (d) very hard water (TH > 180 mg/L) [73].

Permeability index (PI): PI is used to assess and classify the quality of irrigation water. Long-term irrigation can impact soil permeability, which is influenced by the concentrations of Na^+, Ca^{2+}, Mg^{2+}, and HCO_3^- in the soil. Doneen [63,64] proposed Equation (22) (with all ion concentrations expressed in meq/L) to calculate PI:

$$PI = \frac{Na^+ + \sqrt{HCO_3^-}}{Ca^{2+} + Mg^{2+} + Na^+} \times 100 \quad (22)$$

PI values can be categorized into three classes: (a) class I—suitable for irrigation (PI > 75%), (b) class II—good for irrigation (PI = 25–75%), and (c) class III—unsuitable for irrigation (PI < 25%).

Irrigation water quality index (IWQI): IWQI was calculated based on five key water quality parameters: EC, SAR, Na^+, Cl^-, and HCO_3^- following the methodology of

Meireles et al. [74]. First, the concentration units of the water samples were converted from mg/L to meq/L. In this section, the IQWI was assessed by calculating the water quality parameter values (q_i) and the cumulative weight (W_i). Table 1 summarizes the irrigation water quality parameters and their proposed limiting values [74]. The q_i values for the five water quality parameters (qEC, qSAR, qNa^+, qCl^-, and $qHCO_{3^-}$) were determined using Equation (23). The upper limits of the parameter ranges indicated in Table 1 were used as the highest value of the observed sample values to evaluate x_{imap}.

$$IWQI = Q_{max} - \left(\frac{\left(x_{ij} - x_{inf} \right) \times q_{imap}}{x_{amp}} \right) \tag{23}$$

where Q_{max} is the upper value of the q_i class of, X_{ij} represents the observed values for each parameter (Table 1), X_{inf} is the lower limit of the class to which the observed parameter belongs, q_{imap} represents the class amplitude for qi classes, and x_{imap} corresponds to class amplitude for the specific parameter.

Table 1. Parameter-limiting values of IWQI for quality measurement (q_i) calculation [74].

q_i	EC (μS/cm)	SAR $(meq/L)^{1/2}$	Na^+ (meq/L)	Cl^- (meq/L)	HCO_3^- (meq/L)
85–100	200–750	<3	2–3	<4	1–1.5
60–85	750–1500	3–6	3–6	4–7	1.5–4.5
35–60	1500–3000	6–12	6–9	7–10	4.5–8.5
0–35	<200 or >3000	>12	<2 or >9	>10	<1 or >8.5

Finally, the overall IWQI was determined using Equation (24):

$$IWQI = \sum_{1}^{n} q_i w_i \tag{24}$$

where *n* is the number of parameters (in this case a total of 5 parameters); q_i values from Table 1 are multiplied by the corresponding weight (w_i) of each parameter, as listed in Table 2 [74]. It is important to note that, in calculating the final IWQI value, we applied four classification categories of limiting values for the five parameters (Table 1), following the methodological framework outlined by Meireles et al. [70], and incorporating the five distinct categories of restrictions and recommendations for water usage presented in Table 3.

Table 2. The weights of the IWQI parameters [74].

Parameters	Wi
EC	0.211
Na^+	0.204
HCO_{3^-}	0.202
Cl^-	0.194
SAR	0.189
Total	1.000

Total dissolved solids (TDS): TDS measures the total amount of dissolved solids in the groundwater, expressed in mg/L; TDS is a crucial measure for determining the presence and concentration of pollutants. Generally, TDS indicates the amount of dissolved salts in irrigation water, with higher levels reducing a plant's ability to absorb nutrients efficiently due to increased salinity [75]. TDS value was computed by adding the major cations and anions determined in each of the water samples, as indicated by the following equation (Equation (25)):

$$TDS = \sum Major\ cations + \sum Major\ anions \tag{25}$$

TDS values > 500 mg/L are considered the highest desirable limits according to the World Health Organization (WHO) [76]. While TDS is often used as a general water quality index, in this study, it is specifically treated as an irrigation suitability index due to its significant impact on agricultural water use, particularly in areas with high salinity. In the context of this study, which is located near the sea, high TDS levels in groundwater can affect irrigation practices. Although high TDS levels (up to 1500 mg/L) are generally not harmful to humans [77], they can adversely affect living organisms, causing major cardiovascular and renal disorders [78].

Table 3. IWQI characteristics [74].

IWQI	Water Use Restriction	Recommendation	
		Soil	Plant
85–100	No restriction (NR)	Suitable for most soils, with a low likelihood of causing salinity or sodicity issues. Leaching is recommended during irrigation practices, except for soils with extremely low permeability	No toxicity risk for most plants
70–85	Low restriction (LR)	Recommended for irrigated soils with light texture or moderate permeability, with salt leaching advised. Sodicity risk may occur in heavy-textured soils, so use should be avoided in soils with high clay content (2:1 clay types)	Avoid salt-sensitive plants
55–70	Moderate restriction (MR)	Suitable for soils with moderate to high permeability. Moderate leaching of salts is suggested	Plants with moderate salt tolerance can be cultivated
40–55	High restriction (HR)	Applicable for soils with high permeability and no compact layers. Frequent irrigation is required when EC > 2000 dS/m and SAR > 7	Suitable for plants with moderate to high salt tolerance. Special salinity control practices are recommended, except when Na^+, Cl^-, and $HCO_3{}^-$ values are low
0–40	Severe restriction (SR)	Generally unsuitable for irrigation. May be used occasionally under special conditions. For low-salt, high-SAR water, gypsum application is necessary. For highly saline water, soils should be highly permeable, and excess water must be applied to prevent salt accumulation	Only highly salt-tolerant plants should be irrigated, except with waters that have extremely low Na^+, Cl^-, and $HCO_3{}^-$ levels

PTE Load

The evaluation of groundwater quality for PTE load was performed using the following geo-environmental indices: C_d, HPI, and HEI. Detail description and information for each index is provided below:

Contamination index (C_d): The C_d indicates the extent of contamination or the cumulative effect of various water quality parameters that are considered harmful for domestic use [79]; C_d represents the sum of all contamination factors that exceed permissible limits, calculated using Equation (26).

$$C_d = \sum_{i=1}^{n} C_{fi} \tag{26}$$

where $C_{fi} = \frac{C_{Ai}}{C_{Ni}} - 1$ rerepresents the contamination factor. Herein, C_{Ai} refers to the measured value and C_{Ni} is the upper permissible concentration of the ith parameter. The letter 'N' stands for 'normative value', with C_{Ni} being equivalent to the maximum allowable concentration (MAC). C_d values are typically classified into three categories based on contamination levels: low ($C_d < 1$), medium ($C_d = 1–3$), and high ($C_d > 3$).

Heavy metal pollution index (HPI): The HPI is an effective tool for assessing groundwater pollution, reflecting the combined impact of various PTEs on overall water quality. This methodology, originally developed by Horton [53], assigns weights to each PTE in groundwater samples based on their potential hazards to human health. To calculate HPI, a weight (W_i) is assigned to each variable, with pollutant criteria selected accordingly. The rating ranges from 0 to 1, indicating the relative significance assigned to specific water quality parameters. This rating is determined using an inverse proportionality function, based on the recommended standard (S_i) for each component. The concentration thresholds—where 'S_i' represents the standard value and 'I_i' the ideal value—are based on guidelines set by the WHO [76]. HPI is calculated using Equations (27) and (28):

$$HPI = \frac{\sum_{i=1}^{n} W_i Q_i}{\sum_{i=1}^{n} W_i} \quad (27)$$

where Q_i represents the sub-index of the ith parameter, W_i is the unit weight assigned to the ith parameter, and n is the total number of parameters considered. The sub-index Q_i is calculated as follows (Equation (28)):

$$Q_i = \sum_{i=1}^{n} \frac{|M_i - I_i|}{S_i - I_i} \quad (28)$$

where M_i is the monitored value of the PTE for the ith parameter, I_i is the ideal value (the maximum desirable level for drinking water), and S_i is the standard value (the highest permissible level for drinking water) of the ith parameter. The minus sign (-) in the equation indicates the numerical difference between the two values, disregarding the algebraic sign. Water quality is classified into two categories based on HPI: suitable (HPI < 100) and unsuitable (HPI > 100). The concentration limits, including the highest permissible value (S_i) and the maximum desirable value (I_i) for each parameter, were sourced from the WHO [76].

Heavy metal evaluation index (HEI): HEI is used to assess the overall water quality based on the concentrations of PTEs. Like HPI, it provides insight into water's quality concerning PTE contamination [80]. HEI is calculated using Equation (29):

$$HEI = \sum_{i=1}^{n} \frac{Hc_i}{Hmac_i} \quad (29)$$

where Hc_i represents the measured value of the ith parameter, and H_{maci} refers to the MAC of that parameter. This index is valuable for assessing pollution levels, and its straightforward calculation makes it particularly useful [80,81]. HEI values typically categorize contamination into three levels based on contamination: low (HEI < 10), medium (HEI = 10–20), and high (HEI > 20).

Ionic Ratios

Groundwater quality was evaluated using various ionic ratios, including Ca/Mg, Ca/SO_4, Ca/Na, Cl/NO_3, Cl/HCO_3, and Si/NO_3. Each ionic ratio offers insights into underlying hydrogeochemistry, such as the origin of water from specific geological formations, various geochemical processes, anthropogenic influences, etc.

2.3.2. Statistics

The statistical analysis included descriptive, correlation, and multivariate techniques. Descriptive statistics, such as the mean, minimum, maximum, median, standard deviation, and first and third quartiles, were calculated. For chemical parameters below the detection limit (BDL), values were substituted with the detection limit (DL) to enable further analysis. Spearman's non-parametric rank correlation coefficient (r) [82] was used to assess the association between geo-environmental indices. Correlation coefficients, ranging from −1 to 1, were categorized as very strong (0.8–1), strong (0.6–0.79), moderate (0.4–0.59), weak (0.2–0.39), and very weak (0–0.2) [83]. Two multivariate statistical methods—factor analysis

(FA) and hierarchical cluster analysis (HCA)—were employed. Generally, multivariate statistical methods are commonly applied to independent variables, but they can also be used with dependent variables, depending on the objectives of the analysis [83–85]. To meet the normality assumption, the input data were transformed into normal scores. The FA technique was employed to examine interrelationships among parameters, reduce dimensionality, and identify underlying factors. Principal component analysis (PCA) with Varimax rotation [86] was conducted, retaining only factors with eigenvalues greater than or equal to 1, in accordance with Kaiser's criterion [87]. Factor loadings were classified as strong (0.75–1), moderate (0.5–0.75), or weak (0.3–0.5) [88,89]. The Kaiser–Meyer–Olkin (KMO) test [90] and Bartlett's test [91] were used to assess data adequacy. An HCA approach, using Ward's method [92] and squared Euclidean distances, was performed in R-mode to classify geo-environmental indices based on similarity. All statistical analyses were conducted using IBM SPSS v.22 (IBM Corp., Armonk, NY, USA).

3. Results and Discussion

3.1. Hydrogeochemical Characteristics

The groundwater samples from the study area exhibit alkaline ($pH \geq 7.45$; from 7.45 to 9.64) and oxidizing ($Eh \geq 231$ mV; from 231 mV to 371 mV) geochemical conditions [30]. The EC values range from 316 μS/cm to 3943 μS/cm, while DO values vary from 3.62 mg/L to 13.49 mg/L [30], suggesting the influence of diverse hydrogeochemical processes on groundwater quality. Previous studies have identified factors/processes such as seawater intrusion [28–30] and initial stages of denitrification [30] as potential contributors to these variations. More information about the descriptive statistics of the above-mentioned dataset is presented in the study by Papazotos et al. [30]. The Schinos area consistently showed the lowest pH values, while the highest were observed in the springs of the Gerania Mts. The groundwater was predominantly classified as Mg-HCO^3 water type, which indicates rainwater–ultramafic rock/soil interaction, as confirmed by the Piper plot presented in our previous work [30]. In the Loutraki area and Gerania Mts., Mg^{2+} concentrations among major cations exceeded those of Ca^{2+} or alkali (Na^+ + K^+), whereas Schinos displayed both high Mg^{2+} and alkali levels. Regarding major anions, HCO_3- predominated in the region, followed by Cl^-, with the samples of the Schinos area showing elevated Cl^- concentrations. The main hydrochemical types were Mg-HCO_3, Mg-Cl, and Na-Cl, with samples of the Schinos area suggesting possible seawater intrusion due to increased Na^+ and Cl^-; these results are in agreement with previous studies in the same area (e.g., [28–30,33]). Bivariate diagrams were employed to further explore these hydrogeochemical relationships (Figure 3). The strong linear correlations observed between Na^+ vs. Cl^-, as well as Ca^{2+} vs. ${SO_4}^{2-}$, indicate a shared origin for these ions, evidenced by their alignment along the theoretical 1:1 line associated with halite and gypsum dissolution (Figure 3a,b). The samples from the Schinos area exhibited notably higher groundwater concentrations of dissolved ions. This alignment strongly suggests that seawater intrusion plays a major role as a driving geochemical process in the study area. Additionally, the majority of the samples exhibited a minor deviation, plotting above the 1:1 line in the Ca + Mg vs. Na plot (Figure 3c). This pattern underscores the importance of reverse cation exchange as a significant geochemical process, marked by a decrease in Na^+ and an increase in other major cations, particularly Ca^{2+} and Mg^{2+} [89,93]. In Figure 3d, a slight decrease in the Na^+/Cl^- ratio is observed as EC increases, further supporting the presence of a seawater intrusion regime in the Schinos area, where EC values reached up to 3943 μS/cm [30]. The presence of Mg^{2+} in groundwater is primarily associated with the abundance of Mg-bearing silicate minerals found in ultramafic rocks in the study area [30,31], such as those in the olivine (e.g., forsterite [Mg_2SiO_4]), and pyroxene (e.g., enstatite [$Mg_2Si_2O_6$]) groups, as well as secondary minerals such as the serpentine group [$Mg_3Si_2O_5(OH)_4$] and talc [$Mg_3Si_4O_{10}(OH)_2$]. Additionally, significant contributions of Mg^{2+} result from the dissolution Mg-rich carbonates, including magnesite [$MgCO_3$], hydromagnesite [$Mg_5(CO_3)_4(OH)_2\cdot 4(H_2O)$], huntite [$Mg_3Ca(CO_3)_4$], and pyroaurite [$Mg_6Fe^{3+}{}_2CO_3(OH)_{16}\cdot 4H_2O$] [94], a process that may explain the observed predominance of $HCO_3{}^- + SO_4{}^{2-}$ over $Ca^{2+} + Mg^{2+}$ (Figure 3e). This process can be represented by the following dissolution reactions for magnesite (Equation (30)) and forsterite (Equation (31)), which also apply to other relevant mineral phases:

$$MgCO_3 + CO_2 + H_2O \rightarrow Mg^{2+} + 2HCO_3{}^- \quad (30)$$

$$Mg_2SiO_4 + 4CO_2 + 4H_2O \rightarrow 2Mg^{2+} + 4HCO_3{}^- + H_4SiO_4 \quad (31)$$

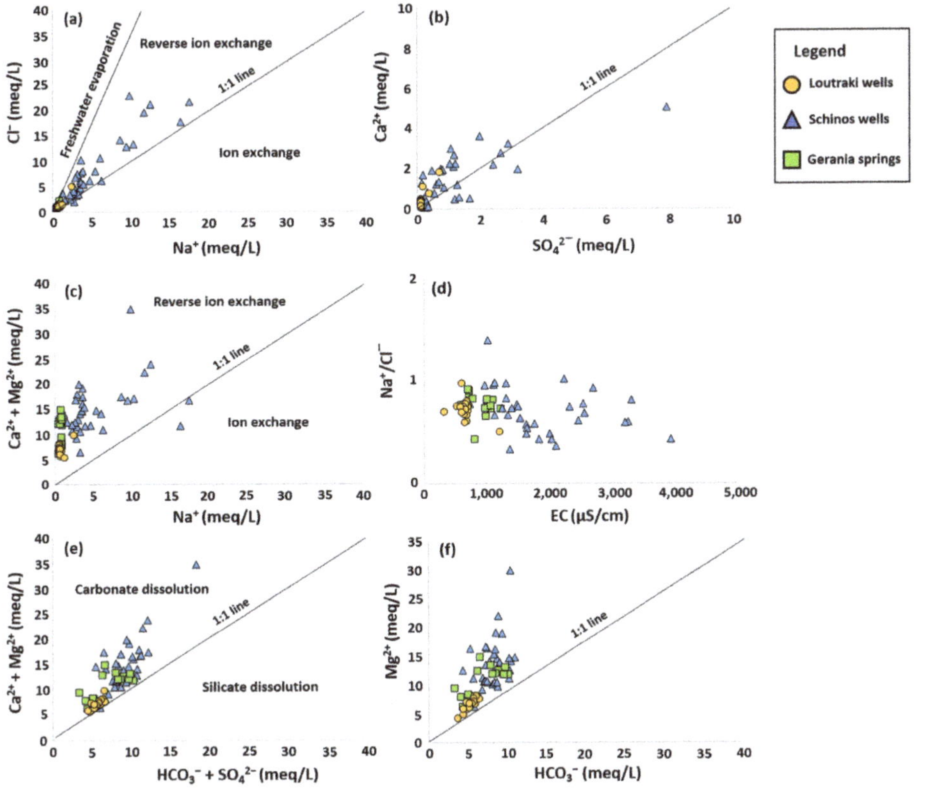

Figure 3. Cross-plots of (**a**) Na^+ vs. Cl^-, (**b**) $SO_4{}^{2-}$ vs. Ca^{2+}, (**c**) Na^+ vs. ($Ca^{2+} + Mg^{2+}$), (**d**) EC vs. Na^+/Cl^-, (**e**) ($HCO_3{}^- + SO_4{}^{2-}$) vs. ($Ca^{2+} + Mg^{2+}$), and (**f**) $HCO_3{}^-$ vs. Mg^{2+} for the 68 groundwater samples from the Loutraki–Schinos–Gerania Mts. region.

However, the higher concentrations of Mg^{2+} compared to $HCO_3{}^-$ (Figure 3f), particularly evident in the Schinos area, suggest an additional source of Mg^{2+} in the groundwater, likely due to seawater intrusion and reverse ion exchange processes.

The Ficklin diagram (Figure 4) demonstrates that although concentrations of PTEs tend to increase with decreasing pH, the overall levels of PTEs in the groundwater remain relatively low based on the diagram's classification [95]. Typically, most PTEs in groundwater exist as cations, and their mobilization requires acidic conditions. As a result, the alkaline conditions prevalent throughout the region generally inhibit their mobility.

However, elevated groundwater concentrations of specific PTEs, such as Cr, Cr(VI), and As, were detected, particularly in the Schinos area [30]. These PTEs form oxyanions, which have a different geochemical mobility compared to cations, allowing them to remain mobile even under alkaline pH conditions.

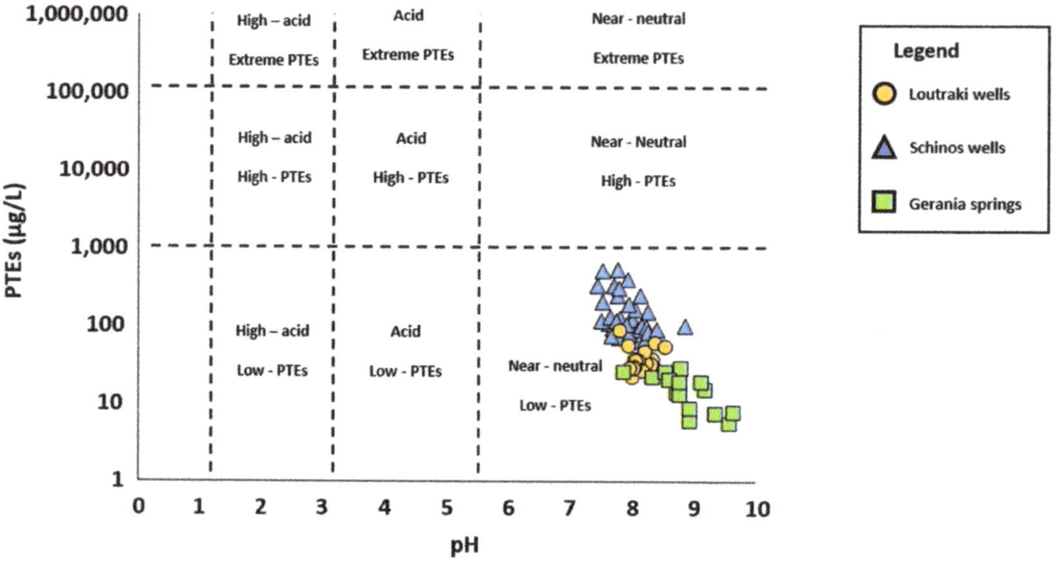

Figure 4. Ficklin diagram [95] of 68 groundwater samples showing the sum of PTEs vs. pH.

It is important to note that the Schinos area dominates a seawater intrusion regime, leading to the salinization of groundwater. This phenomenon is evidenced by elevated groundwater concentrations of Cl^- and Na^+, which are typically abundant in seawater [30]. In contrast, the Gerania Mts. and the Loutraki area are dominated by ultramafic rocks composed of Mg-rich minerals, resulting in high groundwater concentrations of Mg^{2+}, while Ca^{2+} concentrations remain notably low in the same regions.

The absence of significant anthropogenic activities in the study area is confirmed by the low NO_{3^-} groundwater concentrations. However, in the Schinos area, concentrations exceeding 50 mg/L are observed, surpassing the guideline values set by the WHO [76] and Greek legislation (FEK B 3525/25.05.2023). This increase is likely attributable to relatively limited anthropogenic influences on the environment (e.g., small-scale farming, irrigation, low-impact tourism), with fertilizers and septic tanks identified as potential sources of contamination [28,30,33].

Notably, groundwater in the Schinos area exhibits significantly higher concentrations of Cr and Cr(VI), along with notable levels of As. Previous studies link elevated Cr and Cr(VI) levels to synergistic anthropogenic effects, such as fertilizer use and septic tanks, particularly in Cr-rich environments influenced by ultramafic rocks and soils [30,33,89]. In contrast, the occurrence and mobilization of As in groundwater is associated with the prevailing seawater intrusion regime and specific geochemical processes that are related to N-cycling. The combination of seawater intrusion and denitrification processes creates favorable conditions for the mobilization and release of As into the groundwater [30].

3.2. Groundwater Quality Evaluation for Various Purposes

The geo-environmental indices calculated and assessed for groundwater suitability and quality in the study area are categorized into the following four groups, as outlined in Sections 3.2.1–3.2.4:

- Drinking water quality indices
- Irrigation water quality indices
- PTE load indices
- Ionic ratios

Overall, the geo-environmental indices discussed have been widely used to assess groundwater suitability and quality across all continents and countries worldwide, such as Nigeria [22,96], Iran [97], Algeria [98], Morocco [99,100], Ethiopia [101], Brazil [74], Poland [102], Syria [103], China [104], United Arab Emirates [105], Turkey [106], India [107,108], Australia [109], Spain, [110], etc. These indices have proven invaluable in identifying areas where groundwater quality is compromised, revealing patterns and factors that influence water usability for drinking, irrigation, and other purposes. Studies consistently report that poor water quality globally is shaped by a combination of anthropogenic and geogenic factors. Among these, nitrate pollution emerges as a significant contributor. Elevated groundwater concentrations of NO_{3-} often result from agricultural activities, particularly the use of chemical fertilizers, as well as from improper waste disposal, and sewage infiltration. These processes introduce excessive NO_{3-} into groundwater systems, leading to a decline in overall water quality and suitability for human consumption and agricultural use [99,103,108,111–113]. Another critical factor affecting groundwater quality is salinization. This phenomenon is characterized by high levels of Cl^- in groundwater, which can be attributed to several causes, such as seawater intrusion into coastal aquifers, the presence of ancient saline formations, the weathering of salt-rich rocks, and the dissolution of halide minerals. Such conditions not only compromise the usability of groundwater for drinking but also pose challenges for agricultural irrigation and soil health [99,103,113,114]. In addition to these processes, additional human activities have been shown to exacerbate groundwater contamination. Industrial operations, mining activities, and changes in land use often introduce pollutants into aquifers, further reducing water quality. The discharge of industrial effluents and the alteration of natural landscapes can mobilize harmful substances into groundwater systems, making them less suitable for any form of use [23,115–118]. Lastly, geogenic factors play a significant role in shaping groundwater quality. Naturally occurring processes such as the dissolution of host rock minerals release PTEs and major dissolved ions into aquifers. While these processes are natural, their contributions to groundwater chemistry can significantly influence the values of geo-environmental indices, particularly in areas with specific geological conditions [113,119,120]. All the factors mentioned above directly influence the final scores of geo-environmental indices. These scores not only pinpoint areas requiring close monitoring but also emphasize the urgent need for effective management strategies to mitigate contamination from both anthropogenic and geogenic sources.

3.2.1. Drinking Purposes

To assess groundwater suitability for drinking purposes, five geo-environmental indices—NPI, RI, PIG, WQI, and WPI—were calculated, with the results summarized in Table 4. The majority of the samples suggest that groundwater quality is generally good and meets the standards for drinking, based on the indices analyzed. However, specific samples, particularly those evaluated by the NPI and PIG indices, are classified as unsuitable for drinking. These samples, predominantly from the Schinos area, display elevated concentrations of certain major elements (e.g., Cl^-, Na^+, NO_{3-}). Additionally, it is important to emphasize that these geo-environmental indices neglect the influence of determined PTEs, whose toxicity can pose serious risks to human health and other living organisms.

Table 4. The number and percentage of samples for five geo-environmental indices (NPI, RI, PIG, WQI, and WPI), and related to drinking purposes, based on 20 groundwater samples from the Loutraki area, 33 samples from the Schinos area, and 15 samples from the Gerania Mts.

Geo-Environmental Index	Classes	Description	Total		Loutraki Area		Schinos Area		Gerania Mts.	
			N	%	N	%	N	%	N	%
NPI	<0	Clean	33	48.5	17	85	3	9.1	13	86.7
	0–1	Lightly polluted	12	17.6	3	15	7	21.2	2	13.3
	1–2	Moderately polluted	5	7.4	0	0	5	15.2	0	0
	2–3	Significantly polluted	4	5.9	0	0	4	12.1	0	0
	>3	Very significantly polluted	14	20.6	0	0	14	42.4	0	0
RI	<0.5	Unaffected	44	64.7	19	95	11	33.3	14	93.3
	0.5–6.6	Slightly affected	24	35.3	1	5	22	66.7	1	6.7
	>6.6	Strongly affected	0	0	0	0	0	0	0	0
PIG	<1	No significant pollution	39	57.4	20	100	4	12.1	15	100
	1–1.5	Low pollution	17	25.0	0	0	17	51.5	0	0
	1.5–2	Moderate pollution	8	11.8	0	0	8	24.2	0	0
	2–2.5	High pollution	3	4.4	0	0	3	9.1	0	0
	>2.5	Very high pollution	1	1.5	0	0	1	3.0	0	0
WQI	<50	Excellent	46	67.6	20	100	11	33.3	15	100
	50–100	Good	19	27.9	0	0	19	57.6	0	0
	100–200	Poor	3	4.4	0	0	3	9.1	0	0
	200–300	Very poor	0	0.0	0	0	0	0.0	0	0
	>300	Unsuitable	0	0.0	0	0	0	0.0	0	0
WPI	<0.5	Excellent	52	76.5	20	100	16	48.5	15	100
	0.5–0.75	Good	14	20.6	0	0	14	42.4	0	0
	0.75–1	Moderately polluted	3	4.4	0	0	3	9.1	0	0
	>1	Highly polluted	0	0	0	0	0	0	0	0

3.2.2. Irrigation Purposes

To evaluate groundwater suitability for irrigation, indices such as SAR, KR, %Na, PS, MAR, RSC, SSP, TH, PI, IWQI, and TDS were calculated, analyzed, and assessed. According to established classification systems, most indices suggest that groundwater in the study areas is suitable for irrigation, posing minimal health risks and having only minor impacts on soil quality (Table 5). However, some indices, particularly the PS index in certain Schinos samples, indicate moderate water quality. This index, which incorporates Cl^-, showed elevated values in areas experiencing groundwater salinization, such as the Schinos coastal aquifer [28,30]. These findings illustrate that different geo-environmental indices applied to the same water samples within the same area can yield contrasting results, due to variations in the parameters each index evaluates, whether cations, anions, or other chemical elements.

Table 5. The number and percentage of samples for 11 geo-environmental indices (SAR, KR, %Na, PS, MAR, RSC, SSP, TH, PI, IWQI, and TDS), and related to irrigation purposes, based on 20 groundwater samples from the Loutraki area, 33 samples from the Schinos area, and 15 samples from the Gerania Mts.

Geo-Environmental Index	Classes	Description	Total		Loutraki Area		Schinos Area		Gerania Mts.	
			N	%	N	%	N	%	N	%
SAR	<10	Excellent	68	100	20	100	33	100	15	100
	10–18	Good	0	0	0	0	0	0	0	0
	18–26	Doubtful	0	0	0	0	0	0	0	0
	>26	Unsuitable	0	0	0	0	0	0	0	0
KR	<1	Suitable	66	97.1	20	100	31	93.9	15	100
	>1	Unsuitable	2	2.94	0	0	2	6.06	0	0
%Na	<20%	Excellent	50	73.5	20	100	15	45.5	15	100
	20–40%	Good	16	23.5	0	0	16	48.5	0	0
	40–60%	Permissible	2	2.94	0	0	2	6.06	0	0
	60–80%	Doubtful	0	0	0	0	0	0	0	0
	>80%	Unsuitable	0	0	0	0	0	0	0	0
PS	<3	Safe	37	54.4	19	95	3	9.09	15	100
	>3	Unsafe	31	45.6	1	5	30	90.9	0	0
MAR	<50	Suitable	0	0	0	0	0	0	0	0
	>50	Unsuitable	68	100	20	100	33	100	15	100
RSC	<1.25	Good	68	100	20	100	33	100	15	100
	1.25–2.5	Moderate	0	0	0	0	0	0	0	0
	>2.5	Unsuitable	0	0	0	0	0	0	0	0
SSP	<50	Suitable	66	97.1	20	100	31	93.9	15	100
	>50	Unsuitable	2	2.94	0	0	2	6.06	0	0
TH	<60 mg/L	Soft	0	0	0	0	0	0	0	0
	60–120 mg/L	Moderately hard	0	0	0	0	0	0	0	0
	120–180 mg/L	Hard	0	0	0	0	0	0	0	0
	>180 mg/L	Very hard	68	100	20	100	33	100	15	100
PI	>75%	Suitable	0	0	0	0	0	0	0	0
	25–75%	Good	67	98.5	20	100	33	100	14	93.3
	<25%	Unsuitable	1	1.47	0	0	0	0	1	6.67
IWQI	85–100	No restriction	0	0	0	0	0	0	0	0
	70–85	Low restriction	15	22.1	6	30	4	12.1	5	33.3
	55–70	Moderate Restriction	39	57.4	14	70	19	57.6	6	40
	40–55	High restriction	8	11.8	0	0	4	12.1	4	26.7
	0–40	Severe restriction	6	8.82	0	0	6	18.2	0	0
TDS	>500 mg/L	Unsuitable	48	70.6	6	30	33	100	9	60
	<500 mg/L	Suitable	20	29.4	14	70	0	0	6	40

3.2.3. PTE Load Evaluation

The geo-environmental indices applied to assess PTEs in this study include Cd, HPI, and HEI. These indices provide a basis for evaluating water quality concerning trace elements for determining water usability. Elements included in these calculations—such as As, Cr, Cu, Mn, Ni, Pb, Sb, Co, and Zn—represent key PTEs with environmental significance, as highlighted in recent studies [121]. Results indicate that most samples were classified as suitable for use or of good quality/low contamination (Table 6). However, some groundwater samples from the Schinos area displayed moderate to poor quality due to elevated concentrations of specific PTEs, including Cr [as Cr(VI)] and As [28,30,38].

Table 6. The number and percentage of samples for three geo-environmental indices (C_d, HPI, and HEI), and related to PTE load, based on 20 groundwater samples from the Loutraki area, 33 samples from the Schinos area, and 15 samples from the Gerania Mts.

Geo-Environmental Index	Classes	Description	Total		Loutraki Area		Schinos Area		Gerania Mts.	
			N	%	N	%	N	%	N	%
C_d	<1	Low	68	100	20	100	33	100	15	100
	1–3	Medium	0	0	0	0	0	0	0	0
	>3	High	0	0	0	0	0	0	0	0
HPI	<100	Suitable	53	77.9	19	95	19	57.6	15	100
	>100	Unsuitable	15	22.1	1	5	13	39.4	0	0
HEI	<10	Low	66	97.1	20	100	31	93.9	15	100
	10–20	Medium	2	2.94	0	0	2	6.06	0	0
	>20	High	0	0	0	0	0	0	0	0

3.2.4. Ionic Ratios Results

Ionic ratios offer valuable insights into the hydrogeochemical characteristics of groundwater resources, shedding light on factors such as aquifer matrix lithology, salinization processes, and the differentiation between anthropogenic and geogenic influences. The statistical distributions of the calculated ionic ratios (e.g., Ca/Mg, Ca/SO_4, Ca/Na, Cl/NO_3, Cl/HCO_3, and Si/NO_3) are illustrated as boxplots in Figure 5. Notably, groundwater samples from the Schinos area exhibit a broader range in certain ionic ratios, such as Ca/Mg (Figure 5a), Ca/SO_4 (Figure 5b), Ca/Na (Figure 5c), Cl/NO_3 (Figure 5d), and Cl/HCO_3 (Figure 5e), which can be attributed to the intense seawater intrusion regime of this aquifer [28–30], characterized by elevated groundwater concentrations of dissolved ions such as Na^+ and Cl^-. For Ca/Mg ratios (Figure 5a), the low values recorded in the Loutraki area and the Gerania Mts. suggest the predominance of Mg^{2+} over Ca^{2+}, as previously discussed. For Si/NO_3 ratios (Figure 5f), the lower Si/NO_3 values in the groundwater of the Schinos area indicate elevated groundwater NO_3- concentrations, reflecting the influence of specific anthropogenic activities on groundwater quality. In contrast, samples from the Gerania Mts. display a wide range of Si/NO_3 ratios. However, the lower values in this region cannot be attributed to anthropogenic activities, as evidenced by the low NO_3- concentrations (up to 15.6 mg/L) and the lack of significant human impact on this sub-basin. Instead, the low Si/NO_3 ratios are associated with low Si concentrations (median concentration: 2.76 mg/L; minimum concentration <40 μg/L [30]) in the groundwater, which can be explained by the short groundwater residence time within the aquifer, the limited presence of silicate mineral phases, and their inherently low solubility as previously discussed. These findings underscore the importance of combining ionic ratios with elemental concentration data for a comprehensive understanding of groundwater hydrogeochemistry.

3.3. Coupling Geo-Environmental Indices for Assess the Hydrogeochemical Properties of an Area

The geo-environmental indices examined, though relatively easy to apply, are somewhat limited in scope, as they primarily focus on the concentrations of selected elements, neglecting other PTEs crucial to the health of living organisms and ecosystems. Correlation and multivariate statistical analyses, including FA and R-mode HCA, revealed notable patterns between various geo-environmental indices. Figure 6 presents the Spearman correlation coefficients for 25 calculated parameters based on 68 groundwater samples in the study area. The strongest statistically significant correlations are highlighted using a color-coded scale: positive correlations are depicted in shades ranging from orange to red, while negative correlations are represented by light to dark blue (see Figure 6). Correlations selected for further analysis are those with p-values below 0.01 or 0.05.

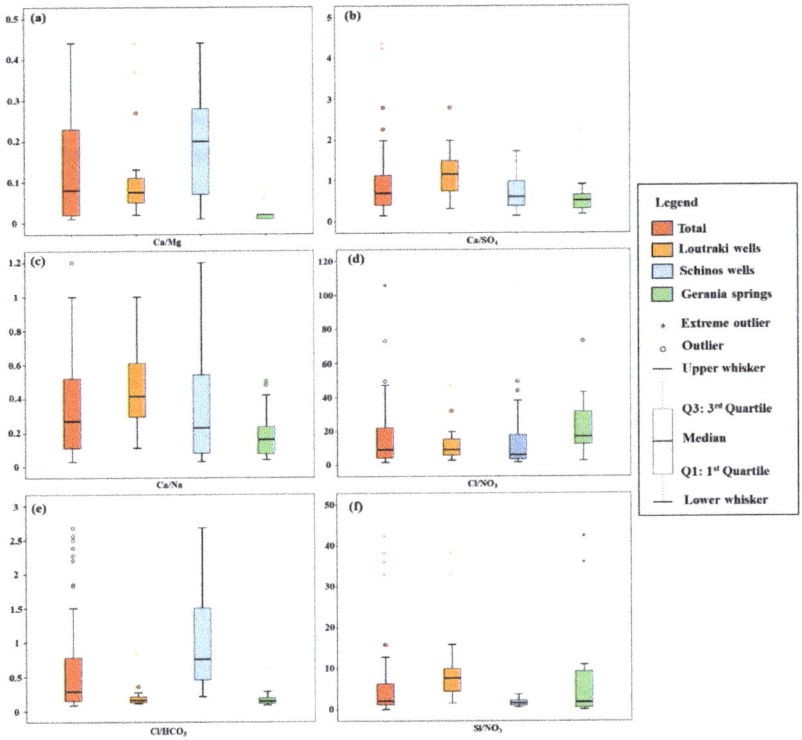

Figure 5. Boxplots of ionic ratios (**a**) Ca/Mg, (**b**) Ca/SO_4, (**c**) Ca/Na, (**d**) Cl/NO_3, (**e**) Cl/HCO_3, and (**f**) Si/NO_3 for the groundwater samples from the Loutraki–Schinos–Gerania Mts. region.

It is common for many geo-environmental indices, particularly those related to water quality for drinking purposes, to exhibit very strong, statistically significant correlations such as PIG and WQI (r = 0.977). However, it is especially challenging to explore the statistical patterns among indices associated with different suitability uses or ionic ratios. This evaluation can be effectively carried out using multivariate statistical tools such as FA via the method of PCA and R-mode HCA.

All 25 variables (NPI, RI, PIG, WQI, WPI, SAR, KR, %Na, PS, MAR, RSC, SSP, TH, PI, IWQI, TDS, C_d, HPI, HEI, Ca/Mg, Ca/SO_4, Ca/Na, Cl/NO_3, Cl/HCO_3, and Si/NO_3) were used to calculate multivariate statistics for the 68 groundwater samples. Geo-environmental indices, calculated using formulas that incorporate determined chemical element data, are inherently dependent variables; however, the literature often treats them as independent variables (e.g., [122,123]). This direction does not constrain our research, as such data are well suited for multivariate statistical analyses [83–85]. Furthermore, the objective of this study is to explore the interrelationships among geo-environmental indices and uncover underlying patterns by grouping these indices to reveal hidden connections. This approach facilitates the identification of key chemical parameters and processes associated with each factor or cluster, enhancing their applicability in environmental science and groundwater pollution studies.

Regarding FA, the scree plot method (Cattell, 1966), depicted in Figure 7, highlights that five components have eigenvalues exceeding 1, meeting Kaiser's criterion [86] for classification as principal components, while the remaining factors are excluded.

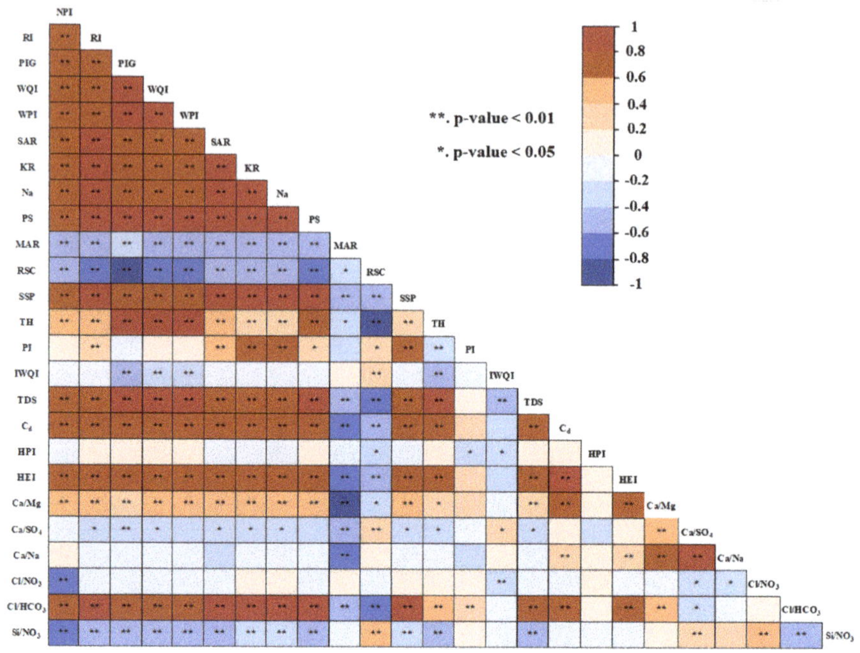

Figure 6. The Spearman correlation matrix, along with significance levels (*p*-values), for the 25 calculated geo-environmental indices in the Loutraki–Schinos–Gerania Mts. region (n = 68 groundwater samples).

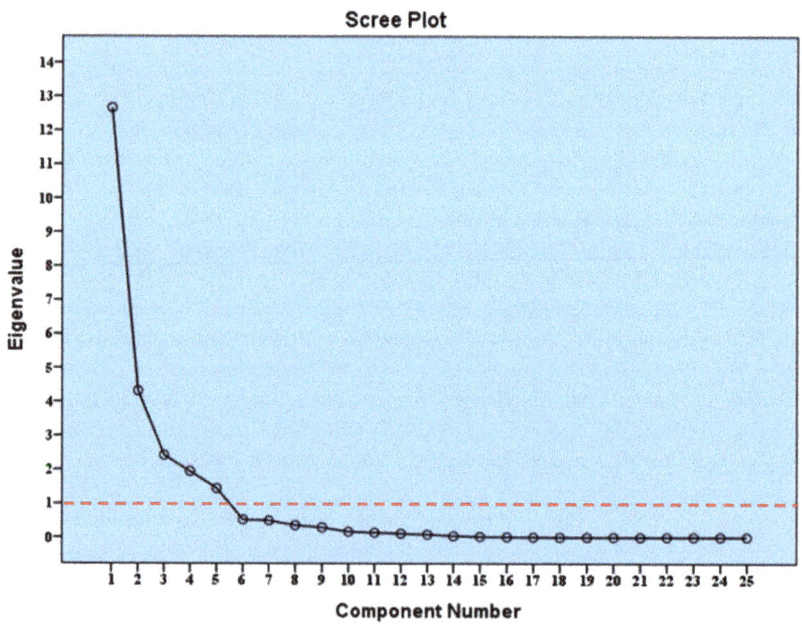

Figure 7. The scree plot of eigenvalues for the components derived from 68 groundwater samples collected from the Loutraki–Schinos–Gerania Mts. region indicates that five components have eigenvalues > 1, signifying their statistical significance within the FA approach.

The five factors explained 91.162% of the total data variance. The KMO value is equal to 0.673, indicating statistically significant results. Additionally, Bartlett's test of sphericity yielded a *p*-value < 0.05, confirming the validity and suitability of the data for FA. These two criteria, KMO and Bartlett's test of sphericity, are widely used in geochemical studies (e.g., [17,21,89]) and are necessary to verify the quality of the multivariate statistical analysis. Table 7 presents the outcomes of the FA, conducted using the PCA method, for the 68 groundwater samples from the Loutraki–Schinos–Gerania Mts. region. The parameter loadings are color-coded: strong loadings are shown in red, moderate loadings in light orange, and weak loadings in light blue. The results of the FA are presented in Table 7 and can be summarized as follows:

- The first factor (FA1) accounts for 50.629% of the total variance. It features strong positive loadings for variables TH (0.97), PIG (0.94), TDS (0.90), WQI (0.87), WPI (0.86), and PS (0.84). Additionally, it includes a strong negative loading for RSC (−0.92); moderate positive loadings for RI (0.71), Cl/HCO_3 (0.71), and SAR (0.52); a moderate negative loading for IWQI (−0.68); and weak positive loadings for KR (0.36), SSP (0.41), %Na (0.41), C_d (0.31), HEI (0.31), Cl/NO_3 (0.46), and NPI (0.37).
- The second factor (FA2) explains 17.251% of the total variance. It shows strong positive loadings for PI (0.95), KR (0.91), SSP (0.89), %Na (0.89), and SAR (0.83). Moderate positive loadings are observed for PS (0.50), RI (0.61), and Cl/HCO_3 (0.61), while weak positive loadings appear for TDS (0.36), WQI (0.37), and WPI (0.36). There is also a weak negative loading for IWQI (−0.39).
- The third factor (FA3) accounts for 9.708% of the total variance, with strong positive loadings for HPI (0.94), C_d (0.87), and HEI (0.87); a weak positive loading for Ca/Mg (0.39); and a weak negative loading for MAR (−0.39).
- The fourth factor (FA4) explains 7.781% of the total variance. It includes strong positive loadings for Ca/Na (0.87), Ca/SO_4 (0.80), and Ca/Mg (0.79); a strong negative loading for MAR (−0.80); and weak positive loadings for NPI (0.34) and Si/NO_3 (0.38).
- The fifth factor (FA5) accounts for 5.793% of the total variance and features a strong negative loading for Cl/NO_3 (−0.79), a moderate positive loading for NPI (0.74), and a moderate negative loading for Si/NO_3 (−0.69).

Table 7. Varimax-rotated principal components of 68 groundwater samples from the Loutraki–Schinos–Gerania Mts. region; strong, moderate, and weak loadings of the parameters are given with red, orange, and blue colors, respectively.

Geo-Environmental Index	Component				
	FA1	FA2	FA3	FA4	FA5
TH	**0.97**	−0.03	0.19	0.03	0.07
PIG	**0.94**	0.25	0.15	0.06	0.14
RSC	**−0.92**	−0.01	−0.29	−0.01	0.08
TDS	**0.90**	0.36	0.10	0.05	0.09
WQI	**0.87**	0.37	0.16	0.12	0.23
WPI	**0.86**	0.36	0.17	0.11	0.24
PS	**0.84**	0.50	0.12	0.01	−0.06
RI	**0.71**	0.61	0.19	−0.01	−0.06
Cl/HCO_3	**0.71**	0.61	0.19	−0.01	−0.06
IWQI	**−0.68**	−0.39	0.01	0.16	0.21
PI	−0.15	**0.95**	−0.14	0.00	−0.02
KR	0.36	**0.91**	0.04	−0.07	0.01
SSP	0.41	**0.89**	0.08	−0.02	0.09
%Na	0.41	**0.89**	0.08	−0.02	0.09
SAR	0.52	**0.83**	0.06	−0.07	−0.01
HPI	0.06	−0.06	**0.94**	0.03	−0.02
C_d	0.31	0.08	**0.87**	0.23	0.19

Table 7. *Cont.*

Geo-Environmental Index	Component				
	FA1	FA2	FA3	FA4	FA5
HEI	0.31	0.08	**0.87**	0.23	0.19
Ca/Na	−0.06	−0.27	0.22	**0.87**	−0.02
Ca/SO_4	−0.20	−0.16	−0.23	**0.80**	−0.02
MAR	−0.28	−0.22	−0.39	**−0.80**	−0.08
Ca/Mg	0.27	0.21	0.39	**0.79**	0.06
Cl/NO_3	0.46	0.10	−0.04	−0.08	**−0.79**
NPI	0.37	0.07	0.21	0.34	**0.74**
Si/NO_3	−0.26	−0.10	−0.12	0.38	**−0.69**
Initial eigenvalues of variances in %	**50.629**	**17.251**	**9.708**	**7.781**	**5.793**
Cumulative % of variance	**50.629**	**67.880**	**77.588**	**85.370**	**91.162**

Bold values indicate significant score loadings.

These five factors collectively explain the majority of the variance in the data, highlighting the most influential variables.

On the other hand, the R-mode HCA employed Ward's method (1963) as the linkage rule, using squared Euclidean distances to measure similarity between variables, a method that is extensively utilized in other geochemical studies (e.g., [101,124,125]). The dendrogram in Figure 8 illustrates the results of the R-mode HCA, depicting 68 groundwater samples collected from the Loutraki–Schinos–Gerania Mts. region. Variables with a linkage distance equal to eight (marked by the red dashed line in Figure 8) and equal to five (indicated by the yellow dashed line in Figure 8) were clustered together, forming distinct groups characterized by similar patterns. As shown in the dendrogram (Figure 8), the variables were partitioned into six clusters at lower linkage distances and three clusters at greater linkage distances. The relationship between the two approaches is that the additional clusters formed by the yellow dashed line represent proximity between specific variables. These variables are grouped together at a higher linkage distance, as indicated by the red dashed line. The three clusters created from a linkage distance equal to eight are as follows:

- Cluster C1: C_d, HEI, HPI, NPI, and Ca/Mg.
- Cluster C2: %Na, SSP, SAR, KR, PI, RI, Cl/HCO_3, PS, WQI, WPI, TDS, PIG, and TH.
- Cluster C3: RSC, IWQI, MAR, Ca/SO_4, Ca/Na, Cl/NO_3, and Si/NO_3

The six clusters created from a linkage distance equal to five are as follows:

- Cluster C1: C_d, HEI, HPI, NPI, and Ca/Mg.
- Cluster C2A: %Na, SSP, SAR, KR, and PI.
- Cluster C2B: RI, Cl/HCO_3, PS, WQI, WPI, TDS, PIG, and TH.
- Cluster C3A: RSC, IWQI, and MAR.
- Cluster C3B: Ca/SO_4 and Ca/Na.
- Cluster C3C: Cl/NO_3 and Si/NO_3.

Considering the results of correlation and multivariate statistical analyses, distinct patterns emerge among the geo-environmental indices. FA1 reflects indices that are strongly influenced by elevated concentrations of dissolved major ions, such as TH, PIG, TDS, WQI, WPI, RI, and Cl/HCO_3. In particular, the strong negative loadings of RSC and IWQI in this factor indicate the influence of alkalinity on other ions in the aqueous solution. Notably, ions like HCO_{3^-} and Cl^-, which occur in higher concentrations, play a key role in defining the geo-environmental indices associated with this factor. Processes that increase dissolved salts in groundwater, such as seawater intrusion, exert a significant influence on these geo-environmental indices. This is further confirmed by the Spearman's correlation coefficients and the R-mode HCA (cluster C2B), which highlight the close relationships among these indices. Additionally, Na-related indices are included in this category due to the strong positive correlation observed between Cl^- and Na^+ in most hydrogeochemical studies (e.g., [89]). FA2 identifies a group of Na-related indices, such as PI, SAR, %Na, KR, and SSP, with their proximity further supported by cluster C2A of the HCA. FA3 points to indices related to the presence of PTEs in groundwater, such as C_d, HEI, and HPI; these indices are grouped within C1 of the HCA. FA4 highlights a group of Ca-related indices, such as Ca/Na, Ca/SO_4, Ca/Mg, and MAR. The negative loading of MAR is linked to the inverse relationship of Ca, where high Ca levels correspond with low MAR values. It is important to note that some of the indices mentioned above belong to different HCA clusters, reflecting different interpretations for each group. For instance, the Ca/Mg ionic ratio is part of C1, indicating its relationship with PTE loadings in groundwater or anthropogenic influences, due to the presence of PTE-related indices and NPI. In contrast, MAR, Ca/SO_4, and Ca/Na are grouped in C3, indicating that this cluster is associated with increased dissolved salts in groundwater. Lastly, FA5 highlights indices such as NPI, Si/NO_3, and Cl/NO_3, which reveal anthropogenic influences in the area. These indices also provide insights into other hydrogeochemical processes, including geogenic contributions, seawater intrusion, and the specific sources of N in the environment.

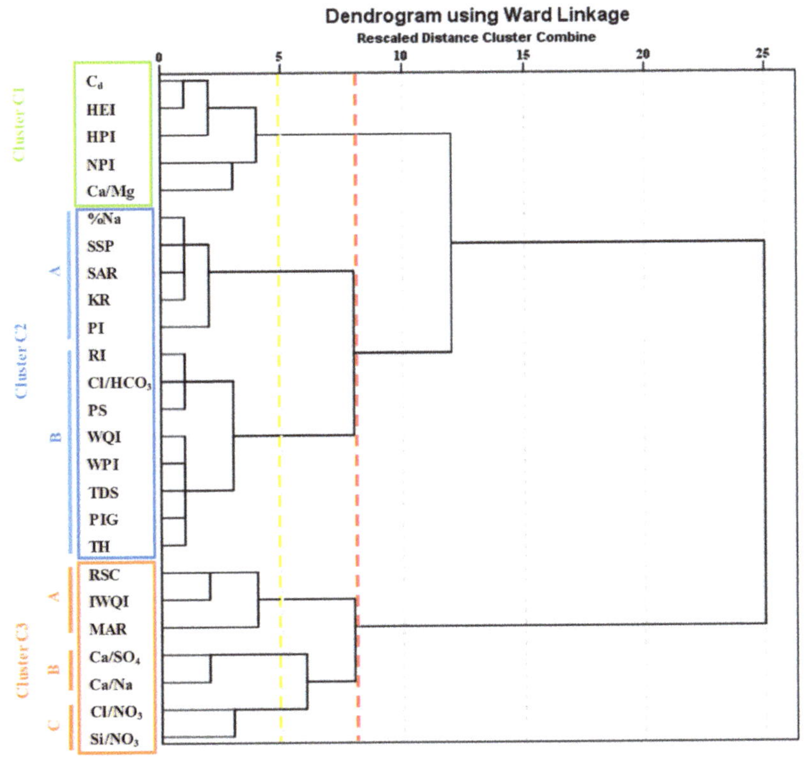

Figure 8. Dendrogram of R-mode HCA for 25 variables calculated in 68 groundwater samples from the Loutraki–Schinos–Gerania Mts. region. The red and yellow dashed lines represent the linkage distances used to create different distinct clusters.

The results of the geo-environmental indices and their statistical analysis indicate that many indices exhibit significant similarities, making it unnecessary to calculate all of them. Specifically, the identified strong, statistically significant correlations suggest that successful groundwater suitability assessments and hydrogeochemical evaluations can be achieved without relying on an extensive number of geo-environmental indices. However, the selection of appropriate geo-environmental indices remains challenging, as it depends on various factors, including geology, land use, anthropogenic activities, proximity to the sea, etc. Therefore, we strongly recommend prioritizing the efficient utilization of selected indices and adopting a more robust methodological framework, both of which are crucial in the scientific disciplines of hydrogeochemistry and groundwater management. Focusing on key geo-environmental indices based on critical parameters/elements (e.g., Ca, Na, PTE, etc.) and the specific requirements of each study (e.g., water use, land use) is preferable. Nonetheless, it is important to recognize that different indices used to assess water suitability for a specific use may occasionally produce contradictory results, potentially leading to incorrect conclusions. According to Tables 4–6, certain indices yield conflicting results regarding the suitability of groundwater samples. While some indices classify all samples as suitable, others consider them entirely unsuitable or unsafe. For instance, the MAR and SAR indices, both used to evaluate irrigation suitability, produce opposing conclusions: the MAR index indicates that all groundwater samples in the Schinos area are unsuitable for irrigation, whereas the SAR index classifies them as suitable. Similar inconsistencies are also observed with other indices. This discrepancy underscores the need for a thorough hydrogeochemical assessment to ensure accurate and holistic interpretation of the data. In the case of the present study area, the presence of ultramafic rocks significantly influences the MAR index values due to the increased Mg^{2+} groundwater concentrations.

Another limitation is that some geo-environmental indices have values influenced by the relative weight of their parameters (e.g., PIG, WQI) or rely on ideal and/or guideline values (e.g., HPI, NPI, PIG, WQI, etc.), a consideration that impacts the final result of the index and is directly associated with the researcher's experience/knowledge and choices. It is likely that different researchers calculating the same index may obtain varying values due to differences in methodology or parameter weighting, which, in some extreme cases, could lead to different interpretations. Therefore, the most crucial principle in such studies is that the geo-environmental indices should not be prioritized over elemental concentrations. Thus, they serve as tools to highlight water suitability or to compare its quality against certain standards. Additionally, some indices are based on identical or nearly identical calculation methods, making their inclusion redundant and unnecessary for further consideration. Some geo-environmental indices show similar results with only minor variations (e.g., %Na-SSP, WQI-PIG, RI-Cl/HCO_3, MAR-Ca/Mg, etc.). This observation is further reinforced by the individual index results and the statistical analysis, which reveals correlation values close to 1, similar factor loadings in the FA, and low linkage distances in the HCA (Figures 6–8, Table 7). Generally, the application of multivariate statistical methods highlights significant similarities among geo-environmental indices, raising concerns about the necessity of calculating all of them individually and underscoring the importance of focusing on the most meaningful indices to enhance analytical efficiency and provide more targeted, insightful results. It is important to note that a water sample may show excellent quality in terms of major element chemistry, yet still contain elevated levels of PTEs. In such cases, the water might be classified as good or excellent and deemed suitable for drinking and irrigation based solely on its major element concentrations. However, an analysis focusing on PTE-related indices could categorize the same water as poor quality. This highlights the importance of conducting a combined analysis using multiple indices, as recommended by this study, to achieve a more comprehensive and accurate assessment of water quality and its suitability for various uses. For instance, coupling two geo-environmental indices with different objectives can yield critical insights into chemometric analysis and water use, ultimately enhancing groundwater management.

This clustering effect stems from the significant influence certain elements exert on the final index values. While each index is typically assessed separately, combining indices from different categories allows for a more robust evaluation of groundwater quality. For example, Figure 9 illustrates a bivariate plot of the WQI, which assesses water suitability for drinking, alongside the HPI, which evaluates water quality based on the occurrence of PTEs. This integrated approach creates six distinct sub-fields, enabling the classification of samples based on their ability to meet both criteria for good water quality. By merging these indices, the assessment of groundwater quality is significantly enhanced, offering a more comprehensive understanding compared to the use of individual indices alone. For example, the coupling of WQI and HPI revealed that 64.7% of the samples fall into the excellent and suitable category, 11.8% into the good and suitable category, 1.47% into the poor and suitable category, 2.94% into the excellent and unsuitable category, 16.2% into the good and unsuitable category, and 2.94% into the poor and unsuitable category. All samples from the Loutraki area and the Gerania Mts. are classified as having excellent water quality based on WQI, with only one sample from the Loutraki area deemed unsuitable according to HPI. In contrast, the majority of samples from the Schinos area exhibit lower water quality when evaluated using both WQI and HPI results (Figure 9).

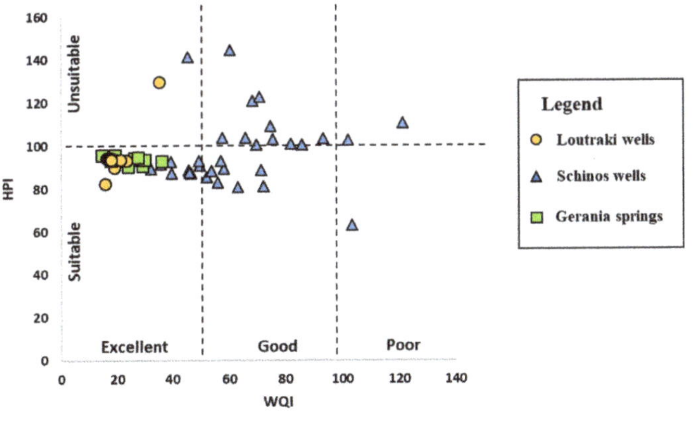

Figure 9. Cross-plot of WQI vs. HPI for the 68 groundwater samples from the Loutraki–Schinos–Gerania Mts. Region.

The combined analysis of ionic ratios is a valuable tool for understanding the occurrence, mobility, and transport of PTEs in the environment. In this investigation, the methodology of Papazotos et al. [21], used to distinguish the sources of PTEs, was followed, an approach that was first applied to various PTEs in the Psachna Basin, Euboea, Greece. Figures 10 and 11 display several ionic ratio diagrams that illustrate the dominant hydrochemical processes in the study area. Figure 10a presents the Cl/HCO_3–Cl and Cl/HCO_3–HCO_3 ratio diagrams for the Loutraki–Schinos–Gerania Mts. region. The majority of the groundwater samples from the study area are plotted in the 'carbonate dissolution' field, indicating that groundwater recharge is the dominant hydrogeological process. However, some samples exhibit a molar ratio greater than 0.5 (and in several cases, >1), reflecting the influence of seawater intrusion in certain parts of the study area. Figure 10b presents the Si/NO_3–Si and Si/NO_3–NO_3 ratio diagrams for the Loutraki–Schinos–Gerania Mts. region. Most groundwater samples fall within the 'silicate dissolution' field, while some from the Schinos area are plotted in the 'nitrate pollution' field. Samples outside these two ranges may be influenced by other factors, such as carbonate mineral dissolution, as indicated in Figure 10a. It is important to emphasize that, in any case, the coexistence of geogenic and anthropogenic activities (such as agricultural practices, sewage discharges, etc.) cannot be ruled out. However, this should be further investigated by analyzing the complete dataset and other qualitative criteria. The concentration of chemical elements is a crucial indicator in environmental geochemical research and should be carefully considered. Neglecting these concentrations can lead to misleading results, as focusing solely on molar ratios may not provide the desired information. For instance, areas unaffected by human activities typically exhibit NO_3- concentrations below 10 mg/L [50]. Therefore, NO_3- concentrations below 0.16 mmol/L should not be attributed to anthropogenic influences, regardless of the Si/NO_3 molar ratio. Conversely, a high or very high Si/NO_3 molar ratio accompanied by low Si concentrations should not be attributed to the dissolution of silicate minerals but rather to other natural factors, such as the dissolution of carbonates, or the influence of seawater or rainwater.

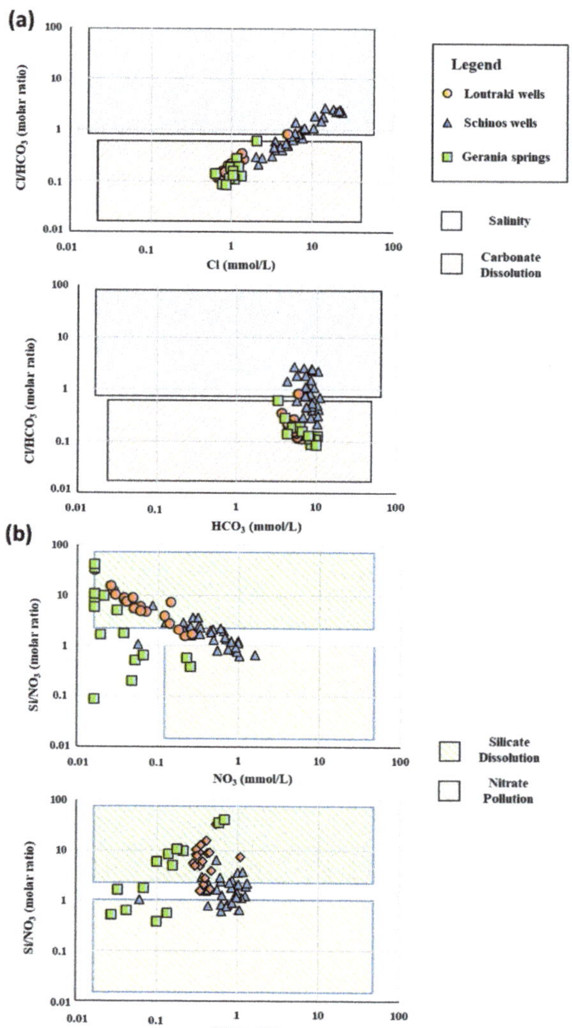

Figure 10. Bivariate diagrams of (**a**) Cl/HCO_3 (molar ratio) vs. Cl (mmol/L) and Cl/HCO_3 (molar ratio) vs. HCO_3 (mmol/L) and (**b**) Si/NO_3 (molar ratio) vs. NO_3 (mmol/L) and Si/NO_3 (molar ratio) vs. Si (mmol/L) for the 68 groundwater samples from the Loutraki–Schinos–Gerania Mts. region.

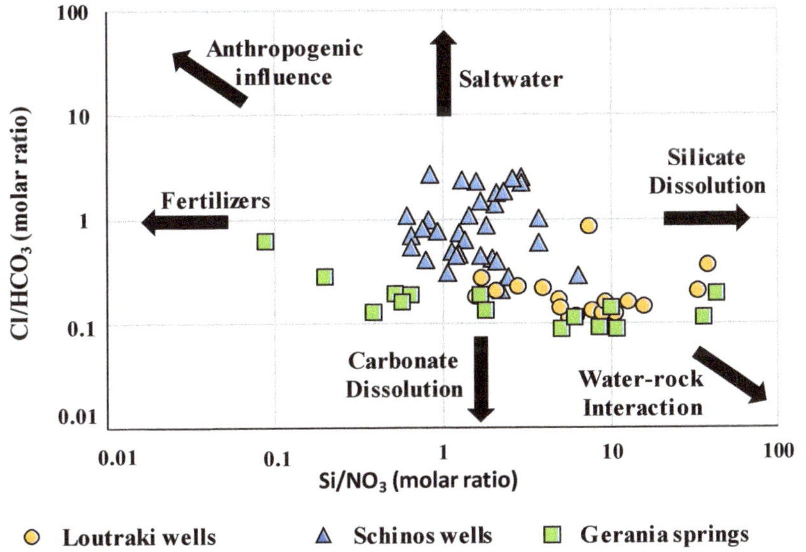

Figure 11. Evaluation of major hydrogeochemical processes affecting water chemistry using Cl/HCO_3 vs. Si/NO_3 diagram.

Finally, combining the information presented, the coupling of the two molar ratios (Cl/HCO_3 and Si/NO_3) can provide valuable insights into the groundwater geochemistry of a complex aquifer system. This approach accounts for processes such as seawater intrusion, agricultural activities, and water–rock/soil interactions, including the involvement of both carbonate and silicate mineral phases (Figure 11). As shown in Figure 11, the samples from the springs of the Gerania Mt. area are characterized by relatively stable Cl/HCO_3 ionic ratios and varying Si/NO_3 ionic ratios. However, the co-evaluation with the diagrams that contain ionic ratios and elemental concentrations (Figure 10) indicates the dominance of geogenic factors in groundwater quality. The same pattern is observed in the samples from the Loutraki area, without the same variability. In contrast, the samples from the Schinos area exhibit an upward and leftward trend, likely due to seawater intrusion and human activities, which contribute to the degradation of the aquifer table.

The high salinity caused by seawater intrusion into the groundwater is linked to the elevated As concentrations observed in the study area, as concentrations tend to increase in the central and upper parts of the plot of Figure 12a. Multivariate statistical analyses (e.g., FA, HCA) and correlation coefficients from our previous study [30] have revealed a statistical association between As and major/trace elements commonly found in seawater, such as Cl^-, Br, and Li. In contrast, Cr exhibits a different pattern, with low to high concentrations trending from the center to the left of the diagram (Figure 12b). The category with the highest Cr concentrations (>Q3) is dominated by low Si/NO_3 ratios, suggesting that agricultural activities are the primary source of Cr. Multivariate statistical analyses (e.g., FA, HCA), correlation coefficients, and spatial distribution maps from other studies [30,89,126–128] further support the role of fertilization in Cr mobilization in the environment. Thus, coupling ionic ratios can reveal valuable insights into PTE geochemistry, and environmental geoscientists are increasingly recommending the use of these tools in their research.

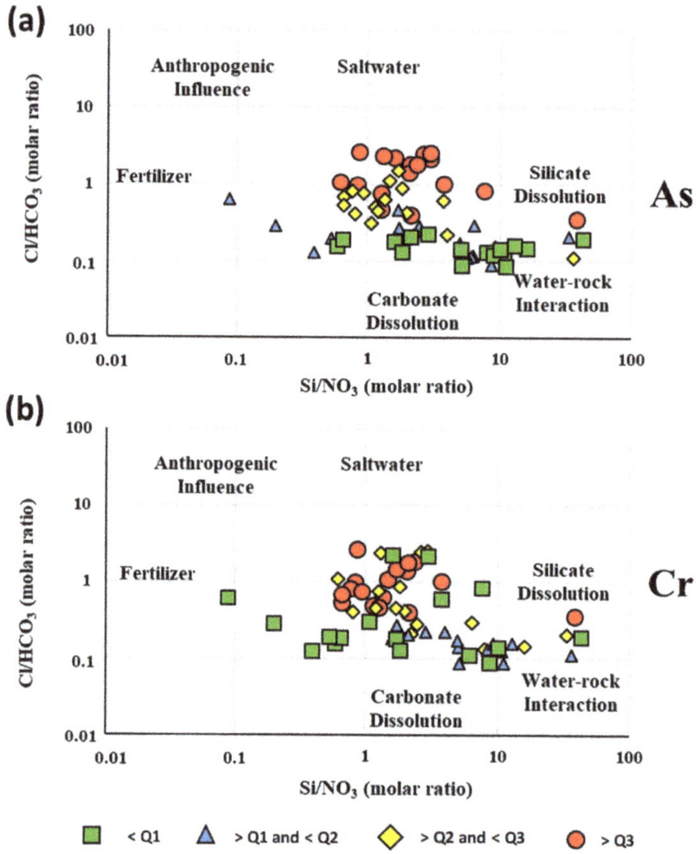

Figure 12. Bivariate Cl/HCO_3 vs. Si/NO_3 diagrams of (**a**) As, and (**b**) Cr (Q1: first quartile; Q2: second quartile or median; Q3: third quartile).

This study suggests the development of a novel more comprehensive geo-environmental index to enhance the accuracy and effectiveness of environmental research. Such an index would integrate both quantitative data, such as concentrations of chemical elements, and qualitative factors, including land use, geology, hydrogeology, and key physicochemical parameters (e.g., pH, Eh, and DO). Future research is encouraged to integrate both qualitative and quantitative groundwater data within this geo-environmental framework, leveraging ML and deep learning (DL) techniques to develop the new index for optimal water quality assessment and management. Furthermore, it is important to note that the new geo-environmental index should replace the outdated and misrepresentation term 'heavy metals' ([129–131] with the modern term 'potentially toxic elements' [121,129]; the established widely used geo-environmental indices require modernization such as HEI and HPI.

This study does not merely offer another groundwater suitability and quality assessment for a study area. It stands apart from other published works by, for the first time, providing a comprehensive exploration of the critical role of geo-environmental indices. The paper offers a detailed evaluation of the strengths and weaknesses of these methodologies and presents specific recommendations for their optimal application, highlighting the novel contributions of this research. Assessing geo-environmental indices and optimizing their application in environmental science can enhance sustainable groundwater management practices and support the achievement of Sustainable Development Goal 6 (SDG-6), particularly in mitigating groundwater pollution. By systematically evaluating geo-environmental indices, pollution sources can be identified, and targeted strategies can be developed to improve water quality, ensuring safe water resources for current and future generations.

4. Conclusions

This paper evaluates the groundwater quality of the NE Peloponnese for various uses by employing geo-environmental indices and ionic ratios. For the first time, data from 68 groundwater samples previously collected in the study area by Papazotos et al. [30] were utilized to calculate these indices and ratios. The analysis also incorporated geological, hydrogeological, geochemical, land/water use, and mineralogical information, applying both classical and multivariate statistical methods. The main findings of this research are summarized as follows:

- Most groundwater samples from the study area are suitable for irrigation purposes. However, some samples from the Schinos area exhibit medium quality due to elevated concentrations of NO_{3^-}, Cl^-, and Na^+. This degradation in water quality is primarily attributed to relatively limited anthropogenic influences on the environment (e.g., small-scale farming, irrigation, low-impact tourism), including the use of septic tanks and fertilizer application, as well as seawater intrusion.
- All samples from the Loutraki area and the Gerania Mts. demonstrate good quality for drinking water. In the Schinos area, while some samples maintain good water quality, others are classified as medium to poor.
- Most samples exhibit good quality regarding PTE concentrations. However, elevated concentrations of Cr and As were detected in some samples from the Loutraki and Schinos areas.
- Based on all indices assessing groundwater suitability for drinking, irrigation, and PTE load, water samples from the Loutraki area and the Gerania Mts. exhibit good to excellent quality, whereas the Schinos area shows comparatively poorer groundwater quality.
- Statistical analysis identified distinct groups of geo-environmental indices based on dominant parameters, suggesting that focusing on the most significant indices can enhance the clarity and relevance of the analysis, reduce redundancy, and improve interpretability, ultimately emphasizing the need for a more robust and systematic methodological framework for assessing groundwater quality rather than relying on numerous geo-environmental indices.
- Some geo-environmental indices designed to evaluate similar water uses yield conflicting results. This issue arises because elemental concentrations are a more critical determinant of water quality than the indices themselves.
- All geo-environmental indices, influenced by parameter weighting (e.g., WQI, HPI, IWQI, etc.), produce results that vary with the researcher's experience/knowledge and methodology, leading to potential differences in interpretation. To minimize inaccuracies in data interpretation, a more robust and standardized scientific approach is required.
- Coupling geo-environmental indices significantly enhances the comprehensive interpretation and evaluation of groundwater quality, optimizing groundwater management.
- Coupling ionic ratios proved valuable in determining the occurrence and mobility of PTEs in the environment. This approach was effectively applied to evaluate As and Cr concentrations in NE Peloponnese, Greece (study area).

In conclusion, further research is recommended to establish regulations governing water use, informed by both usage patterns and epidemiological studies. This research would support policymakers in using multiple-criteria decision analysis (MCDA) and enable stakeholders to enhance public health while optimizing water resource management. Furthermore, this study provides a comprehensive guide for researchers by consolidating the geo-environmental indices used in groundwater quality suitability and hydrogeochemical evaluation. The strategic integration of geo-environmental indices with data from other scientific disciplines (e.g., medicine, biology, chemistry, etc.), could significantly advance environmental sustainability and contribute to achieving the SDGs.

Author Contributions: Conceptualization, P.P.; methodology, P.P., M.V. and D.P.; software, P.P. and MV.; validation, P.P., M.V., D.P. and E.V.; formal analysis, P.P.; investigation, P.P., M.V., D.P. and M.P.; resources, M.P.; data curation, P.P., M.V., D.P., E.V. and M.P.; writing—original draft preparation, P.P.; writing—review and editing, M.V., D.P., E.V. and M.P.; visualization, P.P. and D.P.; supervision, P.P. and M.P.; project administration, M.P.; funding acquisition, E.V. and M.P. All authors have read and agreed to the published version of the manuscript.

Funding: This research received no external funding.

Data Availability Statement: The data presented in this study are available on request from the corresponding author due to privacy restrictions.

Acknowledgments: We would like to thank the three anonymous reviewers for their constructive comments and suggestions, which significantly enhanced the quality of this paper.

Conflicts of Interest: The authors declare no conflicts of interest.

References

1. Priyan, K. Issues and Challenges of Groundwater and Surface Water Management in Semi-Arid Regions. In *Groundwater Resources Development and Planning in the Semi-Arid Region*; Springer: Berlin/Heidelberg, Germany, 2021.
2. Scanlon, B.R.; Fakhreddine, S.; Rateb, A.; de Graaf, I.; Famiglietti, J.; Gleeson, T.; Grafton, R.Q.; Jobbagy, E.; Kebede, S.; Kolusu, S.R.; et al. Global Water Resources and the Role of Groundwater in a Resilient Water Future. *Nat. Rev. Earth Environ.* **2023**, *4*, 87–101. [CrossRef]
3. Qadir, M.; Sharma, B.R.; Bruggeman, A.; Choukr-Allah, R.; Karajeh, F. Non-Conventional Water Resources and Opportunities for Water Augmentation to Achieve Food Security in Water Scarce Countries. *Agric. Water Manag.* **2007**, *87*, 2–22. [CrossRef]
4. Brindha, K.; Schneider, M. Impact of Urbanization on Groundwater Quality. *GIS Geostat. Tech. Groundw. Sci.* **2019**, *2019*, 179–196.
5. Chowdhary, P.; Bharagava, R.N.; Mishra, S.; Khan, N. Role of Industries in Water Scarcity and Its Adverse Effects on Environment and Human Health. In *Environmental Concerns and Sustainable Development*; Springer: Berlin/Heidelberg, Germany, 2020.
6. Lall, U.; Josset, L.; Russo, T. A Snapshot of the World's Groundwater Challenges. *Annu. Rev. Environ. Resour.* **2020**, *45*, 171–194. [CrossRef]
7. Mukherjee, A.; Scanlon, B.R.; Aureli, A.; Langan, S.; Guo, H.; McKenzie, A. Global Groundwater: From Scarcity to Security through Sustainability and Solutions. In *Global Groundwater: Source, Scarcity, Sustainability, Security, and Solutions*; Elsevier: Amsterdam, The Netherlands, 2020.
8. Sharma, R.; Kumar, R.; Agrawal, P.R.; Ittishree; Chankit; Gupta, G. Groundwater Extractions and Climate Change. In *Water Conservation in the Era of Global Climate Change*; Elsevier: Amsterdam, The Netherlands, 2021.
9. Makanda, K.; Nzama, S.; Kanyerere, T. Assessing the Role of Water Resources Protection Practice for Sustainable Water Resources Management: A Review. *Water* **2022**, *14*, 3153. [CrossRef]
10. Chakraborty, T.K.; Islam, M.S.; Ghosh, G.C.; Ghosh, P.; Zaman, S.; Habib, A.; Hossain, M.R.; Bosu, H.; Islam, M.R.; Al Imran, M.; et al. Human Health Risk and Hydro-Geochemical Appraisal of Groundwater in the Southwest Part of Bangladesh Using GIS, Water Quality Indices, and Multivariate Statistical Approaches. *Toxin Rev.* **2023**, *42*, 285–299. [CrossRef]
11. Arumugam, T.; Kinattinkara, S.; Kannithottathil, S.; Velusamy, S.; Krishna, M.; Shanmugamoorthy, M.; Sivakumar, V.; Boobalakrishnan, K.V. Comparative Assessment of Groundwater Quality Indices of Kannur District, Kerala, India Using Multivariate Statistical Approaches and GIS. *Environ. Monit. Assess.* **2023**, *195*, 29. [CrossRef] [PubMed]
12. Kazapoe, R.W.; Addai, M.O.; Amuah, E.E.Y.; Dankwa, P. Hydrogeochemical Characterization of Groundwater in the Wassa Amenfi East and Prestea-Huni Valley Areas of Southern Ghana Using GIS-Based and Multivariate Statistical Techniques. *Sustain. Water Resour. Manag.* **2023**, *9*, 141. [CrossRef]

13. Benyoussef, S.; Arabi, M.; El Yousfi, Y.; Makkaoui, M.; Gueddari, H.; El Ouarghi, H.; Abdaoui, A.; Ghalit, M.; Zegzouti, Y.F.; Azirar, M.; et al. Assessment of Groundwater Quality Using Hydrochemical Process, GIS and Multivariate Statistical Analysis at Central Rif, North Morocco. *Environ. Earth Sci.* **2024**, *83*, 515. [CrossRef]
14. Singh, G.; Singh, J.; Wani, O.A.; Egbueri, J.C.; Agbasi, J.C. Assessment of Groundwater Suitability for Sustainable Irrigation: A Comprehensive Study Using Indexical, Statistical, and Machine Learning Approaches. *Groundw. Sustain. Dev.* **2024**, *24*, 101059. [CrossRef]
15. Egbueri, J.C.; Agbasi, J.C. Data-Driven Soft Computing Modeling of Groundwater Quality Parameters in Southeast Nigeria: Comparing the Performances of Different Algorithms. *Environ. Sci. Pollut. Res.* **2022**, *29*, 38346–38373. [CrossRef] [PubMed]
16. Haggerty, R.; Sun, J.; Yu, H.; Li, Y. Application of Machine Learning in Groundwater Quality Modeling—A Comprehensive Review. *Water Res.* **2023**, *233*, 119745. [CrossRef]
17. Vasileiou, E.; Papazotos, P.; Dimitrakopoulos, D.; Perraki, M. Expounding the Origin of Chromium in Groundwater of the Sarigkiol Basin, Western Macedonia, Greece: A Cohesive Statistical Approach and Hydrochemical Study. *Environ. Monit. Assess.* **2019**, *191*, 509. [CrossRef]
18. Quino-Lima, I.; Ramos-Ramos, O.; Ormachea-Muñoz, M.; Quintanilla-Aguirre, J.; Duwig, C.; Maity, J.P.; Sracek, O.; Bhattacharya, P. Spatial Dependency of Arsenic, Antimony, Boron and Other Trace Elements in the Shallow Groundwater Systems of the Lower Katari Basin, Bolivian Altiplano. *Sci. Total Environ.* **2020**, *719*, 137505. [CrossRef]
19. Ijumulana, J.; Ligate, F.; Bhattacharya, P.; Mtalo, F.; Zhang, C. Spatial Analysis and GIS Mapping of Regional Hotspots and Potential Health Risk of Fluoride Concentrations in Groundwater of Northern Tanzania. *Sci. Total Environ.* **2020**, *735*, 139584. [CrossRef]
20. Verma, P.; Singh, P.K.; Sinha, R.R.; Tiwari, A.K. Assessment of Groundwater Quality Status by Using Water Quality Index (WQI) and Geographic Information System (GIS) Approaches: A Case Study of the Bokaro District, India. *Appl. Water Sci.* **2020**, *10*, 27. [CrossRef]
21. Papazotos, P.; Vasileiou, E.; Vasilakis, S.; Perraki, M. A Novel Hydrogeochemical Approach to Delineate the Origin of Potentially Toxic Elements in Groundwater: Sophisticated Molar Ratios as Environmental Tracers. *Environ. Sci. Pollut. Res.* **2023**, *30*, 74771–74790. [CrossRef]
22. Nwankwoala, H.O.; Okujagu, D.C.; Bolaji, T.A.; Papazotos, P.G.; Ugbenna, K.G. Assessment of Groundwater Quality for Irrigation Suitability: A Case Study of Khana and Gokana LGAs, Rivers State, Nigeria. *Environ. Earth Sci.* **2023**, *82*, 292. [CrossRef]
23. Kumar, M.; Kumari, K.; Ramanathan, A.; Saxena, R. A Comparative Evaluation of Groundwater Suitability for Irrigation and Drinking Purposes in Two Intensively Cultivated Districts of Punjab, India. *Environ. Geol.* **2007**, *53*, 553–574. [CrossRef]
24. Karunanidhi, D.; Aravinthasamy, P.; Deepali, M.; Subramani, T.; Bellows, B.C.; Li, P. Groundwater Quality Evolution Based on Geochemical Modeling and Aptness Testing for Ingestion Using Entropy Water Quality and Total Hazard Indexes in an Urban-Industrial Area (Tiruppur) of Southern India. *Environ. Sci. Pollut. Res.* **2021**, *28*, 18523–18538. [CrossRef]
25. Stamatis, G.; Parpodis, K.; Filintas, A.; Zagana, E. Groundwater Quality, Nitrate Pollution and Irrigation Environmental Management in the Neogene Sediments of an Agricultural Region in Central Thessaly (Greece). *Environ. Earth Sci.* **2011**, *64*, 1081–1105. [CrossRef]
26. Papazotos, P.; Koumantakis, I.; Vasileiou, E. Hydrogeochemical Assessment and Suitability of Groundwater in a Typical Mediterranean Coastal Area: A Case Study of the Marathon Basin, NE Attica, Greece. *HydroResearch* **2019**, *2*, 49–59. [CrossRef]
27. Panagopoulos, G.; Lambrakis, N.; Chalvantzis, C.; Bekiari, V.; Avramidis, P. Assessing the suitability of groundwater for drinking and agricultural uses in the zacharo basin, sw peloponnesus. *Bull. Geol. Soc. Greece* **2017**, *50*, 899–906. [CrossRef]
28. Papadopoulos, K.; Lappas, I. Groundwater quality degradation due to Cr6+ presence in Schinos area, prefecture of Corinth, Central Greece. In Proceedings of the 10th International Hydrogeological Congress of Greece, Thessaloniki, Greece, 8–10 October 2014.
29. Pyrgaki, K.; Argyraki, A.; Kelepertzis, E.; Paraskevopoulou, V.; Botsou, F.; Dassenakis, E.; Mitsis, I.; Skourtsos, E. Occurence of Hexavalent Chromium in the Ophiolite Related Aquifers of Loytraki and Schinos Areas. *Bull. Geol. Soc. Greece* **2017**, *50*, 2261–2270. [CrossRef]
30. Papazotos, P.; Vasileiou, E.; Perraki, M. Elevated Groundwater Concentrations of Arsenic and Chromium in Ultramafic Environments Controlled by Seawater Intrusion, the Nitrogen Cycle, and Anthropogenic Activities: The Case of the Gerania Mountains, NE Peloponnese, Greece. *Appl. Geochem.* **2020**, *121*, 104697. [CrossRef]
31. Pyrgaki, K.; Argyraki, A.; Botsou, F.; Kelepertzis, E.; Paraskevopoulou, V.; Karavoltsos, S.; Mitsis, I.; Dassenakis, E. Hydrogeochemical Investigation of Cr in the Ultramafic Rock-Related Water Bodies of Loutraki Basin, Northeast Peloponnese, Greece. *Env. Earth Sci.* **2021**, *80*, 62. [CrossRef]
32. Kelepertzis, E.; Pyrgaki, K.; Argyraki, A.; Botsou, F.; Boeckx, P.; Karavoltsos, S.; Dassenakis, M. Application of Dual Isotopes (δ 15 N, δ 18 O) to Determine Nitrate Contamination Sources in Cr(VI)-Impacted Groundwater of Central Greece Aquifers. In Proceedings of the 15th International Congress of the Geological Society of Greece Athens, Athens, Greece, 22–24 May 2019.

33. Pyrgaki, K.; Kelepertzis, E.; Argyraki, A.; Boeckx, P.; Botsou, F.; Dassenakis, E. Identification of Sources and Transformations of Nitrate in Cr(VI)-Impacted Alluvial Aquifers by a Hydrogeochemical and Δ15N-NO3− and Δ18O-NO3− Isotopes Approach. *Environ. Sci. Pollut. Res.* **2022**, *29*, 57703–57719. [CrossRef] [PubMed]
34. Bornovas, J.; Eleutheriou, A.; Gaitanakis, P. 1970. Geological Map of Greece, 1: 50000, Kaparellion Sheet, IGME.
35. Bornovas, J.; Lalechos, N.; Filippakis, N. 1970. Geological Map of Greece, 1:50000, Korinthos Sheet, IGME.
36. Bornovas, J.; Lalechos, N.; Filippakis, N. *Geological Map of Greece, 1:50000, Korinthos Sheet*; Institute of Geology and Mineral Exploration (IGME): Athens, Greece, 1969.
37. Voudouris, K.; Stamatis, G. Distribution and Fractal Analysis of Parameters of the Upper Part of Unsaturated Zone of the Alluvial Aquifer of Loutraki Basin. *Min. Metall. Ann.* **2002**, *12*, 39–54.
38. Pyrgaki, K.; Argyraki, A.; Kelepertzis, E.; Botsou, F.; Megremi, I.; Karavoltsos, S.; Dassenakis, E.; Mpouras, T.; Dermatas, D. A DPSIR Approach to Selected Cr(VI) Impacted Groundwater Bodies of Central Greece. *Bull. Environ. Contam. Toxicol.* **2021**, *106*, 446–452. [CrossRef] [PubMed]
39. Stamatis, G.; Voudouris, K. Delineation of Protection Zones According to Hydrogeological Criteria: The Case Study of Loutraki Alluvial Aquifer. *Miner. Wealth* **2000**, *2000*, 13–36.
40. Dotsika, E.; Poutoukis, D.; Raco, B. Fluid Geochemistry of the Methana Peninsula and Loutraki Geothermal Area, Greece. *J. Geochem. Explor.* **2010**, *104*, 97–104. [CrossRef]
41. D'Alessandro, W.; Calabrese, S.; Bellomo, S.; Brusca, L.; Daskalopoulou, K.; Li Vigni, L.; Randazzo, L.; Kyriakopoulos, K. Impact of Hydrothermal Alteration Processes on Element Mobility and Potential Environmental Implications at the Sousaki Solfataric Field (Corinthia—Greece). *J. Volcanol. Geotherm. Res.* **2020**, *407*, 107121. [CrossRef]
42. Giannakopoulou, P.P.; Petrounias, P.; Rogkala, A.; Tsikouras, B.; Stamatis, P.M.; Pomonis, P.; Hatzipanagiotou, K. The Influence of the Mineralogical Composition of Ultramafic Rocks on Their Engineering Performance: A Case Study from the Veria-Naousa and Gerania Ophiolite Complexes (Greece). *Geosciences* **2018**, *8*, 251. [CrossRef]
43. Vassilakis, E.; Royden, L.; Papanikolaou, D. Kinematic Links between Subduction along the Hellenic Trench and Extension in the Gulf of Corinth, Greece: A Multidisciplinary Analysis. *Earth Planet. Sci. Lett.* **2011**, *303*, 108–120. [CrossRef]
44. Kaplanis, A.; Koukouvelas, I.; Xypolias, P.; Kokkalas, S. Kinematics and Ophiolite Obduction in the Gerania and Helicon Mountains, Central Greece. *Tectonophysics* **2013**, *595–596*, 215–234. [CrossRef]
45. ELSTAT (2021). Hellenic Statistical Authority. Available online: https://www.statistics.gr/en/home (accessed on 15 October 2024).
46. HNMS. Available online: https://emy.gr/hnms-stations (accessed on 20 October 2024).
47. European Environment Agency (EEA) CORINE Land Cover (CLC) 2018, Version 2020_20u1. Derived by Integrating the Data of Land Cover Changes between the Years 2012–2018 (Available at: CLC 2018—Copernicus Land Monitoring Service). Available online: https://land.copernicus.eu/en/products/corine-land-cover/clc2018 (accessed on 20 October 2024).
48. *UNE-EN ISO/IEC/17025:2005*; General Requirements for the Competence of Testing and Calibration Laboratories. 2005.
49. Mutewekil, M.; Awawdeh, M.; Abu, F.; Al-Ajlouni, A. An Innovative Nitrate Pollution Index and Multivariate Statistical Investigations of Groundwater Chemical Quality of Umm Rijam Aquifer (B4), North Yarmouk River Basin, Jordan. In *Water Quality Monitoring and Assessment*; InTech: Rijeka, Croatia, 2012.
50. Panno, S.V.; Kelly, W.R.; Martinsek, A.T.; Hackley, K.C. Estimating Background and Threshold Nitrate Concentrations Using Probability Graphs. *Ground Water* **2006**, *44*, 697–709. [CrossRef]
51. Revelle, R. Criteria for Recognition of the Sea Water in Ground-waters. *Eos Trans. Am. Geophys. Union* **1941**, *22*, 593–597. [CrossRef]
52. Subba Rao, N. PIG: A Numerical Index for Dissemination of Groundwater Contamination Zones. *Hydrol. Process* **2012**, *26*, 3344–3350. [CrossRef]
53. Horton, R.K. An Index Number System for Rating Water Quality. *J. Water Pollut. Control Fed.* **1965**, *37*, 300–306.
54. Brown, M.; McClelland, N.I.; Deininger, R.A.; Tozer, R.G. A Water Quality Index: Do We Dare? *Water Sew. Work.* **1970**, *117*, 339–343.
55. Hossain, M.; Patra, P.K. Water Pollution Index—A New Integrated Approach to Rank Water Quality. *Ecol. Indic.* **2020**, *117*, 106668. [CrossRef]
56. Richards, L.A. Diagnosis and Improvement of Saline and Alkali Soils. *Soil Sci.* **1954**, *78*, 154. [CrossRef]
57. Todd, D.K. *Groundwater Hydrology*, 2nd ed.; Wiley: New York, NY, USA, 1980.
58. Kelley, W.P. Permissible Composition and Concentration of Irrigation Water. *Trans. Am. Soc. Civ. Eng.* **1941**, *106*, 849–855. [CrossRef]
59. Kelley, W.P. *Alkali Soils—Their Formation Properties and Reclamation*; Reinhold Publishing Co.: New York, NY, USA, 1951.
60. Wilcox, L.V. *Classification and Use of Irrigation Waters*; United States Department of Agriculture: Washington, DC, USA, 1955.
61. Ayers, R.S.; Westcot, D.W. *Water Quality for Agriculture*; FAO United Nations: Rome, Italy, 1985.
62. Collins, R.; Jenkins, A. The Impact of Agricultural Land Use on Stream Chemistry in the Middle Hills of the Himalayas, Nepal. *J. Hydrol.* **1996**, *185*, 71–86. [CrossRef]

63. Doneen, L.D. The Influence of Crop and Soil on Percolating Waters. In Proceedings of the 1961 Biennial Conference Groundwater Recharge, Berkeley, CA, USA, 26–27 June 1962; pp. 156–163.
64. Doneen, L.D. *Notes on Water Quality in Agriculture*; Department of Water Science and Engineering, University of California: Oakland, CA, USA, 1964.
65. Rawat, K.S.; Singh, S.K.; Gautam, S.K. Assessment of Groundwater Quality for Irrigation Use: A Peninsular Case Study. *Appl. Water Sci.* **2018**, *8*, 233. [CrossRef]
66. Szabolcs, I.; Darab, C. The Influences of Irrigation Water of High Sodium Carbonate Contents on Soils. In Proceedings of the 8th International Congress Soil Science Sodic Soils, Research Institute for Soil Science and Agricultural Chemistry of the Hungarian Academy of Sciences, Izmir, Turkey, 15–17 May 1964; ISSS Trans II: Tsukuba, Japan, 1964; pp. 802–812.
67. Paliwal, K.V. *Irrigation with Saline Water*; Water Technology Centre, Indian Agriculture Research Institute: New Delhi, India, 1972.
68. Hem, J.D. *Study and Interpretation of the Chemical Characteristics of Natural Water*; US Geological Survey Water-Supply Paper; United States Government Printing Office: Washington, DC, USA, 1985; Volume 2254.
69. Giggenbach, W.F. Geothermal Solute Equilibria. Derivation of Na-K-Mg-Ca Geoindicators. *Geochim. Cosmochim. Acta* **1988**, *52*, 2749–2765. [CrossRef]
70. Eaton, F.M. Significance of Carbonates in Irrigation Waters. *Soil Sci.* **1950**, *69*, 123–134. [CrossRef]
71. Boyd, C.E.; Tucker, C.S. *Handbook for Aquaculture Water Quality*; Craftmaster Printers, Inc.: Auburn, AL, USA, 2014.
72. Sengupta, P. Potential Health Impacts of Hard Water. *Int. J. Prev. Med.* **2013**, *4*, 866–875.
73. Sawyer, G.N.; McCarthy, D.L. *Chemistry of Sanitary Engineers*; McGraw Hill: New York, NY, USA, 1967; p. 518.
74. Meireles, A.C.M.; de Andrade, E.M.; Chaves, L.C.G.; Frischkorn, H.; Crisostomo, L.A. A New Proposal of the Classification of Irrigation Water. *Rev. Ciência Agronômica* **2010**, *41*, 349–357. [CrossRef]
75. Mkilima, T. Groundwater Salinity and Irrigation Suitability in Low-Lying Coastal Areas. A Case of Dar Es Salaam, Tanzania. *Watershed Ecol. Environ.* **2023**, *5*, 173–185. [CrossRef]
76. World Health Organization (WHO). *Guidelines for Drinking Water Quality*; WHO: Geneva, Switzerland, 2017.
77. Subramani, T.; Elango, L.; Damodarasamy, S.R. Groundwater Quality and Its Suitability for Drinking and Agricultural Use in Chithar River Basin, Tamil Nadu, India. *Environ. Geol.* **2005**, *47*, 1099–1110. [CrossRef]
78. Dore, M.P.; Parodi, G.; Portoghese, M.; Errigo, A.; Pes, G.M. Water Quality and Mortality from Coronary Artery Disease in Sardinia: A Geospatial Analysis. *Nutrients* **2021**, *13*, 2858. [CrossRef] [PubMed]
79. Backman, B.; Bodiš, D.; Lahermo, P.; Rapant, S.; Tarvainen, T. Application of a Groundwater Contamination Index in Finland and Slovakia. *Environ. Geol.* **1998**, *36*, 55–64. [CrossRef]
80. Edet, A.E.; Offiong, O.E. Evaluation of Water Quality Pollution Indices for Heavy Metal Contamination Monitoring. A Study Case from Akpabuyo-Odukpani Area, Lower Cross River Basin (Southeastern Nigeria). *GeoJournal* **2002**, *57*, 295–304. [CrossRef]
81. Prasanna, M.V.; Praveena, S.M.; Chidambaram, S.; Nagarajan, R.; Elayaraja, A. Evaluation of Water Quality Pollution Indices for Heavy Metal Contamination Monitoring: A Case Study from Curtin Lake, Miri City, East Malaysia. *Environ. Earth Sci.* **2012**, *67*, 1987–2001. [CrossRef]
82. Spearman, C. The Proof and Measurement of Association between Two Things. *Am. J. Psychol.* **1904**, *15*, 72–101. [CrossRef]
83. Evans, J. *Straightforward Statistics for the Behavioral Sciences*; Brooks/Cole Publishing Company: Pacific Grove, CA, USA, 1996.
84. Shrestha, N. Factor Analysis as a Tool for Survey Analysis. *Am. J. Appl. Math. Stat.* **2021**, *9*, 4–11. [CrossRef]
85. Dziuban, C.D.; Shirkey, E.C. When Is a Correlation Matrix Appropriate for Factor Analysis? Some Decision Rules. *Psychol. Bull.* **1974**, *81*, 358–361. [CrossRef]
86. Kaiser, H.F. The Varimax Criterion for Analytic Rotation in Factor Analysis. *Psychometrika* **1958**, *23*, 187–200. [CrossRef]
87. Kaiser, H.F. The Application of Electronic Computers to Factor Analysis. *Educ. Psychol. Meas.* **1960**, *20*, 141–151. [CrossRef]
88. Liu, C.W.; Lin, K.H.; Kuo, Y.M. Application of Factor Analysis in the Assessment of Groundwater Quality in a Blackfoot Disease Area in Taiwan. *Sci. Total Environ.* **2003**, *313*, 77–89. [CrossRef]
89. Papazotos, P.; Vasileiou, E.; Perraki, M. The Synergistic Role of Agricultural Activities in Groundwater Quality in Ultramafic Environments: The Case of the Psachna Basin, Central Euboea, Greece. *Environ. Monit. Assess.* **2019**, *191*, 317. [CrossRef] [PubMed]
90. Kaiser, H.F. A Second Generation Little Jiffy. *Psychometrika* **1970**, *35*, 401–415. [CrossRef]
91. Bartlett, M.S. Tests of Significance in Factor Analysis. *Br. J. Stat. Psychol.* **1950**, *3*, 77–85. [CrossRef]
92. Ward, J.H. Hierarchical Grouping to Optimize an Objective Function. *J. Am. Stat. Assoc.* **1963**, *58*, 236–244. [CrossRef]
93. Meybeck, M. Global Chemical Weathering of Surficial Rocks Estimated from River Dissolved Loads. *Am. J. Sci.* **1987**, *287*, 401–428. [CrossRef]
94. Stamatakis, M.G.; Mitsis, I. The Occurrences of Mg-Hydroxycarbonates in Serpentinites of the Western Section of the South Aegean Volcanic Arc (West Attica Peninsula-Northeastern Argolis Peninsula), Greece. *Bull. Geol. Soc. Greece* **2013**, *47*, 427–437. [CrossRef]

95. Ficklin, W.H.; Plumlee, G.S.; Smith, K.S.; McHugh, J.B. Geochemical Classification of Mine Drainages and Natural Drainages in Mineralized Areas. In Proceedings of the International Symposium on Water-Rock Interaction, Park City, UT, USA, 13–18 July 1992; pp. 381–384.
96. Egbueri, J.C.; Mgbenu, C.N.; Digwo, D.C.; Nnyigide, C.S. A Multi-Criteria Water Quality Evaluation for Human Consumption, Irrigation and Industrial Purposes in Umunya Area, Southeastern Nigeria. *Int. J. Environ. Anal. Chem.* **2023**, *103*, 3351–3375. [CrossRef]
97. Rezaei, A.; Hassani, H.; Jabbari, N. Evaluation of Groundwater Quality and Assessment of Pollution Indices for Heavy Metals in North of Isfahan Province, Iran. *Sustain. Water Resour. Manag.* **2019**, *5*, 491–512. [CrossRef]
98. Bouselsal, B.; Satouh, A.; Egbueri, J.C. Evaluating Water Quality, Mineralization Mechanisms, and Potential Health Risks of Nitrate Contamination in the Continental Intercalaire Aquifer of Reggane, Algeria. *Environ. Earth Sci.* **2024**, *83*, 539. [CrossRef]
99. El Mountassir, O.; Bahir, M.; Ouazar, D.; Chehbouni, A.; Carreira, P.M. Temporal and Spatial Assessment of Groundwater Contamination with Nitrate Using Nitrate Pollution Index (NPI), Groundwater Pollution Index (GPI), and GIS (Case Study: Essaouira Basin, Morocco). *Environ. Sci. Pollut. Res.* **2022**, *29*, 17132–17149. [CrossRef] [PubMed]
100. Sanad, H.; Mouhir, L.; Zouahri, A.; Moussadek, R.; El Azhari, H.; Yachou, H.; Ghanimi, A.; Oueld Lhaj, M.; Dakak, H. Assessment of Groundwater Quality Using the Pollution Index of Groundwater (PIG), Nitrate Pollution Index (NPI), Water Quality Index (WQI), Multivariate Statistical Analysis (MSA), and GIS Approaches: A Case Study of the Mnasra Region, Gharb Plain, Morocco. *Water* **2024**, *16*, 1263. [CrossRef]
101. Demelash, A.; Atlabachew, A.; Jothimani, M.; Abebe, A. Hydrogeochemical Characterization and Appraisal of Groundwater Quality in Yisr River Catchment, Blue Nile River Basin, Ethiopia, by Using the GIS, WQI, and Statistical Techniques. *J. Chem.* **2023**, *2023*, 8199000. [CrossRef]
102. Witkowski, A.J.; Dąbrowska, D.; Wróbel, J. Groundwater Quality Assessment in the Area of the Zinc Smelter in Miasteczko Śląskie (Poland) Using Selected Metal Indices. *Water* **2024**, *16*, 279. [CrossRef]
103. Abou Zakhem, B.; Hafez, R. Heavy Metal Pollution Index for Groundwater Quality Assessment in Damascus Oasis, Syria. *Environ. Earth Sci.* **2015**, *73*, 6591–6600. [CrossRef]
104. Zhang, Q.; Qian, H.; Xu, P.; Hou, K.; Yang, F. Groundwater Quality Assessment Using a New Integrated-Weight Water Quality Index (IWQI) and Driver Analysis in the Jiaokou Irrigation District, China. *Ecotoxicol. Environ. Saf.* **2021**, *212*, 111992. [CrossRef] [PubMed]
105. Batarseh, M.; Imreizeeq, E.; Tilev, S.; Al Alaween, M.; Suleiman, W.; Al Remeithi, A.M.; Al Tamimi, M.K.; Al Alawneh, M. Assessment of Groundwater Quality for Irrigation in the Arid Regions Using Irrigation Water Quality Index (IWQI) and GIS-Zoning Maps: Case Study from Abu Dhabi Emirate, UAE. *Groundw. Sustain. Dev.* **2021**, *14*, 100611. [CrossRef]
106. Varol, S.; Davraz, A. Evaluation of the Groundwater Quality with WQI (Water Quality Index) and Multivariate Analysis: A Case Study of the Tefenni Plain (Burdur/Turkey). *Environ. Earth Sci.* **2015**, *73*, 1725–1744. [CrossRef]
107. Chaurasia, A.K.; Pandey, H.K.; Tiwari, S.K.; Prakash, R.; Pandey, P.; Ram, A. Groundwater Quality Assessment Using Water Quality Index (WQI) in Parts of Varanasi District, Uttar Pradesh, India. *J. Geol. Soc. India* **2018**, *92*, 76–82. [CrossRef]
108. Panneerselvam, B.; Karuppannan, S.; Muniraj, K. Evaluation of Drinking and Irrigation Suitability of Groundwater with Special Emphasizing the Health Risk Posed by Nitrate Contamination Using Nitrate Pollution Index (NPI) and Human Health Risk Assessment (HHRA). *Hum. Ecol. Risk Assess.* **2020**, *27*, 1324–1348. [CrossRef]
109. Jahan, S.; Strezov, V. Water Quality Assessment of Australian Ports Using Water Quality Evaluation Indices. *PLoS ONE* **2017**, *12*, e0189284. [CrossRef] [PubMed]
110. Pascual Aguilar, J.A.; Campo, J.; Meneu, S.N.; Gimeno-García, E.; Andreu, V. Analysis of Existing Water Information for the Applicability of Water Quality Indices in the Fluvial-Littoral Area of Turia and Jucar Rivers, Valencia, Spain. *Appl. Geogr.* **2019**, *111*, 102062. [CrossRef]
111. Al-Aizari, H.S.; Aslaou, F.; Al-Aizari, A.R.; Al-Odayni, A.B.; Al-Aizari, A.J.M. Evaluation of Groundwater Quality and Contamination Using the Groundwater Pollution Index (GPI), Nitrate Pollution Index (NPI), and GIS. *Water* **2023**, *15*, 3701. [CrossRef]
112. Bahrami, M.; Zarei, A.R.; Rostami, F. Temporal and Spatial Assessment of Groundwater Contamination with Nitrate by Nitrate Pollution Index (NPI) and GIS (Case Study: Fasarud Plain, Southern Iran). *Environ. Geochem. Health* **2020**, *42*, 3119–3130. [CrossRef] [PubMed]
113. Boualem, B.; Egbueri, J.C. Graphical, Statistical and Index-Based Techniques Integrated for Identifying the Hydrochemical Fingerprints and Groundwater Quality of In Salah, Algerian Sahara. *Environ. Geochem. Health* **2024**, *46*, 158. [CrossRef] [PubMed]
114. Ravindra, B.; Subba Rao, N.; Dhanamjaya Rao, E.N. Groundwater Quality Monitoring for Assessment of Pollution Levels and Potability Using WPI and WQI Methods from a Part of Guntur District, Andhra Pradesh, India. *Environ. Dev. Sustain.* **2023**, *25*, 14785–14815. [CrossRef]
115. Shooshtarian, M.R.; Dehghani, M.; Margherita, F.; Gea, O.C.; Mortezazadeh, S. Land Use Change and Conversion Effects on Ground Water Quality Trends: An Integration of Land Change Modeler in GIS and a New Ground Water Quality Index Developed by Fuzzy Multi-Criteria Group Decision-Making Models. *Food Chem. Toxicol.* **2018**, *114*, 204–214. [CrossRef] [PubMed]

116. Kumar, A.; Krishna, A.P. Groundwater Quality Assessment Using Geospatial Technique Based Water Quality Index (WQI) Approach in a Coal Mining Region of India. *Arab. J. Geosci.* **2021**, *14*, 1126. [CrossRef]
117. Abugu, H.O.; Egbueri, J.C.; Agbasi, J.C.; Ezugwu, A.L.; Omeka, M.E.; Ucheana, I.A.; Aralu, C.C. Hydrochemical Characterization of Ground and Surface Water for Irrigation Application in Nigeria: A Review of Progress. *Chem. Afr.* **2024**, *7*, 3011–3036. [CrossRef]
118. Petropoulou, I.; Frousiou, M.-S.; Vasileiou, E. An Overview of Quality Assessment Methods for Water and Soil in Mining Regions. *Mater. Proc.* **2023**, *15*, 31. [CrossRef]
119. Arifullah; Changsheng, H.; Akram, W.; Rashid, A.; Ullah, Z.; Shah, M.; Alrefaei, A.F.; Kamel, M.; Aleya, L.; Abdel-Daim, M.M. Quality Assessment of Groundwater Based on Geochemical Modelling and Water Quality Index (WQI). *Water* **2022**, *14*, 3888. [CrossRef]
120. Bouteraa, O.; Mebarki, A.; Bouaicha, F.; Nouaceur, Z.; Laignel, B. Groundwater Quality Assessment Using Multivariate Analysis, Geostatistical Modeling, and Water Quality Index (WQI): A Case of Study in the Boumerzoug-El Khroub Valley of Northeast Algeria. *Acta Geochim.* **2019**, *38*, 796–814. [CrossRef]
121. Papazotos, P. Potentially Toxic Elements in Groundwater: A Hotspot Research Topic in Environmental Science and Pollution Research. *Environ. Sci. Pollut. Res.* **2021**, *28*, 47825–47837. [CrossRef] [PubMed]
122. Shaw, S.K.; Sharma, A. Assessment of Groundwater Quality and Suitability for Irrigation Purpose Using Irrigation Indices, Remote Sensing and GIS Approach. *Groundw. Sustain. Dev.* **2024**, *26*, 101297. [CrossRef]
123. Khan, R.A.; Khan, N.A.; El Morabet, R.; Alsubih, M.; Qadir, A.; Bokhari, A.; Mubashir, M.; Asif, S.; Cheah, W.Y.; Manickam, S.; et al. Geospatial Distribution and Health Risk Assessment of Groundwater Contaminated within the Industrial Areas: An Environmental Sustainability Perspective. *Chemosphere* **2022**, *303*, 134749. [CrossRef]
124. Tziritis, E.; Sachsamanoglou, E.; Güler, C. Evaluating Spatiotemporal Groundwater Quality Changes in the Rhodope Coastal Aquifer System (NE Greece) Employing a GIS-Assisted Hybrid Approach of Multivariate Statistics and Inverse Geochemical Modeling. *Sci. Total Environ.* **2024**, *947*, 174676. [CrossRef]
125. Agbasi, J.C.; Ayejoto, D.A.; Egbueri, J.C.; Khan, N.; Abba, S.I.; Ahmad, V.; Abuzinadah, M.F. HERisk and Statistical Clustering Integrated for Health Risk Modelling of PTEs in Natural Water Resources for Drinking and Sanitary Uses. *Toxin Rev.* **2024**, *43*, 513–539. [CrossRef]
126. Vasileiou, E.; Perraki, M.; Stamatis, G.; Gartzos, E. The Effects of Water Rock Interaction and the Human Activities on the Occurrence of Hexavalent Chromium in Waters. The Case Study of the Psachna Basin, Central Euboea, Greece. *EGU Gen. Assem. Conf. Abstr.* **2014**, *16*, 15467.
127. Megremi, I.; Vasilatos, C.; Vassilakis, E.; Economou-Eliopoulos, M. Spatial Diversity of Cr Distribution in Soil and Groundwater Sites in Relation with Land Use Management in a Mediterranean Region: The Case of C. Evia and Assopos-Thiva Basins, Greece. *Sci. Total Environ.* **2019**, *651*, 656–667. [CrossRef] [PubMed]
128. Remoundaki, E.; Vasileiou, E.; Philippou, A.; Perraki, M.; Kousi, P.; Hatzikioseyian, A.; Stamatis, G. Groundwater Deterioration: The Simultaneous Effects of Intense Agricultural Activity and Heavy Metals in Soil. *Procedia Eng.* **2016**, *162*, 545–552. [CrossRef]
129. Pourret, O. On the Emergence of Tortured Phrases: A Threat to Scientific Integrity—The Example of "Heavy Metal". *Eur. Sci. Ed.* **2024**, *50*.
130. Pourret, O.; Hursthouse, A. It's Time to Replace the Term "Heavy Metals" with "Potentially Toxic Elements" When Reporting Environmental Research. *Int. J. Environ. Res. Public Health* **2019**, *16*, 4446. [CrossRef]
131. Pourret, O.; Bollinger, J.C.; Hursthouse, A. Heavy Metal: A Misused Term? *Acta Geochim.* **2021**, *40*, 466–471. [CrossRef]

environments

MDPI

Article

Seasonality of Pharmaceuticals and Personal Care Products in Shallow Lakes, Florida, USA—Part A

Elzbieta Bialkowska-Jelinska [1], Philip van Beynen [1,*] and Laurent Calcul [2]

[1] School of Geosciences, University of South Florida, Tampa, FL 33620, USA; elzbieta@usf.edu

[2] Chemical Purification Analysis and Screening Core Facility (CPAS), University of South Florida, Tampa, FL 33620, USA; calcul@usf.edu

* Correspondence: vanbeyne@usf.edu

Abstract

Shallow lakes are highly vulnerable to pollution due to their small water volume. Those that receive effluent from the drainfields of onsite wastewater treatment systems (septic tanks) may contain pharmaceuticals and personal care products (PPCPs) that escaped removal during treatment. This study examined the effects of seasonal rainfall variability on the assemblages and concentrations of fourteen PPCPs in two shallow lakes in West–Central Florida, USA: one surrounded by residents equipped with septic tanks and the other located within a nature preserve. Water samples were collected weekly during an 18-week interval from April to August 2021. Liquid chromatography–mass spectrometry analyses revealed the omnipresence of five PPCPs: theophylline, caffeine, cotinine, DEET, and testosterone, although acetaminophen, ibuprofen, and sulfamethoxazole were also common. Of all the PPCPs detected, theophylline, DEET, and acetaminophen concentrations were higher during the wet season in the septic tank-influenced lake, while caffeine, cotinine, and testosterone concentrations decreased. In the lake located in the nature preserve, theophylline, caffeine, and acetaminophen levels increased in the wet season. In contrast, cotinine, DEET, and testosterone levels decreased. Overall, more compounds were detected during the wet season, with highly hydrophobic PPCPs (fluoxetine, atorvastatin, and octocrylene) only present during this period.

Keywords: pharmaceuticals; personal care products; seasonality; septic tanks

1. Introduction

Pharmaceuticals and personal care products (PPCPs) use has risen with a growing and aging population [1]. New PPCPs are developed each year, adding to the already thousands of chemicals that pollute our water bodies, and their toxicity has already led to acute exposure to aquatic organisms [2,3]). PPCPs are ubiquitous, being found in the atmosphere [4], polar ice sheets [5], oceans [6], rivers and lakes [3,7], aquifers [8], and soils [9]. They bioaccumulate in flora [10] and fauna [3]. Their introduction to the environment is mainly through the release of treated water from wastewater treatment plants (WWTPs) and onsite wastewater treatment systems (OWTS), more commonly known as septic tanks. These waters are released into rivers, lakes, and drainfields. Because of the inability of these treatment systems to remove all PPCPs from the effluent, these contaminants can be incorporated into the tissue of organisms that interact with this polluted water [7].

There are 15,000 WTTPs in the USA discharging 129^6 m^3 of wastewater per day [11]. These facilities release treated water into water bodies through several practices. First is the intentional release into rivers and lakes, where dilution overcomes the potentially high levels of PPCPs in the effluent. Second, reclaimed water produced by tertiary-level processing of wastewater is used for municipal irrigation. Overwatering with reclaimed water, a common practice in residential areas, leads to urban runoff, which enters community stormwater ponds, local streams, and groundwater [12]. Reclaimed water for irrigation is also commonplace in the agricultural setting [13]. Finally, there is unintentional discharge when the WWTPs are overwhelmed by major storm events [14]. These facilities tend to be regional and contribute significant amounts of contaminated water to the environment. At a local level, the drainfields of OWTS are designed to slowly release human waste, including PPCPs, into the surrounding soils. OWTS are the largest contributors of wastewater entering groundwater [15]. Increased precipitation leads to effluent entering the groundwater or overland runoff [16]. The USA has over 21 million OWTS, each discharging 160 to 200 L/day per capita [17]. Assuming a minimum of two individuals per system, OWTS contribute at least 3.36 million m^3 of effluent per day.

Human residences equipped with OWTS, located along lake shorelines, can discharge untreated PPCPs and contribute to water pollution. Shallow lakes play an important role in biodiversity as they provide habitats for fish, invertebrates, and waterfowl and support ecosystem services like carbon sequestration, nutrient cycling, and aquaculture [18]. However, their relatively shallow depth makes them highly sensitive to disturbances such as tropical storms, fluctuating water levels, and human activities like agriculture and urbanization, impairing their water quality and availability [19].

This study examines how seasonal changes in precipitation affect the mobilization of PPCPs from OWTS drainfields located upgradient of a residential lake. For comparison, we include a lake located in a nature preserve that has most of the same physical characteristics as the residential lake, albeit with surrounding OWTS. Specifically, we examine how the onset of the wet season in West–Central Florida will influence both the concentrations and assemblages of PPCPs in both lakes. The novelty of our study relates to the following: (1) the two lakes are small enclosed basins with no significant inputs of PPCPs from other sources besides OWTS; (2) there are few studies in karst regions with a similar sampling resolution; (3) the lakes have very similar physical characteristics, which reduces complications of different climate, geology, and soil type; and (4) there is clear climatic delineation between the dry season and the wet season.

2. Conceptual Model of Study

Our conceptual model of PPCP mobilization is illustrated in Figure 1. During the dry season, PPCPs are primarily retained in unsaturated drainfield soils through hygroscopic behavior via absorption or adsorption, thereby limiting their movement into the lake [20] (Figure 1a). With the onset of the wet season, soil saturation increases, flushing out more hydrophilic PPCPs from the drainfields/soils and helping initiate the mobilization of those with more hydrophobic properties (Figure 1b).

Transferring this concept to our study area, specifically the residential lake, two different scenarios are evident (Figure 1a,b). For the dry season, little, if any, PPCPs reach the lake. It should be noted that the homes surrounding the lake are supplied by wells that tap the Floridan Aquifer System (FAS). The aquifer is the main source of water for the lake for this season (Figure 1c). With the onset of the wet season, heavy rainfall saturates the drainfield soils, flushing PPCPs out into the lake, overwhelming the contribution of the aquifer water. Increased groundwater flow mobilizes both stored and newly introduced PPCPs, leading to higher concentrations. While the initial precipitation may temporarily reduce PPCP levels through dilution, concentrations rise as mobilized PPCPs enter the lake. In concert, more hydrophobic PPCPs may be flushed from the drainfields and soils.

Finally, although the piezometric surface of the Intermediate Aquifer System (IAS), which lies between the lake and FAS, may rise, its contribution to the lake is unlikely due to the lake's greater hydraulic head as lake volume increases (Figure 1d). Consequently, the underlying aquifer is unlikely to contribute PPCPs or impact lake volume significantly. Previous studies [21–23] have noted increased PPCP concentrations in water bodies during periods of elevated rainfall due to the reduced removal efficiency of OWTS and higher mobilization rates.

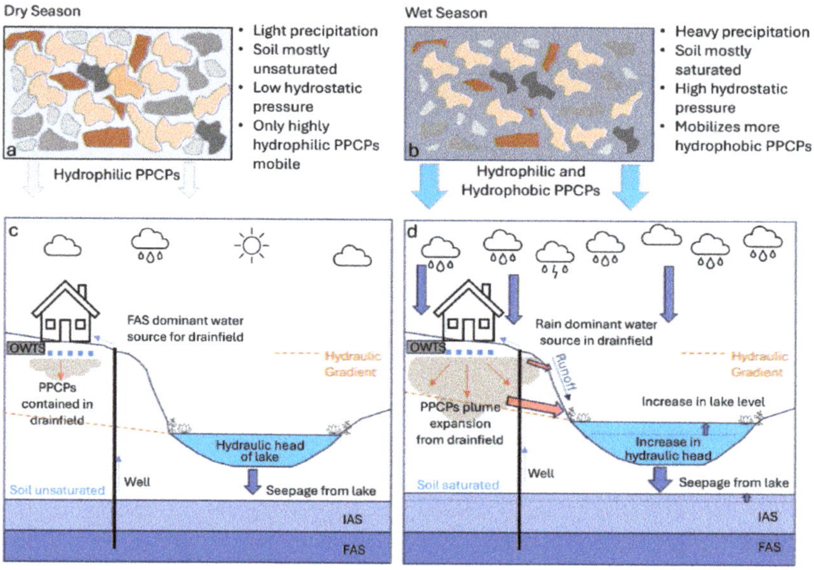

Figure 1. Conceptual model of PPCP movement in the soil during the dry and wet seasons: (**a**) hygroscopic water surrounding soil particles is dark blue; (**b**) due to saturation, all soil pores are filled with water; mobilization of PPCPs from OWTS drainfields into a residential lake in the dry (**c**) and wet (**d**) season. The two aquifers identified are the Floridan Aquifer System (FAS) and the Intermediate Aquifer System (IAS).

3. Materials and Methods

3.1. Study Area

This study was conducted in West–Central Florida, in the Alafia River watershed (Figure 2). The residential lake (RL) is internally drained, 120 m wide, with a maximum depth of 5.95 m. It is surrounded by 18 residential homes equipped with OWTS for wastewater treatment and water wells that draw water from the Upper Floridan Aquifer (UFA). The natural lake (NL), which serves as a control site, is located 1800 m northwest of RL. It is 100 m wide and 2.95 m at its deepest point (Figure 2a). The study area's climate, represented by the nearest town, Riverview, is classified as humid subtropical, characterized by distinct wet and dry seasons. Annual precipitation averages 764 mm in the wet season and 348 mm in the dry season. The mean annual temperature for the region is 22.2 °C.

The UFA represents the upper aquifer of the FAS. FAS is subdivided into three units: UFA, the Middle Confining Unit, and the Lower Floridan Aquifer. In the hydrogeological framework of Florida, the FAS, the IAS, and the Surficial Aquifer are the main aquifers. Each of these aquifer systems is exposed at the land surface in various regions of Florida. Within the study area, the FAS is geologically confined by IAS, while the Surficial Aquifer is present at the surface to the east of the study site [24] (Figure 2).

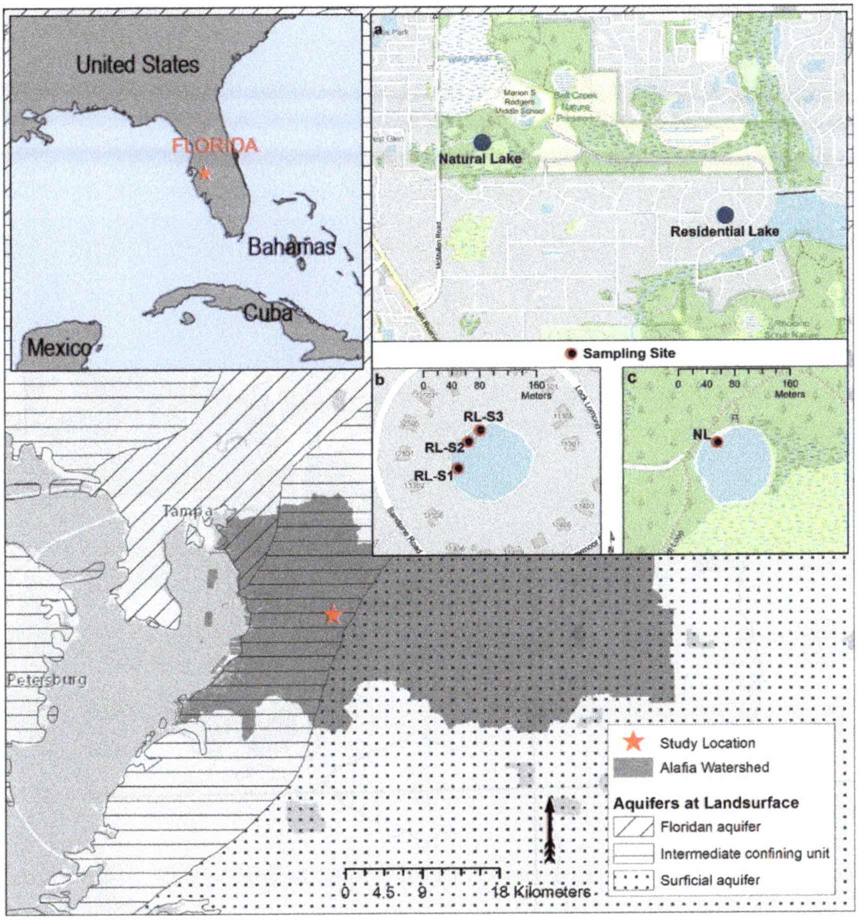

Figure 2. Study location in Florida, Alafia River watershed, and the major aquifers found at the land surface; (**a**) residential (27°50′09.7″ N 82°16′46.6″ W) and natural (27°50′28.8″ N 82°17′54.4″ W) lakes; (**b**) sampling sites RL-S1, RL-S2, and RL-S3 at RL; (**c**) sampling site at NL.

Our aim was to determine how the onset of the wet season altered the PPCPs present in the lake water. Consequently, starting sampling at the end of the dry season allows us to determine a baseline before the hydraulic regime changes in the area. As such, weekly water samples were collected from 30 April to 30 August 2021 (18 weeks). Within RL, sampling was conducted at three sites: Site 1 (RL-S1), Site 2 (RL-S2), and Site 3 (RL-S3), and at one site in NL (Figure 2b,c). The three sites from RL were selected because of their similarities: (1) distance of their OWTS from the lake, (2) hydraulic gradients, and (3) width of the shoreline vegetation. Additionally, only these three property owners allowed us access to the lake.

3.2. Water Samples

Water samples were collected in the littoral zone of RL and NL at approximately 10 am each sampling day. Sixty-nine (69) water samples were collected, including 51 from the residential and 18 from the natural lakes. Due to the accessibility issues at RL-S2, this site was not sampled on the last three dates (13, 20, 30 August). For each site, 1 L of water was collected at a depth of 35 cm below the lake surface in amber glass bottles after rinsing with the sample water twice before sample collection. The samples were placed in a cooler and transported to the Chemical Purification Analysis and Screening (CPAS) Lab at the University of South Florida (USF) in Tampa. Samples were filtered twice through vacuum-assisted Whatman 0.7 µm glass microfiber fiber (GF/F) filters, stored at 4 °C, and processed within 24 h.

The physiochemical analyses of the water samples were performed right before the water collection. YSI Pro20 Dissolved Oxygen Meter was used to measure dissolved oxygen (DO). pH and temperature were measured with Hanna Instruments H198127 pH-Temperature tester. Total dissolved solids (TDS) and electric conductivity (EC) were measured with Hanna Dissolved Solid Tester (DiST).

3.3. Precipitation Data

Precipitation data for the study were obtained from the Southwest Florida Water Management District [25] using the NexRad extraction tool. This tool provides detailed hourly rainfall data. To facilitate analysis, the hourly rainfall information was aggregated to daily totals, giving a comprehensive view of daily precipitation levels for the study area.

3.4. Chemicals, Reagents, and Materials

Fourteen compounds were selected for this study based on their documented occurrence in aquatic environments near OWTS [26–31]. These PPCPs represent a broad spectrum of substances, including antidepressants, nonsteroidal anti-inflammatory drugs, lipid regulators, anticonvulsants, antibiotics, respiratory drugs, nicotine metabolites, preservatives, stimulants, insect repellents, and sunscreen ingredients. Compounds such as octocrylene, atorvastatin, fluoxetine, testosterone, propylparaben, carbamazepine, N,N-diethyl-meta-toluamide (DEET), methylparaben, and cotinine were sourced from Neugen Labs (Tampa, FL). Other compounds, including ibuprofen, sulfamethoxazole, acetaminophen, caffeine, and theophylline, were obtained from Sigma Aldrich (St. Louis, MO, USA). Additionally, theophylline-d6 was procured from Cayman Chemical Company (Ann Arbor, MI, USA). All analytical standards and solvents used in this study were of ≥98% purity. Methanol, acetonitrile, formic acid, and ammonium fluoride were supplied by Thermo Fisher Scientific (Pittsburgh, PA, USA), and Milli-Q water from CPAS was used in the analyses.

Table 1 provides the complete list of target PPCPs, detailing their hydrophobicity, molecular size, formula, CAS number, and classification.

Table 1. List of target compounds.

Compound	Log K_{ow}	Molar Mass ($gmol^{-1}$)	CAS	Molecular Formula	PPCP Class
Octocrylene (OCT)	6.88	361.48	61-97-30-4	$C_{24}H_{27}NO_2$	Sunscreen component
Atorvastatin (ATV)	6.36	558.64	134523-00-5	$C_{33}H_{35}FN_2O_5$	Lipid regulator
Fluoxetine (FLX)	4.05	309.33	54910-89-3	$C_{17}H_{18}F_3NO$	Antidepressant
Ibuprofen (IBU)	3.97	206.3	15687-27-1	$C_{13}H_{18}O_2$	NSAID
Testosterone (TST)	3.32	288.43	58-22-0	$C_{19}H_{28}O_2$	Hormone
Propylparaben (PPB)	3.04	180.2	94-13-3	$C_{10}H_{12}O_3$	Preservative
Carbamazepine (CBZ)	2.45	236.27	298-46-4	$C_{15}H_{12}N_2O$	Anti-convulsant
DEET	2.02	191.27	134-62-3	$C_{12}H_{17}NO$	Insect repellant

Table 1. *Cont.*

Compound	Log K_{ow}	Molar Mass ($gmol^{-1}$)	CAS	Molecular Formula	PPCP Class
Methylparaben (MPB)	1.96	152.15	99-76-3	$C_8H_8O_3$	Preservative
Sulfamethoxazole (SMX)	0.89	253.28	723-46-6	$C_{10}H_{11}N_3O_3S$	Antibiotic
Acetaminophen (APAP)	0.46	151.16	103-90-2	$C_8H_9NO_2$	NSAID
Cotinine (COT)	0.07	176.22	486-56-6	$C_{10}H_{12}N_2O$	Metabolite of nicotine
Caffeine (CAF)	−0.07	194.19	58-08-2	$C_8H_{10}N_4O_2$	Stimulant
Theophylline (THE)	−0.02	180.6	58-55-9	$C_7H_8N_4O_2$	Bronchodilator

Source of log K_{ow} values PubChem (open chemistry database) at National Institutes of Health (NIH) https://pubchem.ncbi.nlm.nih.gov/ (accessed on 23 June 2025).

3.5. *Sample Extraction and Analysis*

3.5.1. Sample Extraction

Solid-phase extraction (SPE) of PPCPs from water was conducted using Oasis hydrophilic–lipophilic balance (HLB) cartridges (500 mg) from Waters Corporation (Milford, MA, USA). HLB cartridges were used due to their dual hydrophilic–lipophilic balance, which is advantageous for extracting a broad range of acidic, basic, and neutral chemicals from various matrices. Previous research has demonstrated the efficacy of HLB Oasis in extracting PPCPs from environmental water samples [32]. Elution of PPCPs was performed using methanol, which has proven effective in solid-phase extraction (SPE) in previous studies [33–36].

HLB cartridges (500 mg) were preconditioned with 6 mL of MeOH followed by 6 mL of H_2O at a flow rate of 1 mL per min. A total of 1 L of sample was loaded into the cartridges at a rate of 6 mL/min and dried under vacuum. Subsequently, the analytes were eluted with 8 mL of MeOH and dried under nitrogen. Dried samples were stored in a freezer at −20 °C until the sampling campaign was completed. All samples were analyzed at the same time.

Dried extracts were reconstituted in 0.5 mL of 80:20 H_2O:MeOH (*v*/*v*) and filtered through 0.45 µm polyvinylidene fluoride (PVDF) filters and transferred to vials. A total of 200 ng/mL of Theophylline-d6 (internal standard) was added to vials to monitor instrument precision. All samples were analyzed using liquid chromatography–tandem mass spectrometry (LC-MS/MS).

3.5.2. Liquid Chromatography–Mass Spectrometry

Fragment ions of each PPCP were optimized via direct infusion of each compound (0.5 mL/min) into the mass spectrometer at a flowrate of 500 µL/min. The two most abundant product ions were selected, except for octocrylene and testosterone, where only one product ion was observed.

The liquid chromatography was performed using Agilent 1260 infinity high-performance liquid chromatography (HPLC) coupled to a 6460 Triple Quadrupole Mass Spectrometer (QqQ MS) with Agilent Jet Stream Technology (AJS) with an electro spray ionization (ESI) source. The presence of contaminants was determined using a Zorbax Eclipse reversed-phase column (100 mm × 2.1 mm, 1.8 um particle size) and tandem mass spectrometry (MS/MS) with the multiple reaction monitoring (MRM) method. The data acquisition was performed in positive (ESI+) and negative (ESI−) ionization polarity modes. The injection volume was 10 µL. The mobile phase flow rate was 200 µL/min, and the column was maintained at 30 °C. In the positive mode, solvent A contained Milli-Q water with 0.1% (*v*/*v*) formic acid at 95% and acetonitrile with 0.1% (*v*/*v*) formic acid as solvent B at 5%. Solvent B increased to 100% at

12 min and held until 15 min, then returned to 5%. A post-time of 5 min was added to allow the column to re-calibrate before the next analysis, resulting in a total run-time of 21 min.

In the negative mode, solvent A contained Milli-Q water with 1 mM of ammonium fluoride at 40%, and solvent B contained 1 mM of ammonium fluoride with 50:50 (v/v) water: methanol at 60%. Solvent B increased to 100% to 2 min and held until 15 min, then returned to 0%. The injection volume was 10 µL. A post-time of 5 min was added to enable conditioning.

3.5.3. Quantification Analysis

The limit of detection (LOD) and limit of quantification (LOQ) were calculated based on the signal-to-noise ratios of 3 and 10, respectively. According to the International Committee on Harmonization [37],

$$LOD = \frac{3.3d}{S} \quad (1)$$

$$LOQ = \frac{10d}{S} \quad (2)$$

where δ is the standard deviation of the response, and S is the slope of the calibration curve.

External calibration method using 7 points was used to quantify the concentrations of PPCPs. Details on PPCPs and chromatography information are presented in Supplementary Materials Tables S1–S4.

3.5.4. Quality Assurance and Quality Control

Instrument blanks were used to check for possible contamination. DEET was detected in blank samples at concentrations of 0.075 µg/L. The smallest concentration of DEET in the water samples was 0.5 µg/L. DEET was detected in blank water samples from several other studies [38,39], and there are known challenges with analyses for DEET, including the co-occurrence of chemically similar substances in the environment or in the solvents used for the analysis, which could result in analytical bias [39].

3.6. Statistical Analysis

Pearson correlation analysis was performed using SPSS software (version 29.0) to establish if there is a correlation between PPCP concentrations at the sampling sites in the residential lake, between PPCP concentrations in the residential and natural lake, and between PPCP concentrations and water physicochemical characteristics. The Mann–Whitney U test was used to determine if there are statistically significant differences in PPCP concentrations between the dry and wet seasons.

Three key assumptions were made in the statistical analyses. First, a PPCP was considered present if it was detected in more than 50% of the water samples ($\geq$9 occurrences). For missing data points, the limit of detection (LOD) value was assigned. Second, only data from RL-2 and RL-3 were used to calculate the average RL values, as these sites exhibited strong correlation, allowing for a representative mean (RL-avg). Third, RL-S1 was excluded from this analysis because its concentrations significantly deviated from those of RL-2 and RL-3. Including RL-S1 would have skewed the results, creating a misleading impression of higher PPCP concentrations in the RL.

4. Results

4.1. Onset of Wet Season

The transition from the dry season to the wet season was noticeable. During the last month of the dry season, there was very little precipitation. From the end of May to mid-July 2021, regular rainfall events occurred, with most of them exceeding 10 mm, with a particularly heavy rain during the first week of July. In August, although the frequency of rain events remained similar to that of the initial six weeks of the wet season, the amounts were mostly below 10 mm (Figure 3).

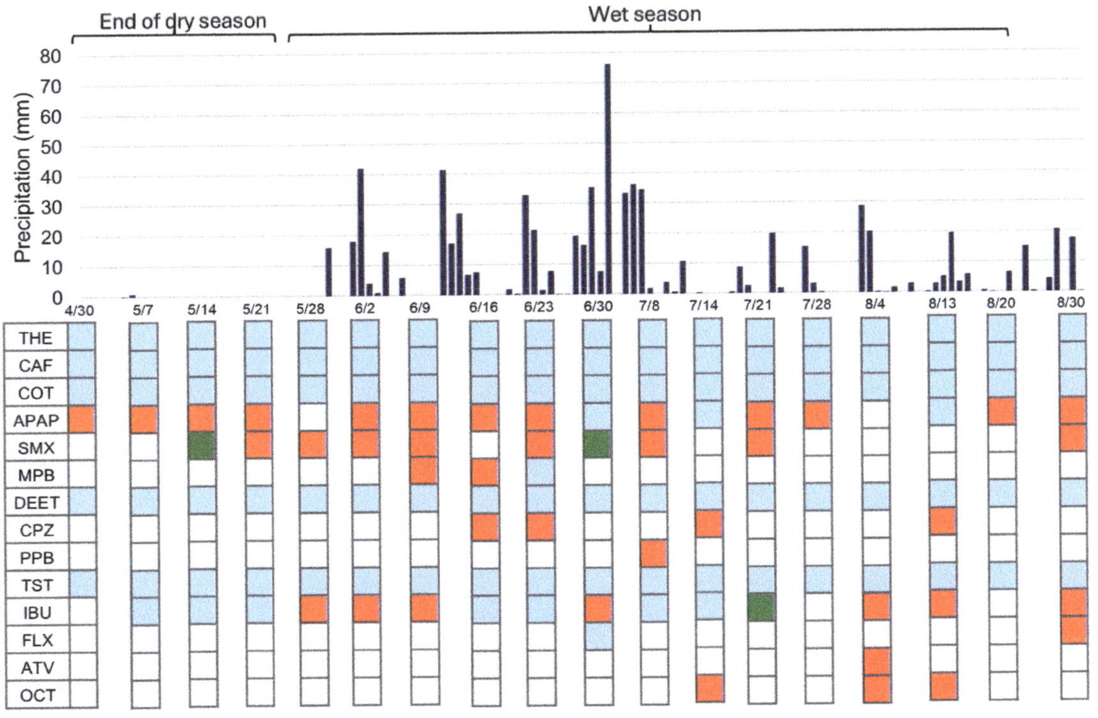

Figure 3. Sampling campaign dates and precipitation values. Presence of PPCPs in the residential and natural lakes. The representation of PPCPs in water samples is color-coded: blue indicates presence in both lakes (in residential lake—at least one detection at RL-S1, RL-S2, or RL-S3), red represents presence only in RL (detection at RL-S1, RL-S2, or RL-S3), and green indicates presence in the NL. Theophylline (THE), caffeine (CAF), cotinine (COT), acetaminophen (APAP), sulfamethoxazole (SMX), methylparaben (MPB), N,N-Diethyl-meta-toluamide (DEET), carbamazepine (CBZ), propylparaben (PPB), testosterone (TST), ibuprofen (IBU), fluoxetine (FLX), atorvastatin (ATV), and octocrylene (OCT).

4.2. PPCPs in RL and NL

All fourteen target PPCPs were identified at least once throughout the study period (Figure 3). To investigate the impact of the OWTS on lake water quality, we compared PPCP values of the residential lake (RL-avg) with the nature preserve lake (NL). Mean concentrations of cotinine and acetaminophen were higher in the RL-avg than the NL by 5.63 μg/L and 0.26 μg/L, respectively. Conversely, theophylline, caffeine, DEET, and testosterone had higher concentrations in NL compared to RL-avg. Theophylline and caffeine had an average concentration of 24.21 μg/L and 14.82 μg/L at RL-avg and 61.79 μg/L and 95.58 μg/L at NL. The average DEET concentration at RL-avg was 8.61 μg/L, and that at NL was 23.36 μg/L. This high average at NL was mainly caused by two samples with concentrations of 182.5 μg/L (14 May) and 76.9 μg/L (20 August). Similarly, the higher average concentration of testosterone at NL was due to two sampling events, first on 30 April and second on 28 July. The testosterone average was 6.39 μg/L at NL and 1.41 μg/L at

RL-avg. Maximum and minimum concentrations and the frequency of occurrences for all identified PPCPs can be found in Table S5.

4.3. *Influence of Precipitation on PPCP Concentrations*

When depicting the relationship, if any, between PPCP concentrations and precipitation, only those compounds that occurred throughout the entire sampling period were used (Figure 4). Also, the values of RL-avg represent the concentrations of the PPCPs in the residential lake (see Section 3). There is no consistent relationship between any of the PPCP concentrations and the onset of the wet season. Of the six PPCPs, only DEET increases in concentration throughout the wet season. For caffeine and theophylline, the onset of the wet season appears to decrease their presence in RL initially, but their concentrations gradually increase toward the latter half of this interval. Surprisingly, cotinine, testosterone, and acetaminophen concentrations vary throughout the entire sampling period.

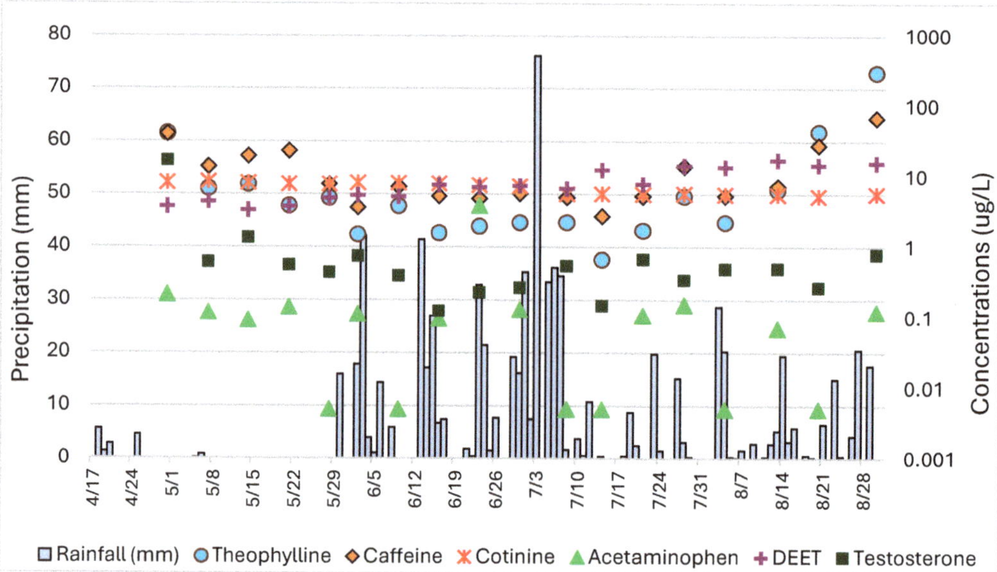

Figure 4. PPCP concentrations for RL-avg. Rain precipitation for the study area is presented as a bar graph representing daily precipitation in mm on the left *Y*-axis, and PPCP concentrations are presented on the right *Y*-axis.

At NL, theophylline, caffeine, and acetaminophen concentrations increased during the wet season, while DEET, cotinine, and testosterone decreased. The only statistically significant difference between the seasons was found for cotinine (Mann–Whitney U = 0.5, $z = -2.939$, $p < 0.001$). Fluoxetine was only detected in the wet season. Methylparaben, carbamazepine, propylparaben, octocrylene, and atorvastatin were not detected in NL.

Seasonal variations in the detection of PPCPs across all four sampling sites show that more PPCPs were detected in the wet season when compared to the dry season (Figure 5). For example, the number of detected PPCPs increased from 7 to 13 in RL-S1, from 8 to 10 in RL-S2, and from 7 to 12 in RL-S3. The natural lake also showed an increase, from 7 PPCPs in the dry season to 10 in the wet season.

Caffeine and theophylline dominated the PPCP profiles across all four sites and seasons. Cotinine, acetaminophen, and ibuprofen were also frequently present but in lower concentrations. Compounds such as fluoxetine, testosterone, and methyl- and propylparaben appeared more frequently during the wet season.

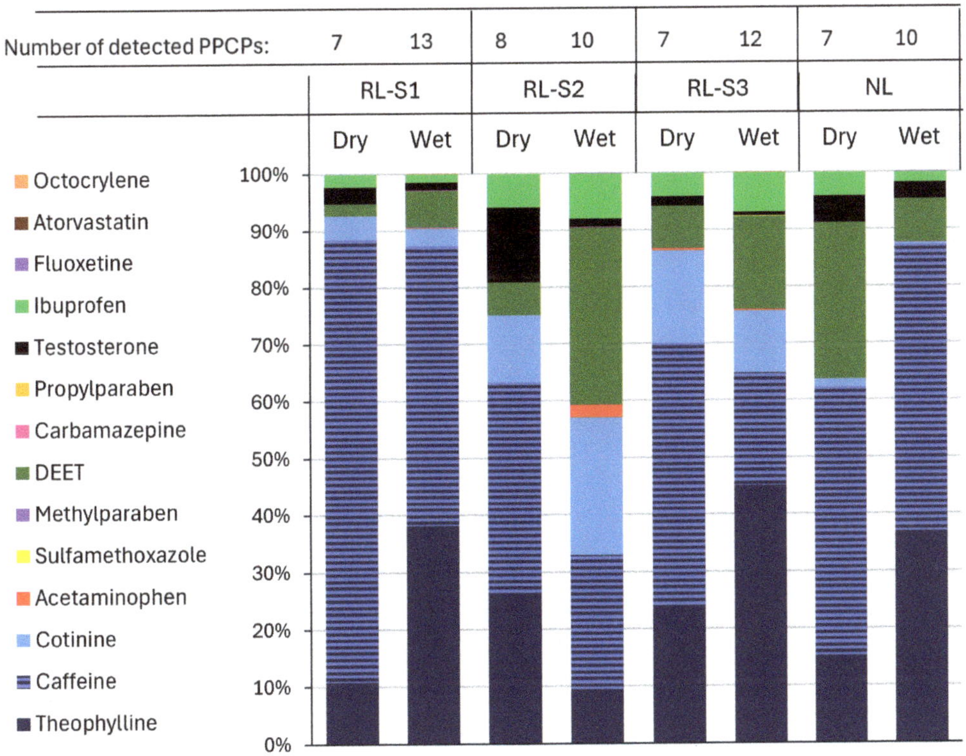

Figure 5. Dry and wet season variation in number of detected PPCPs.

RL showed more diverse and abundant PPCP profiles than the natural lake, particularly during the wet season. RL-S1 and RL-S3 wet season samples had the richest PPCP assemblages, including once-off detections of carbamazepine, atorvastatin, and methylparaben, which were absent in their dry season.

4.4. Water Quality

Various water quality measurements were conducted concurrently with the water collection for the PPCPs (Figure 6). This was performed to determine whether changes in the water quality could explain the variability in the concentrations and assemblages of the PPCPs.

RL's TDS values at each sampling site peaked at 144 ppm, 145 ppm, and 142 ppm just before the wet season onset and then decreased slightly. In contrast, NL maintained stable TDS levels of ~100 ppm. RL exhibited pH ranging from 6.29 to 7.73, and EC varied between 221 µS/cm and 289 µS/cm. Conversely, NL exhibited pH levels ranging from 4.97 µS/cm to 6.38 µS/cm. All the water quality measurements were higher in RL than in NL. The average temperature at RL was similar, 29.3 °C, 29.4 °C, and 28.7 °C at RL-S1, RL-S2, and RL-S3, respectively. The lowest temperature occurred on 21 May. The temperature in NL was higher than in RL. With the lowest temperature of 26.6 °C and the highest of 32.2 °C. In RL, DO levels fluctuated between 0.04 mg/L and 2.82 mg/L, while in the NL, DO ranged from 0.7 mg/L to 5.25 mg/L. Generally, lower levels of DO with slightly lower temperatures were observed at RL.

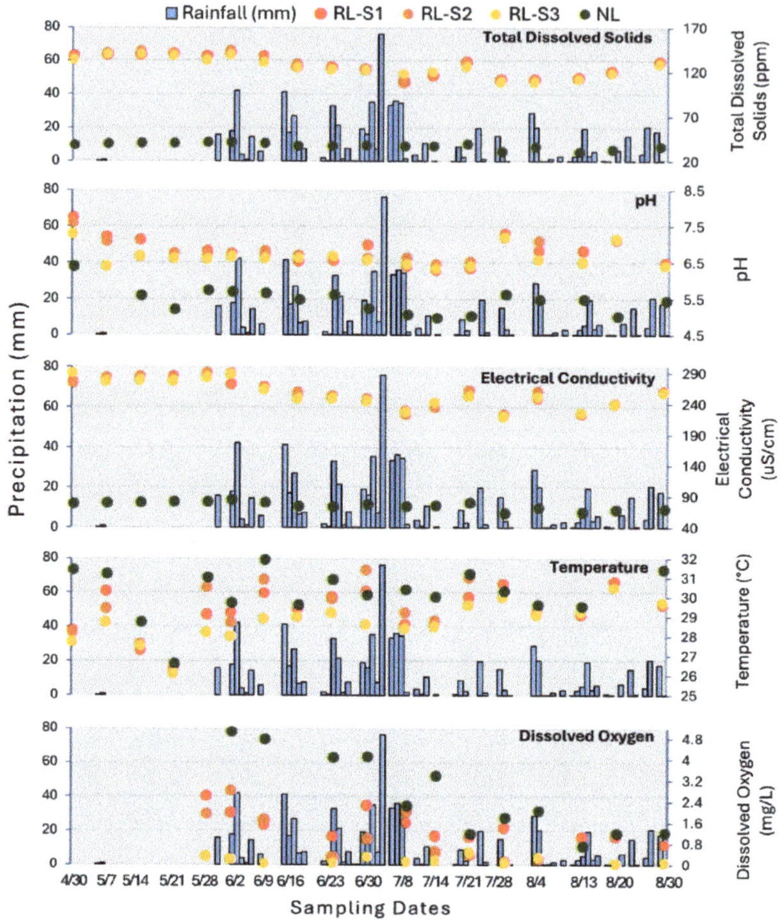

Figure 6. Water quality parameters and precipitation for the residential lake sites and the natural lake.

Lake water physicochemical parameters, including pH, EC, and TDS, were quite homogeneous at all three sites in the lake. A positive correlation was observed between TDS and EC (Table 2). This concurrent change most probably reflects the dilution of the lake water. While there are some statistically significant correlations between the water quality parameters and certain PPCPs, there is little consistency in their relationships. For example, there are positive correlations between testosterone and pH and between DEET and temperature. Conversely, there is a negative correlation between DEET and TDS and DEET and EC. TDS and electrical EC had a positive correlation with cotinine, but neither pH nor DO was correlated with PPCPs.

Table 2. Pearson correlation between physicochemical characteristics of the lakes and PPCP concentrations.

	pH-RL	TDS-RL	EC-RL	T-RL	pH-NL	TDS-NL	EC-NL	T-NL
Pearson	−0.131	0.021	0.031	0.151				
Sig.	0.605	0.934	0.904	0.549				
Pearson	0.631 **	0.185	0.309	−0.288				
Sig.	0.005	0.461	0.212	0.247				

Table 2. *Cont.*

	pH-RL	TDS-RL	EC-RL	T-RL	pH-NL	TDS-NL	EC-NL	T-NL
Pearson	−0.055	−0.717 **	−0.714 **	0.503 *				
Sig.	0.828	0.001	0.001	0.033				
Pearson	0.127	0.756 **	0.767 **	−0.462				
Sig.	0.614	0.000	0.000	0.053				
Pearson	0.236	0.143	0.184	−0.088				
Sig.	0.345	0.572	0.465	0.729				
Pearson					−0.229	0.114	0.055	0.277
Sig.					0.378	0.653	0.829	0.266
Pearson					0.484 *	−0.291	−0.365	0.159
Sig.					0.049	0.241	0.136	0.530
Pearson					−0.066	0.120	0.093	−0.236
Sig.					0.800	0.635	0.715	0.346
Pearson					0.737 **	0.301	0.363	0.064
Sig.					<0.001	0.224	0.138	0.800
Pearson					−0.204	0.349	0.303	0.046
Sig.					0.432	0.156	0.221	0.855

**—Correlation is significant at the 0.01 level (2-tailed). *—Correlation is significant at the 0.05 level (2-tailed).

5. Discussion

5.1. PPCP Concentrations

The concentrations of some of our PPCPs are quite high compared to most studies. Given our unique study area, a small, enclosed lake that receives effluent from 18 OWTS, we found it impossible to find data from comparable settings. However, there are several that have similar concentrations. Ramage et al. [40] sampled the waters of a low-flow stream adjacent to several households with septic tanks. They detected cotinine and caffeine at 31 μg/L and 200 μg/L, respectively. Richards et al. [41] investigated water downstream from two water tanks, one used by two individuals and the other by five, and reported caffeine concentrations ranging from 0.48 to 5 μg/L. Similarly, Tran et al. [42] documented caffeine levels in surface waters reaching up to 14 μg/L. Daneshvar et al. [43] and Diwan et al. [44] detected 60–174 μg/L caffeine levels in Ambazari, Futala, and Gandhi Sagar lakes. Tran et al. [45] investigated concentrations in surface water samples from the sewer-fed catchment. The caffeine concentrations ranged from 14 ng/L to 144.18 μg/L, acetaminophen from below detection (<5.0 ng/L) to 4.82 μg/L, and DEET from 1.4 ng/L to 6.23 μ/L Carbamazepine concentrations ranged from below detection to 53.3 ng/L. DEET with μg/L level concentrations has been measured in waters receiving WWTP effluent. For example, Lee et al. [46] detected levels at 3.7 μg/L in stream water, while concentrations up to 15 μg/L have been found in WWTP effluent [47,48]. Regarding theophylline, only effluents from hospitals (max. 8 μg/L) [49] and PPCP manufacturing plants (max. 33.1 μg/L) [50] have similar values to our maximum concentrations [51].

5.2. PPCP Concentrations in the Dry and Wet Seasons

The results show an increase in average concentrations in the wet season for theophylline, acetaminophen, and DEET, with only the mean for DEET being statistically significant (at significance level 0.05, $p < 0.001$). Conversely, the average caffeine, cotinine, and testosterone concentrations decreased, with all means being statistically significant (significance level 0.05, *p*-values: 0.035, 0.008, and 0.018, respectively). Such results are similar to the findings of Sodré et al. [52], who found higher mean concentrations of caffeine in the dry season. As mentioned in the Introduction, we expected that the onset of the wet season would initially dilute PPCP concentrations in the lake, followed by the arrival of the PPCPs mobilized from the septic tank drainfields. This would be accompanied by decreases in both TDS and specific conductance. However, as none of the PPCPs that adhered to this supposition, except DEET, there must be other complicating factors in play.

5.2.1. Complications at the Source

An important assumption that is required to support our initial supposition is that the OWTS would be operating as expected. If the OWTS has not been properly maintained, this may change its operation regarding the removal efficiency of certain PPCPs in the tank and drainfield [53]. The PPCP concentrations in the environment also depend on how fast the compounds are removed in the OWTS. Removal and transformation of PPCPs in OWTS relies on vadose zone processes, sorption, and microbial degradation [54]. Another factor influencing the presence of PPCPs may be the age of the OWTS. Older systems may not process the effluent as well as newer ones. The pipes of the drainfield may have become clogged or crushed over time. Consequently, this diminished performance of the OWTS could complicate the delivery of PPCPs to the lake. Without a detailed examination of the OWTS (tanks and drainfields), it is impossible to tell the state of operation of these systems. However, if they were not properly maintained, then their efficiency of PPCP removal would be reduced, thereby leading to more PPCPs entering the groundwater and, subsequently, the lake.

5.2.2. Contribution of Aquifer Water

Another factor to consider is the contribution of well water to the runoff entering the lake. The study area has aquifer-fed wells as the only option for residential water. This aquifer water would enter the lake via drainfield effluent and irrigation. Aquifers can act as storage reservoirs for contaminants, including PPCPs [29]. Changes in groundwater flow patterns due to fluctuations in precipitation or pumping from wells can remobilize stored contaminants and release them into the lake [55].

To investigate the aquifers' influence, we measured well water at site one in RL (Table 3). Only those detected at the well are included in the table. First, most of the PPCPs detected in the lake are not present in the aquifer (well) water. This suggests that most of the PPCPs originate from the OWTS. Second, besides DEET's first sample, all PPCP levels are at least one order of magnitude greater than those from the FAS. Consequently, while there are PPCPs in the well water, the aquifer's contribution can be considered minimal.

Cotinine was detected in the well water only during the first sampling. During this period of the dry season, irrigation water is sourced from the well, which taps the aquifer. If there was a short pulse of water contaminated with cotinine passing through the aquifer at the time of pumping, it would be in both the well water and the residential sampling sites. The levels in the residential sites could be higher due to the evaporation of some of the irrigation water. This may represent a single contamination event in an area upgradient from the wells. Dumping of septic tank effluent in a recharge area for the aquifer could explain a temporary high level of cotinine. Other studies have found similar episodic presence of certain PPCPs in aquifer water [56,57].

Table 3. PPCP concentrations of well water at RL-S1 and lake water samples at RL-S1.

Date	Location	PPCP Concentrations (μg/L)				
		Theophylline	**Testosterone**	**DEET**	**Cotinine**	**Caffeine**
30 April 2021	Well	0.28	2.80	4.51	5.73	4.91
	RL-1	17.70	0.19	3.52	8.38	19.68
	RL-2	53.02	31.94	3.70	7.8	51.22
	RL-3	28.81	1.77	3.70	8.43	29.13
30 June 2021	Well	0.17	0.72	0.62	Not detected	0.12
	RL-1	24.44	0.38	8.22	7.54	56.78
	RL-2	1.12	0.35	7.32	7.25	3.76
	RL-3	3.30	0.19	7.38	7.35	7.65
13 August 2021	Well	0.15	0.36	1.9	Not detected	2.2
	RL-1	7.05	0.70	13.7	5.27	9.42
	RL-2	**	**	**	**	**
	RL-3	6.47	0.50	15.7	5.57	7.28

** water samples not collected.

As for the comparatively high DEET levels on the first sampling day, there is no clear explanation. Sampling protocols were identical for all water collections. Additionally, the tap was run at full capacity for approximately one minute to flush "old" water from the system before sample collection began. DEET concentrations could also be a result of residential use.

Another possible source of PPCPs from the aquifer is the upward movement of water through the base of the lake from the Intermediate Aquifer System (IAS), which overlies the FAS. However, this source is unlikely, as local well logs (Figure 7) show that the IAS is over five meters below the base of the lake. Additionally, as this is an internally drained sinkhole lake [58], the excess precipitation must be removed from the watershed through seepage. Consequently, it is unlikely that the upward flow of the IAS into the RL is a major source of the PPCPs.

The high TDS values of the RL may provide an indication of the potential of flocculation to remove PPCPs from the water column. This process may explain the negative correlation between DEET and TDS/EC. Üstün-Odabaşı et al. [59] found a strong negative correlation between total PPCP concentrations and TDS.

The temperature and pH of the lake water impact the degradation of PPCPs. Warmer water temperatures accelerate chemical reactions, which would promote the degradation of PPCPs in both lake water and in soils [60]. Consequently, temporal changes in temperature and pH may impact the presence of PPCPs in the lake. In both RL and NL, neither temperature nor pH has consistent correlations with the concentrations of PPCPs. Such a result fits the conclusions of Du et al. [61], who investigated the role of temperature and pH, amongst others, in the attenuation of PPCPs in lake water.

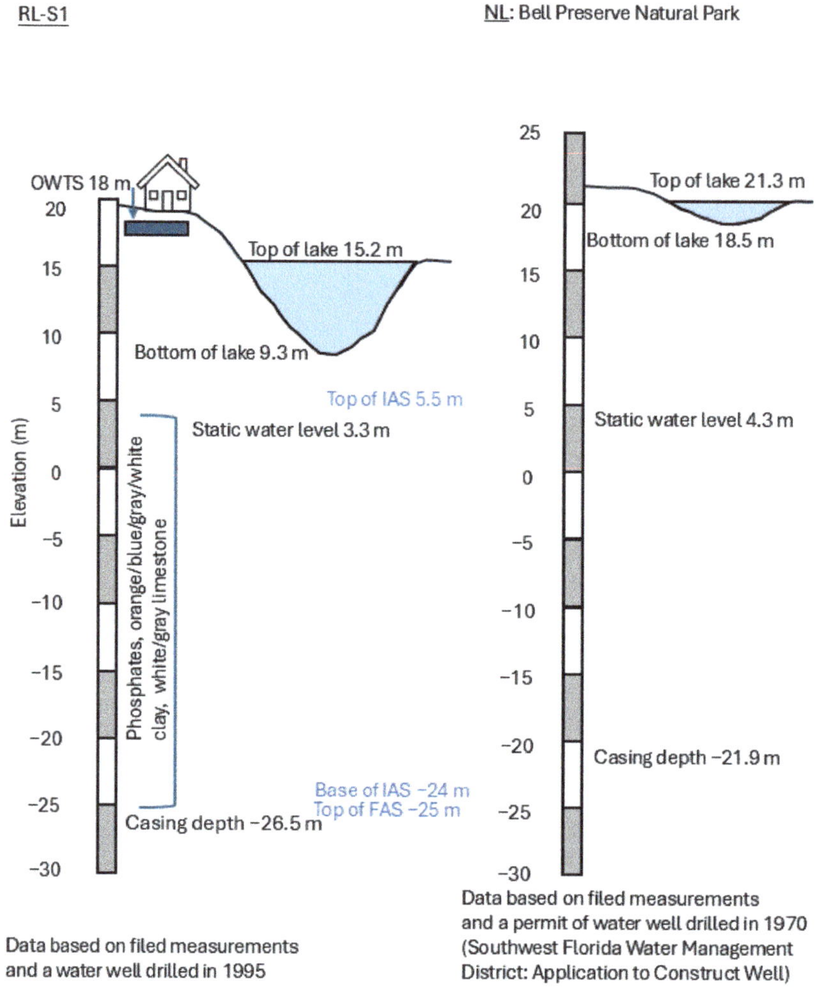

Figure 7. Hydrogeological profiles for both lakes from well log data.

5.3. PPCP Assemblages in the Dry and Wet Seasons

A second supposition of this study is that the onset of the wet season leads to an increase in more hydrophobic PPCPs in the residential lake. During the dry season, eight PPCPs were detected in the lake water: theophylline, caffeine, cotinine, acetaminophen, sulfamethoxazole, DEET, testosterone, and ibuprofen. In the wet season, along with the above PPCPs, atorvastatin, carbamazepine, methylparaben, propylparaben, fluoxetine, and octocrylene were detected. It is highly likely that the physicochemical properties of these latter PPCPs influenced their presence in the lake water [62]. A common trait of organic contaminants is that they may partition (adsorb) to soil. This accumulation can be assessed by the logarithm of octanol–water partition coefficient (log K_{ow}), which measures the distribution of a chemical between water and an organic solvent (typically octanol). The hydrophobic PPCPs fluoxetine, atorvastatin, and octocrylene are hygroscopic in nature, which favors their absorption or adsorption onto soil particles [20]. During the wet season, these PPCPs with an affinity for soil particles (hydrophobic) are less able to sorb onto these particles due to soil saturation [63].

The PPCPs in the residential lake were classified as being hydrophilic (log $K_{OW} < 2.5$), moderately hydrophobic (log K_{OW} between 2.5 and 4.0), and highly hydrophobic (log $K_{OW} > 4$). PPCPs measures that fell within the hydrophilic categories were carbamazepine, DEET, methylparaben, sulfamethoxazole, acetaminophen, cotinine, caffeine, and theophylline, in log K_{OW} order from highest to lowest. These compounds have low sorption potential and would therefore be expected to have high occurrences in water. PPCPs found in the moderately hydrophobic classification are ibuprofen, testosterone, and propylparaben (log K_{OW} order as described above). These PPCPs have medium sorption potential and would be expected to be present in the lake water but also remain in the OWTS drainfields and soils surrounding the lake. The final category, those of high hydrophobic class, are octocrylene, atorvastatin, and fluoxetine (the same order as above). Normally, these PPCPs remain in the soils due to their high sorption potential [64]. Consequently, as their presence in our lake water during only the wet season suggests, their sorption potential was exceeded due to the increase in both the soil saturation and hydrostatic pressure. Dai et al. [63] found that the absorption of more hydrophobic PPCPs was negligible when the soils were saturated compared to unsaturated soils. They reasoned that in unsaturated soils, the greater complexity of the flow path for the PPCPs increases the possibility of contaminant sorption as they travel along the water films surrounding the soil particles. In contrast, saturated soil allows for greater ease of travel for the PPCPs, thereby explaining their presence in the lake during the wet season. This adheres to our conceptual model (Figure 1) that the PPCPs with the highest hydrophobicity are only present when soils are saturated.

5.4. PPCPs in the Residential and the Natural Lakes

We included the NL as a control because it was not surrounded by OWTS but otherwise had similar physical properties. Unfortunately, we could only sample at one location due to the difficulty of access. While we were not surprised to find PPCPs, as they are environmentally ambiguous, the elevated concentrations of those present were unexpected. This result suggests that the NL's source of contaminants is sourced from residential drainfields. Of the PPCPs present in the NL at certain sampling intervals, those that were at higher concentrations than the RL are also frequently detected in other environments [52,65]. Caffeine and theophylline are highly hydrophilic, explaining their presence in surface waters at high concentrations [66,67]. Caffeine in a lake located in a natural preserve is not unique since it was found in high quantities in the Arctic [68] and remote rivers of the Rocky Mountains National Park [38]. DEET, testosterone, and ibuprofen are also commonly found in aquatic environments [30,69,70]. Most studies have attributed the source of the contaminants to be WTTPs.

The pH values in NL were lower, and dissolved oxygen (DO) values were higher compared to the RL. However, as mentioned above, both lakes, for the most part, had similar trends in the PPCPs' seasonal variability. Consequently, it can be assumed that the water quality parameters do not have a major impact on the chemical properties of the PPCP concentrations. This supposition agrees with the findings of Edwards (2014) [71], who found that the concentration of caffeine did not show any significant correlation with hydrologic parameters such as surface water temperature, pH, or dissolved oxygen, indicating its stability and slow pace of degradation. Hence, despite clear differences in the water quality parameters between the two lakes, they can be disregarded as significant influences on PPCP dissimilarities.

While the water quality parameters measured above may not explain discrepancies in the PPCP assemblages and concentrations, they may shed light on their provenance. Both pH and conductivity values for the RL are very similar to the aquifer-fed (FAS) lake in the same county, whereas NL is closer to a rain-fed lake, also in the same area (Figure 8). The RL is partially supplied by the FAS via well water, whereas there is no evidence for this source at the NL. An investigation of water levels in a groundwater well on the NL property found that the aquifer is 17.6 m below the base of the lake. Consequently, source waters are quite different for each lake, thereby suggesting that the FAS could not be the source of the PPCPs in NL. The contribution of FAS to the RL should not be overstated. The wet season RL lake volume is ~19,000 m^3, and for the 123-day period of observation, ~7100 m^3 of precipitation occurred over the lake. Consequently, precipitation is a significant contributor to seasonal changes in lake volume.

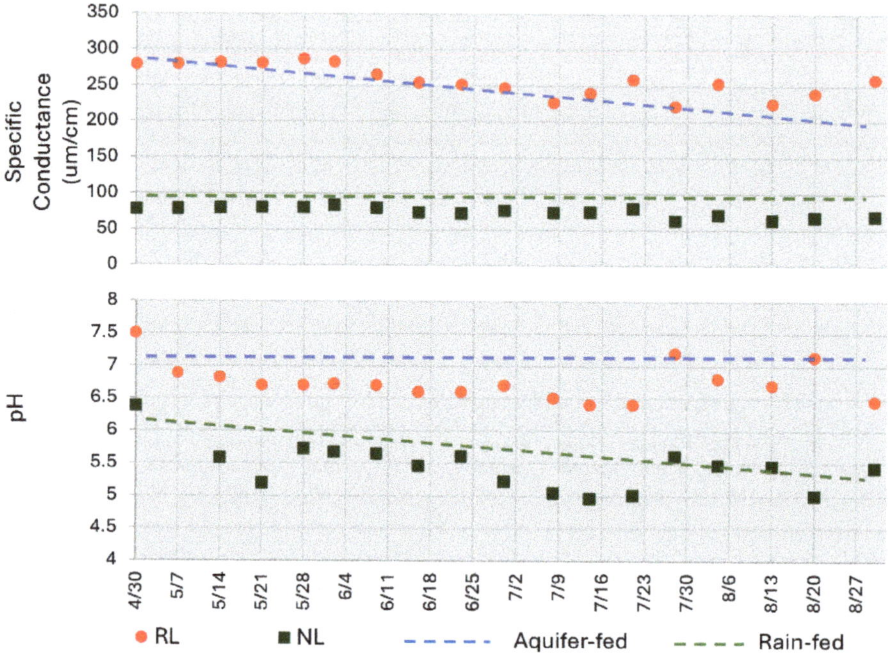

Figure 8. Comparison of natural/residential lakes with aquifer-augmented and rainfed lakes from NW Hillsborough County.

The most likely explanation for the PPCPs in NL is the residential area south of the preserve. The small residential area is approximately 200 m from the lake. These homes use OWTS and are at a higher elevation than the lake itself. This differential creates a shallow hydraulic gradient sloping toward the lake. Consequently, the same process occurs as in RL. The high permeability of the Chandler soil unit, which covers the entire study area [72], would help promote the somewhat free movement of groundwater advection. If OWTS are indeed the primary source of PPCPs for NL, then the concentration of these contaminants should mirror one another. As shown in the results, the trends for most PPCPs are very similar for both lakes. Of course, it is unlikely that both residential homes use the same PPCPs, which would explain the differences between the lakes.

The higher average DEET concentration in NL may provide additional evidence on how the specific location of the two study sites may influence the concentrations. The Bell Creek Preserve has stagnant marshes south and east of the lake. This environment could be a source of a greater abundance of mosquitoes, which would call for a greater application of DEET by park visitors compared to RL, which lacks a similar marsh.

5.5. Implications for Management of Shallow Lakes

Given the results of this study, we can suggest some potential solutions to decreasing the concentrations of PPCPs in these shallow lakes. The first pertains to improving the effectiveness of PPCP removal in the residential OWTS, although these measures may be cost-prohibitive at such a small scale. These technologies include advanced oxidation processes, membrane filtration, and activated carbon adsorption [73]. The simplest option, which is currently being implemented in some areas, is to connect residential areas to the municipal sewer system. However, for neighborhoods that are more isolated, one possible option could be directing these homes to decentralized, small-scale treatment plants that utilize some of the above technologies. It is beyond the scope of this paper to calculate the costs of the alternative approaches.

From a policy/regulation perspective, there are a number of approaches that can be considered, not necessarily being adopted exclusively from one another. Monitoring lake water quality regarding PPCPs is essential to determine the presence of elevated levels and potential sources of such contaminants [74]. Increasing public awareness is a cost-effective approach, which involves both informing the residents of proper disposal of PPCPs and how their actions impact their local environment [75,76]. Next, developing site-specific risk assessments helps identify PPCPs of concern and whether they pose any ecological risks to lake organisms [77]. Finally, a more holistic approach is adopting the practice of Integrated Watershed Management [73], which incorporates various aspects such as land use, sewage infrastructure, and catchment hydrology.

6. Conclusions

This pilot study investigated how precipitation, with the onset of the rainy season, influences the concentrations and assemblages of 14 PPCPs in two lakes, the first surrounded by OWTS and the second located in a nature preserve. Our conceptual model of the influence of the wet season on PPCP concentrations in the RL was not supported by the statistical analyses. Sampling soil water may allow for a more definitive explanation as to why. This sampling would shed light on the spatial and temporal transition of the contaminants between the OWTS and the lake.

In addition to the first hypothesis, we posited that more hydrophobic PPCPs would appear in the lake due to greater soil saturation as precipitation increased. We found that the assemblage of PPCPs did change, as those with more hydrophobic properties were detected only in the wet season. The saturation of the soil overcomes the propensity of these PPCPs to adhere to soil particles, thereby allowing the release into the lake.

The contribution of variable water sources influencing the PPCP concentrations in RL was not initially considered important. However, as the residential homes are supplied by well water from the FAS, which is included for irrigation in the dry season, aquifer water is an important component of the study. The RL does share similarities with other aquifer-fed lakes. However, well water collected from a residential well showed that PPCP concentrations were very small compared to the lake water. Consequently, the OWTS surrounding the lake are the main source of PPCPs in the RL. The OWTS are also the main contributor of PPCPs in the NL, despite being located away from a residential area.

There were several limitations in the study. Only two lakes were analyzed, and including more in future research would allow for more definitive conclusions. Second, the entire dry season was not sampled. However, our interest was the onset of the wet season, not the full hydraulic year. Finally, more sites could have been sampled at both lakes. While three sites seemed reasonable for the RL as well as sampling at the same depth and time of day, it appears that the influence of the total depth of the water column may have been more significant than we first envisioned. Consequently, several more sites at both lakes should be included for the next phase of this study. However, this was the first study on shallow lakes in a karst setting and provides new foci for future research.

Supplementary Materials: The following supporting information can be downloaded at: https://www.mdpi.com/article/10.3390/environments12070219/s1.

Author Contributions: Conceptualization, P.v.B. and E.B.-J.; methodology, E.B.-J., L.C., and P.v.B.; software, E.B.-J. and L.C.; validation, E.B.-J., P.v.B., and L.C.; formal Analysis, E.B.-J. and L.C.; investigation, P.v.B. and E.B.-J.; resources, P.v.B., L.C., and E.B.-J.; data curation, E.B.-J.; writing—original draft preparation, E.B.-J. and P.v.B.; writing—review and editing, E.B.-J. and P.v.B.; visualization, E.B.-J. and P.v.B.; supervision, P.v.B.; project administration, P.v.B. All authors have read and agreed to the published version of the manuscript.

Funding: This research received no external funding.

Data Availability Statement: The original contributions presented in this study are included in the article/Supplementary Material. Further inquiries can be directed to the corresponding author(s).

Acknowledgments: We thank the homeowners for granting access to the residential lake and Rae and Jenna for helping with sample collection.

Conflicts of Interest: The authors declare no conflicts of interest.

References

1. Gao, J.; O'Brien, J.; Du, P.; Li, X.; Ort, C.; Mueller, J.F.; Thai, P.K. Measuring Selected PPCPs in Wastewater to Estimate the Population in Different Cities in China. *Sci. Total Environ.* **2016**, *568*, 164–170. [CrossRef] [PubMed]
2. Junaid, M.; Wang, Y.; Hamid, N.; Deng, S.; Li, W. Prioritizing Selected PPCPs on the Basis of Environmental and Toxicogenetic Concerns: A Toxicity Estimation to Confirmation Approach. *J. Hazard. Mater.* **2019**, *380*, 120828. [CrossRef]
3. Brausch, J.M.; Connors, K.; Brooks, B.W.; Rand, G.M. Human Pharmaceuticals in the Aquatic Environment: A Review of Recent Toxicological Studies and Considerations for Toxicity Testing. In *Reviews of Environmental Contamination and Toxicology*; Whitacre, D.M., Ed.; Springer: Boston, MA, USA, 2012; pp. 1–99.
4. Duong, H.T.; Kadokami, K.; Nguyen, D.T.; Trinh, H.T.; Doan, N.H.; Mizukawa, H.; Takahashi, S. Occurrence, Potential Sources, and Risk Assessment of Pharmaceuticals and Personal Care Products in Atmospheric Particulate Matter in Hanoi, Vietnam. *Environ. Sci. Pollut. Res.* **2023**, *30*, 34814–34826. [CrossRef] [PubMed]
5. Xie, Z.; Zhang, P.; Wu, Z.; Zhang, S.; Wei, L.; Mi, L.; Kuester, A.; Gandrass, J.; Ebinghaus, R.; Yang, R.; et al. Legacy and Emerging Organic Contaminants in the Polar Regions. *Sci. Total Environ.* **2022**, *835*, 155376. [CrossRef]
6. Arpin-Pont, L.; Martínez-Bueno, M.J.; Gomez, E.; Fenet, H. Occurrence of PPCPs in the Marine Environment: A Review. *Environ. Sci. Pollut. Res.* **2016**, *23*, 4978–4991. [CrossRef] [PubMed]
7. Keerthanan, S.; Jayasinghe, C.; Biswas, J.K.; Vithanage, M. Pharmaceutical and Personal Care Products (PPCPs) in the Environment: Plant Uptake, Translocation, Bioaccumulation, and Human Health Risks. *Crit. Rev. Environ. Sci. Technol.* **2021**, *51*, 1221–1258. [CrossRef]
8. Silori, R.; Shrivastava, V.; Singh, A.; Sharma, P.; Aouad, M.; Mahlknecht, J.; Kumar, M. Global Groundwater Vulnerability for Pharmaceutical and Personal Care Products (PPCPs): The Scenario of Second Decade of 21st Century. *J. Environ. Manag.* **2022**, *320*, 115703. [CrossRef]
9. Xu, J.; Wu, L.; Chang, A.C. Degradation and Adsorption of Selected Pharmaceuticals and Personal Care Products (PPCPs) in Agricultural Soils. *Chemosphere* **2009**, *77*, 1299–1305. [CrossRef]
10. Wu, X.; Ernst, F.; Conkle, J.L.; Gan, J. Comparative Uptake and Translocation of Pharmaceutical and Personal Care Products (PPCPs) by Common Vegetables. *Environ. Int.* **2013**, *60*, 15–22. [CrossRef]
11. Center for Sustainable Systems. *U.S. Wastewater Treatment Factsheet*; Pub. No. CSS04-14; Center for Sustainable Systems: Ann Arbor, MI, USA, 2024.
12. Karnjanapiboonwong, A.; Suski, J.G.; Shah, A.A.; Cai, Q.; Morse, A.N.; Anderson, T.A. Occurrence of PPCPs at a Wastewater Treatment Plant and in Soil and Groundwater at a Land Application Site. *Water. Air. Soil Pollut.* **2011**, *216*, 257–273. [CrossRef]
13. Boxall, A.B.A. Fate and Transport of Veterinary Medicines in the Soil Environment. In *Fate of Pharmaceuticals in the Environment and in Water Treatment Systems*; Aga, D.S., Ed.; CRC Press: Boca Raton, FL, USA, 2008; pp. 123–137.
14. Aga, D.S.; Samara, F.; Dronjak, L.; Kanan, S.; Mortula, M.; Vahapoglu, L. Rising Water, Rising Risks: The Hidden Dangers of Emerging Contaminants in Climate-Intensified Storms. *ACS ES&T Water* **2024**, *4*, 2785–2788.
15. Perkins, R. Septic Tanks, Lot Size and Pollution of Water Table Aquifers. *J. Environ. Health* **1984**, *46*, 298–304.

16. Gao, Q.; Blum, K.M.; Gago-Ferrero, P.; Wiberg, K.; Ahrens, L.; Andersson, P.L. Impact of On-Site Wastewater Infiltration Systems on Organic Contaminants in Groundwater and Recipient Waters. *Sci. Total Environ.* **2019**, *651*, 1670–1679. [CrossRef] [PubMed]
17. U.S. Environmental Protection Agency. *Design Manual: Onsite Wastewater Treatment and Disposal Systems*; U.S. Environmental Protection Agency: Washington, DC, USA, 1980.
18. Meerhoff, M.; Beklioğlu, M. Shallow Lakes and Ponds. In *Limnology*; Academic Press: Cambridge, MA, USA, 2024; pp. 859–892.
19. Beklioğlu, M.; Meerhoff, M.; Davidson, T.A.; Ger, K.A.; Havens, K.; Moss, B. Preface: Shallow Lakes in a Fast Changing World: The 8th International Shallow Lakes Conference. *Hydrobiologia* **2016**, *778*, 9–11. [CrossRef]
20. Ng, A.; Weerakoon, D.; Lim, E.; Padhye, L.P. Fate of Environmental Pollutants. *Water Environ. Res.* **2019**, *91*, 1294–1325. [CrossRef]
21. Ma, R.; Wang, B.; Yin, L.; Zhang, Y.; Deng, S.; Huang, J.; Wang, Y.; Yu, G. Characterization of Pharmaceutically Active Compounds in Beijing, China: Occurrence Pattern, Spatiotemporal Distribution and Its Environmental Implication. *J. Hazard. Mater.* **2017**, *323*, 147–155. [CrossRef]
22. Corada-Fernández, C.; Candela, L.; Torres-Fuentes, N.; Pintado-Herrera, M.G.; Paniw, M.; González-Mazo, E. Effects of Extreme Rainfall Events on the Distribution of Selected Emerging Contaminants in Surface and Groundwater: The Guadalete River Basin (SW, Spain). *Sci. Total Environ.* **2017**, *605–606*, 770–783. [CrossRef]
23. Yu, X.; Sui, Q.; Lyu, S.; Zhao, W.; Wu, D.; Yu, G.; Barcelo, D. Rainfall Influences Occurrence of Pharmaceutical and Personal Care Products in Landfill Leachates: Evidence from Seasonal Variations and Extreme Rainfall Episodes. *Environ. Sci. Technol.* **2021**, *55*, 4822–4830. [CrossRef]
24. Upchurch, S.; Scott, T.M.; Alfieri, M.C. Hydrogeology of Florida. In *The Karst Systems of Florida: Understanding Karst in Geologically Young Terrain*; LaMoreaux, J., Ed.; Springer: Berlin/Heidelberg, Germany, 2019; pp. 93–145.
25. SWFWMD NexRad RADAR Rainfall Estimates. Available online: https://edis.ifas.ufl.edu/publication/AE517 (accessed on 2 February 2022).
26. Yang, Y.Y.; Toor, G.S.; Wilson, P.C.; Williams, C.F. Micropollutants in Groundwater from Septic Systems: Transformations, Transport Mechanisms, and Human Health Risk Assessment. *Water Res.* **2017**, *123*, 258–267. [CrossRef]
27. Schaider, L.A.; Ackerman, J.M.; Rudel, R.A. Septic Systems as Sources of Organic Wastewater Compounds in Domestic Drinking Water Wells in a Shallow Sand and Gravel Aquifer. *Sci. Total Environ.* **2016**, *547*, 470–481. [CrossRef]
28. Conn, K.E.; Siegrist, R.L.; Barber, L.B.; Meyer, M.T. Fate of Trace Organic Compounds during Vadose Zone Soil Treatment in an Onsite Wastewater System. *Environ. Toxicol. Chem.* **2010**, *29*, 285–293. [CrossRef]
29. Katz, B.G.; Griffin, D.W.; McMahon, P.B.; Harden, H.S.; Wade, E.; Hicks, R.W.; Chanton, J.P. Fate of Effluent-Borne Contaminants beneath Septic Tank Drainfields Overlying a Karst Aquifer. *J. Environ. Qual.* **2010**, *39*, 1181. [CrossRef]
30. Del Rosario, K.L.; Mitra, S.; Humphrey, C.P.; O'Driscoll, M.A. Detection of Pharmaceuticals and Other Personal Care Products in Groundwater beneath and Adjacent to Onsite Wastewater Treatment Systems in a Coastal Plain Shallow Aquifer. *Sci. Total Environ.* **2014**, *487*, 216–223. [CrossRef] [PubMed]
31. Hinkle, S.R.; Weick, R.J.; Johnson, J.M.; Cahill, J.D.; Smith, S.G.; Rich, B.J. *Organic Wastewater Compounds, Pharmaceuticals, and Coliphage in Ground Water Receiving Discharge from Onsite Wastewater Treatment Systems near La Pine, Oregon: Occurrence and Implications for Transport*; Usgs Sir 2005-5055; U.S. Geological Survey: Reston, VA, USA, 2005; 98p.
32. Gilart, N.; Marcé, R.M.; Borrull, F.; Fontanals, N. Determination of Pharmaceuticals in Wastewaters Using Solid-Phase Extraction-Liquid Chromatography-Tandem Mass Spectrometry. *J. Sep. Sci.* **2012**, *35*, 875–882. [CrossRef] [PubMed]
33. Petrie, B.; Youdan, J.; Barden, R.; Kasprzyk-Hordern, B. Multi-Residue Analysis of 90 Emerging Contaminants in Liquid and Solid Environmental Matrices by Ultra-High-Performance Liquid Chromatography Tandem Mass Spectrometry. *J. Chromatogr. A* **2016**, *1431*, 64–78. [CrossRef] [PubMed]
34. Althakafy, J.T.; Kulsing, C.; Grace, M.R.; Marriott, P.J. Liquid Chromatography—Quadrupole Orbitrap Mass Spectrometry Method for Selected Pharmaceuticals in Water Samples. *J. Chromatogr. A* **2017**, *1515*, 164–171. [CrossRef]
35. Archer, E.; Petrie, B.; Kasprzyk-Hordern, B.; Wolfaardt, G.M. The Fate of Pharmaceuticals and Personal Care Products (PPCPs), Endocrine Disrupting Contaminants (EDCs), Metabolites and Illicit Drugs in a WWTW and Environmental Waters. *Chemosphere* **2017**, *174*, 437–446. [CrossRef]
36. Styszko, K.; Proctor, K.; Castrignanò, E.; Kasprzyk-Hordern, B. Occurrence of Pharmaceutical Residues, Personal Care Products, Lifestyle Chemicals, Illicit Drugs and Metabolites in Wastewater and Receiving Surface Waters of Krakow Agglomeration in South Poland. *Sci. Total Environ.* **2021**, *768*, 144360. [CrossRef]
37. Center for Drug Evaluation and Research. *Validation of Analytical Procedures*; Center for Drug Evaluation and Research: Silver Spring, MD, USA, 2024.
38. Battaglin, W.A.; Bradley, P.M.; Iwanowicz, L.; Journey, C.A.; Walsh, H.L.; Blazer, V.S. Pharmaceuticals, Hormones, Pesticides, and Other Bioactive Contaminants in Water, Sediment, and Tissue from Rocky Mountain National Park, 2012–2013. *Sci. Total Environ.* **2018**, *643*, 651–673. [CrossRef]
39. Anumol, T.; Merel, S.; Clarke, B.O.; Snyder, S.A. Ultra High Performance Liquid Chromatography Tandem Mass Spectrometry for Rapid Analysis of Trace Organic Contaminants in Water. *Chem. Cent. J.* **2013**, *7*, 104. [CrossRef]

40. Ramage, S.; Camacho-Muñoz, D.; Petrie, B. Enantioselective LC-MS/MS for Anthropogenic Markers of Septic Tank Discharge. *Chemosphere* **2019**, *219*, 191–201. [CrossRef]
41. Richards, S.; Paterson, E.; Withers, P.J.A.; Stutter, M. Septic Tank Discharges as Multi-Pollutant Hotspots in Catchments. *Sci. Total Environ.* **2016**, *542*, 854–863. [CrossRef] [PubMed]
42. Tran, N.H.; Hu, J.; Ong, S.L. Simultaneous Determination of PPCPs, EDCs, and Artificial Sweeteners in Environmental Water Samples Using a Single-Step SPE Coupled with HPLC-MS/MS and Isotope Dilution. *Talanta* **2013**, *113*, 82–92. [CrossRef] [PubMed]
43. Daneshvar, A.; Aboulfadl, K.; Viglino, L.; Broséus, R.; Sauvé, S.; Madoux-humery, A.; Weyhenmeyer, G.A.; Prévost, M. Chemosphere Evaluating Pharmaceuticals and Caffeine as Indicators of Fecal Contamination in Drinking Water Sources of the Greater Montreal Region. *Chemosphere* **2012**, *88*, 131–139. [CrossRef] [PubMed]
44. Diwan, V.; Lundborg, C.; Tamhankar, A.J. Seasonal and Temporal Variation in Release of Antibiotics in Hospital Wastewater: Estimation Using Continuous and Grab Sampling. *PLoS ONE* **2013**, *8*, e68715. [CrossRef]
45. Tran, N.H.; Li, J.; Hu, J.; Ong, S.L. Occurrence and Suitability of Pharmaceuticals and Personal Care Products as Molecular Markers for Raw Wastewater Contamination in Surface Water and Groundwater. *Environ. Sci. Pollut. Res.* **2014**, *21*, 4727–4740. [CrossRef]
46. Lee, C.J.; Mau, D.P.; Rasmussen, T.J. *Effects of Nonpoint and Selected Point Contaminant Sources on Stream-Water Quality and Relation to Land Use in Johnson County, Northeastern Kansas, October 2002 Through June 2004*; U.S. Geological Survey: Reston, VA, USA, 2005.
47. Focazio, M.J.; Kolpin, D.W.; Barnes, K.K.; Furlong, E.T.; Meyer, M.T.; Zaugg, S.D.; Barber, L.B.; Thurman, M.E. A National Reconnaissance for Pharmaceuticals and Other Organic Wastewater Contaminants in the United States—II) Untreated Drinking Water Sources. *Sci. Total Environ.* **2008**, *402*, 201–216. [CrossRef]
48. Phillips, P.J.; Schubert, C.; Argue, D.; Fisher, I.; Furlong, E.T.; Foreman, W.; Gray, J.; Chalmers, A. Concentrations of Hormones, Pharmaceuticals and Other Micropollutants in Groundwater Affected by Septic Systems in New England and New York. *Sci. Total Environ.* **2015**, *512–513*, 43–54. [CrossRef]
49. Azuma, T.; Arima, N.; Tsukada, A.; Hirami, S.; Matsuoka, R.; Moriwake, R.; Ishiuchi, H.; Inoyama, T.; Teranishi, Y.; Yamaoka, M.; et al. Detection of Pharmaceuticals and Phytochemicals Together with Their Metabolites in Hospital Effluents in Japan, and Their Contribution to Sewage Treatment Plant Influents. *Sci. Total Environ.* **2016**, *548–549*, 189–197. [CrossRef]
50. Kleywegt, S.; Payne, M.; Ng, F.; Fletcher, T. Environmental Loadings of Active Pharmaceutical Ingredients from Manufacturing Facilities in Canada. *Sci. Total Environ.* **2019**, *646*, 257–264. [CrossRef]
51. Wilschnack, M.; Cartmell, E.; Yates, K.; Petrie, B. Septic Tanks as a Pathway for Emerging Contaminants to the Aquatic Environment–Need for Alternative Rural Wastewater Treatment? *Environ. Pollut.* **2024**, *362*, 124988. [CrossRef] [PubMed]
52. Sodré, F.F.; Santana, J.S.; Sampaio, T.R.; Brandão, C.C.S. Seasonal and Spatial Distribution of Caffeine, Atrazine, Atenolol and Deet in Surface and Drinking Waters from the Brazilian Federal District. *J. Braz. Chem. Soc.* **2018**, *29*, 1854–1865. [CrossRef]
53. Schaider, L.A.; Rodgers, K.M.; Rudel, R.A. Review of Organic Wastewater Compound Concentrations and Removal in Onsite Wastewater Treatment Systems. *Environ. Sci. Technol.* **2017**, *51*, 7304–7317. [CrossRef]
54. Yang, Y.Y.; Toor, G.S.; Wilson, P.C.; Williams, C.F. Septic Systems as Hot-Spots of Pollutants in the Environment: Fate and Mass Balance of Micropollutants in Septic Drainfields. *Sci. Total Environ.* **2016**, *566–567*, 1535–1544. [CrossRef]
55. McEachran, A.D.; Shea, D.; Nichols, E.G. Pharmaceuticals in a Temperate Forest-Water Reuse System. *Sci. Total Environ.* **2017**, *581–582*, 705–714. [CrossRef] [PubMed]
56. Hrkal, Z.; Pastuszek, F. Behaviour of PPCP Substances in a Fluvial Aquifer after Infiltration of Treated Wastewater. *Appl. Sci.* **2023**, *13*, 9348. [CrossRef]
57. Dodgen, L.K.; Kelly, W.R.; Panno, S.V.; Taylor, S.J.; Armstrong, D.L.; Wiles, K.N.; Zhang, Y.; Zheng, W. Characterizing Pharmaceutical, Personal Care Product, and Hormone Contamination in a Karst Aquifer of Southwestern Illinois, USA, Using Water Quality and Stream Flow Parameters. *Sci. Total Environ.* **2017**, *578*, 281–289. [CrossRef]
58. Haber, J.D.; Mayfield, G.; Loper, J.E. Sinkhole Formation at Lake Grady, Florida. In *Karst Studies in West Central Florida: USF Seminar in Karst Environments*; The University of South Florida and the Southwest Florida Water Management District: Tampa, FL, USA, 2003; pp. 53–64.
59. Üstün-Odabaşı, S.; Maryam, B.; Özdemir, N.; Büyükgüngör, H. Occurrence and Seasonal Variations of Pharmaceuticals and Personal Care Products in Drinking Water and Wastewater Treatment Plants in Samsun, Turkey. *Environ. Earth Sci.* **2020**, *79*, 311. [CrossRef]
60. Bertin, S.; Yates, K.; Petrie, B. Enantiospecific Behaviour of Chiral Drugs in Soil. *Environ. Pollut.* **2020**, *262*, 114364. [CrossRef]
61. Du, R.; Duan, L.; Zhang, Q.; Wang, B.; Huang, J.; Deng, S.; Yu, G. Analysis on the Attenuation Characteristics of PPCPs in Surface Water and Their Influencing Factors Based on a Compilation of Literature Data. *Water Res.* **2023**, *242*, 120203. [CrossRef]
62. Lapworth, D.J.; Baran, N.; Stuart, M.E.; Ward, R.S. Emerging Organic Contaminants in Groundwater: A Review of Sources, Fate and Occurrence. *Environ. Pollut.* **2012**, *163*, 287–303. [CrossRef]

63. Dai, Y.; Zhuang, J.; Chen, X. Synergistic Effects of Unsaturated Flow and Soil Organic Matter on Retention and Transport of PPCPs in Soils. *Environ. Res.* **2020**, *191*, 110135. [CrossRef]
64. Xu, X.; Xu, Y.; Xu, N.; Pan, B.; Ni, J. Pharmaceuticals and Personal Care Products (PPCPs) in Water, Sediment and Freshwater Mollusks of the Dongting Lake Downstream the Three Gorges Dam. *Chemosphere* **2022**, *301*, 134721. [CrossRef]
65. Katsikaros, A.G.; Chrysikopoulos, C.V. Occurrence and Distribution of Pharmaceuticals and Personal Care Products (PPCPs) Detected in Lakes around the World—A Review. *Environ. Adv.* **2021**, *6*, 100131. [CrossRef]
66. Gonzalez-Rey, M.; Tapie, N.; Le Menach, K.; Dévier, M.H.; Budzinski, H.; Bebianno, M.J. Occurrence of Pharmaceutical Compounds and Pesticides in Aquatic Systems. *Mar. Pollut. Bull.* **2015**, *96*, 384–400. [CrossRef]
67. Picó, Y.; Alvarez-Ruiz, R.; Alfarhan, A.H.; El-Sheikh, M.A.; Alshahrani, H.O.; Barceló, D. Pharmaceuticals, Pesticides, Personal Care Products and Microplastics Contamination Assessment of Al-Hassa Irrigation Network (Saudi Arabia) and Its Shallow Lakes. *Sci. Total Environ.* **2020**, *701*, 135021. [CrossRef] [PubMed]
68. Kallenborn, R.; Brorström-Lundén, E.; Reiersen, L.O.; Wilson, S. Pharmaceuticals and Personal Care Products (PPCPs) in Arctic Environments: Indicator Contaminants for Assessing Local and Remote Anthropogenic Sources in a Pristine Ecosystem in Change. *Environ. Sci. Pollut. Res.* **2018**, *25*, 33001–33013. [CrossRef] [PubMed]
69. Conkle, J.L.; Gan, J.; Anderson, M.A. Degradation and Sorption of Commonly Detected PPCPs in Wetland Sediments under Aerobic and Anaerobic Conditions. *J. Soils Sediments* **2012**, *12*, 1164–1173. [CrossRef]
70. Ng, B.; Quinete, N.; Maldonado, S.; Lugo, K.; Purrinos, J.; Briceño, H.; Gardinali, P. Understanding the Occurrence and Distribution of Emerging Pollutants and Endocrine Disruptors in Sensitive Coastal South Florida Ecosystems. *Sci. Total Environ.* **2021**, *757*, 143720. [CrossRef]
71. Edwards, Q.A.; Kulikov, S.M.; Garner-O'Neale, L.D. Caffeine in Surface and Wastewaters in Barbados, West Indies. *Springerplus* **2015**, *4*, 57. [CrossRef]
72. USDA. *Soil Survey of Hillsborough County, Florida*; USDA: Washington, DC, USA, 1989.
73. Wang, J.; Wang, S. Removal of Pharmaceuticals and Personal Care Products (PPCPs) from Wastewater: A Review. *J. Environ. Manag.* **2016**, *182*, 620–640. [CrossRef] [PubMed]
74. aus der Beek, T.; Weber, F.A.; Bergmann, A.; Hickmann, S.; Ebert, I.; Hein, A.; Küster, A. Pharmaceuticals in the Environment-Global Occurrences and Perspectives. *Environ. Toxicol. Chem.* **2016**, *35*, 823–835. [CrossRef] [PubMed]
75. Tong, A.Y.C.; Peake, B.M.; Braund, R. Disposal Practices for Unused Medications around the World. *Environ. Int.* **2011**, *37*, 292–298. [CrossRef] [PubMed]
76. Boxall, A.B.A. Pharmaceuticals and Personal Care Products (PPCPs): What Are the Big Questions? *Environ. Health Perspect.* **2012**, *120*, 1221–1229. [CrossRef]
77. Bagnis, S.; Fitzsimons, M.F.; Snape, J.; Tappin, A.; Comber, S. Processes of Distribution of Pharmaceuticals in Surface Freshwaters: Implications for Risk Assessment. *Environ. Chem. Lett.* **2018**, *16*, 1193–1216. [CrossRef]

environments

MDPI

Article

Assessing Environmental Risk Posed by Pharmaceuticals and Personal Care Products in Shallow Lakes, Florida, USA—Part B

Elzbieta Bialkowska-Jelinska [1], Philip van Beynen [1,*] and Laurent Calcul [2]

[1] School of Geosciences, University of South Florida, Tampa, FL 33620, USA; elzbieta@usf.edu
[2] Chemical Purification Analysis and Screening Core Facility (CPAS), University of South Florida, Tampa, FL 33620, USA; calcul@usf.edu
* Correspondence: vanbeyne@usf.edu

Abstract

The use of pharmaceuticals and personal care products (PPCPs) is steadily growing as the world's population both increases and ages. Many of these products are released into the environment via municipal wastewater treatment plants and onsite wastewater treatment systems (septic tanks). Consequently, it is essential to ascertain whether these contaminants pose any risk to aquatic organisms who live in the water bodies receiving this waste. Risk quotients (RQ) are a commonly used method to do so. For our pilot study, we undertook such analysis for three trophic levels: algae, crustaceans, and fish from two small lakes, one fed by septic tanks and the other not. This research was conducted in 2021 from the end of the dry season and through most of the wet season in west central Florida, USA. Of the 14 PPCPs measured, six had RQs that posed a risk to all three trophic levels. This risk increased during the wet season. Both lakes, regardless of whether they directly received PPCPs from septic tanks or not, had some level of risk. However, the lake without septic tanks had a smaller risk, both in elevated RQs and the occurrence to the various species. Of the PPCPs measured, DEET, caffeine, and theophylline posed the greatest risk.

Keywords: pharmaceuticals; personal care products; risk quotient; seasonality; lake

1. Introduction

Pharmaceuticals and personal care products (PPCPs) are a group of chemicals widely used by a growing and aging population [1]. Most PPCPs can induce physiological effects even at low concentrations, making them highly potent compounds that can disrupt biological processes or lead to acute exposure to aquatic organisms [2,3]. One of the sources of PPCPs in the environment is through the release of treated water from onsite wastewater treatment systems (OWTS), more commonly known as septic tanks. These waters are released into rivers, lakes, and drainfields. Because of the inability of these treatment systems to remove all PPCPs from the effluent, these contaminants can be incorporated into the tissue of organisms that interact with this polluted water [4].

A widely used method to assess the potential risk of exposure from contaminants is risk quotients (RQ) [5], a ratio of maximum measured environmental concentration to the predicted no-effect concentration [6]. This method has been successfully applied to aquatic environments such as lakes [7], rivers [8], coastal waters [9], and groundwater [10]. Large lakes tend to have lower risk due to the sheer volume of water, whereas rivers and small lakes are at greater risk [8].

Our pilot study investigates the impact of OWTS on aquatic species in a small, shallow, enclosed lake in a residential area. To serve as a control, we also include a nearby, similar

https://doi.org/10.3390/environments12070231

lake located within a nature preserve. The form of impact is PPCPs released from the OWTS surrounding the residential lake. We propose that the concentrations of PPCPs in the residential lake, as measured by the risk quotient (RQ), are at levels not detrimental to aquatic organisms. Additionally, we evaluate how environmental risk assessment varies between dry and wet seasons and whether using maximum versus mean concentration levels influences risk estimation. As this study does not include an entire year's data and is limited to two lakes, we consider our results to be preliminary in nature.

2. Materials and Methods

2.1. Study Area

Our study was conducted in the littoral zone of two lakes in West-Central Florida. The residential lake (RL) is internally drained, 120 m wide, with a maximum depth of 5.95 m. It is surrounded by 18 residential homes equipped with OWTS for wastewater treatment. Potable water for the homes comes from wells that draw water from the Upper Floridan Aquifer (UFA). Natural lake (NL), which serves as a control site, is located 1800 m northwest of RL. It is 100 m wide and 2.8 m at its deepest point. The two lakes are similar in their geomorphology and climate setting, albeit the control lake being smaller and shallower. Within RL, sampling was conducted at three sites—Site 1 (RL-S1), Site 2 (RL-S2), and Site 3 (RL-S3)—and at one site at NL. The three sites from RL were selected because of their similarities: (1) distance of their OWTS from the lake; (2) hydraulic gradients; and (3) width of the shoreline vegetation (Figure 1).

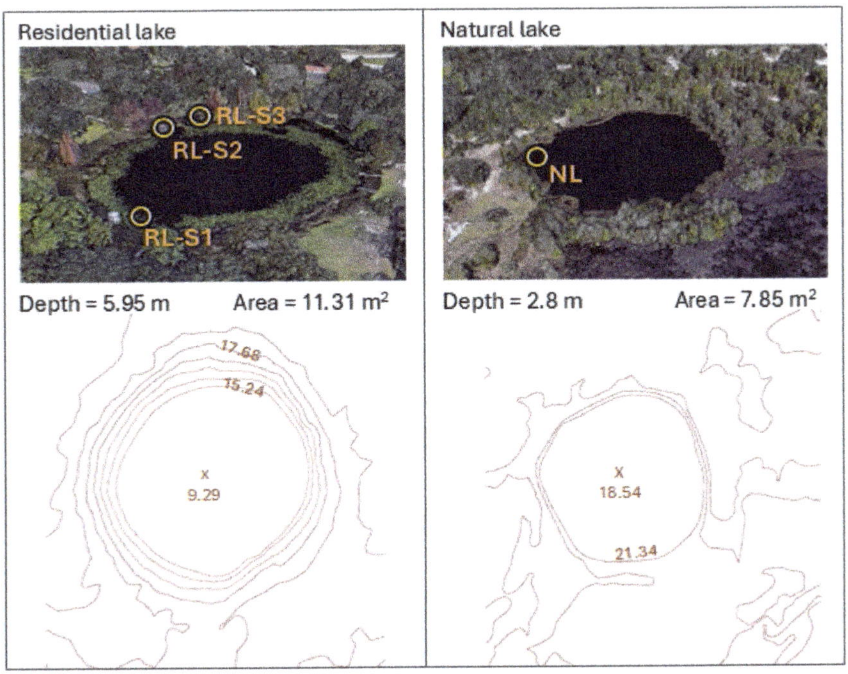

Figure 1. Location of sampling sites within residential and natural lakes, with associated lake depth and surface area.

2.2. Water Sampling and Analyses

Water samples were collected weekly from the end of the dry season (late April 2021) through to the height of the wet season (late August 2021) using 1L amber bottles.

After filtration and solid phase extraction, the concentration of the PPCPs was determined through liquid chromatography mass spectrometry using multiple reaction monitoring (MRM) method, which was employed to detect and quantify the PPCP concentrations. For more details on the study area, chromatographic methods [11–18], and materials, see Bialkowska-Jelinska et al. [19] and Supplementary Materials (Tables S1–S6).

2.3. Ecological Risk Assessment

The ecological risk assessment in this study was conducted using the risk quotient (RQ) method to evaluate the potential effects of PPCPs on aquatic organisms across three trophic levels: fish, crustacean (daphnia), and algae (green algae). The RQ was calculated using the maximum PPCP concentrations detected throughout the study, including the mean and maximum values for both the dry and wet seasons.

The RQ is a widely used metric to assess the ecological risk of contaminants in aquatic ecosystems [1,20,21]. It is calculated as the ratio between the measured environmental concentration (MEC) of a compound and its predicted no-effect concentration (PNEC):

$$RQ = \frac{MEC}{PNEC} \quad (1)$$

$$PNEC = \frac{ChV}{AF} \quad (2)$$

where RQ is a risk quotient; MEC is the maximum and mean concentration of the PPCP detected in surface water; PNEC is predicted-no-effect-concentration values of PPCPs; ChV represents the chronic toxicity values; and AF stands for assessment factor and equals 100.

PNEC can also be calculated using LC50 values, which indicate the lethal concentration for 50% of a target population. ChV represents the concentration required to produce chronic effects over extended exposure. If the chronic toxicity value is not exceeded, no toxic effects are expected [22]. ChVs of analyzed PPCPs were obtained from the 'Organic Module Report' generated using the US EPA's Ecological Structure-Activity Relationship (ECOSAR v2.2) software [23] and are presented in Table 1.

Table 1. Chronic toxicity values for fish, crustacean, and algae.

PPCP	Chronic Toxicity Value (mg/L)		
	Fish	Crustacean	Algae
Octocrylene (OCT)	0.00086	0.0081	0.042
Atorvastatin (ATV)	0.055	0.257	1.61
Fluoxetine (FLX)	0.025	0.019	0.033
Ibuprofen (IBU)	4.94	4.31	15.6
Testosterone (TST)	5.84	0.637	2.43
Propylparaben (PPB)	0.406	2	1.69
Carbamazepine (CBZ)	1.05	1.17	0.096
N,N-Diethyl-meta-toluamide (DEET)	0.49	5.72	3.21
Methylparaben (MPR)	2.06	0.99	3.89
Sulfamethoxazole (SMX)	5	0.07	11.14
Acetaminophen (APAP)	0.124	0.189	0.352
Cotinine (COT)	6.13	109	19.8
Caffeine (CAF)	0.914	2.8	2.63
Theophylline (THE)	1.49	5.55	4.13

Values extracted from ECOSAR v2.2.

Final RQ values were categorized as follows: low risk (RQ < 0.1); medium risk (RQ between 0.1 and 1); and high risk (RQ $\geq$ 1.0) [7,21,24]. The RQ values were determined for the following scenarios:

- The maximum PPCP concentrations detected throughout the sampling campaign (RQ_{max}).
- The maximum PPCP concentrations recorded during the dry and wet seasons ($RQ_{max\text{-}dry}$ and $RQ_{max\text{-}wet}$).
- The mean PPCP concentrations observed during the dry and wet seasons ($RQ_{mean\text{-}dry}$ and $RQ_{mean\text{-}wet}$).

2.4. Statistical Analysis

An ANOVA test was conducted to evaluate whether the risk quotients for PPCPs differed significantly between the dry and wet seasons across four lake sites (Tables S7–S10).

3. Results

The maximum PPCP concentrations for Site RL-S1, RL-S2, RL-S3, and NL and mean concentrations of the six PPCPs are presented in Tables 2 and 3, respectively. Not all PPCPs were detected throughout the sampling interval, with propyl paraben and atorvastatin only being present once. The highest maximum values during the entire sampling period were for caffeine, DEET, and theophylline. For the mean values, only those that were detected more than nine times (50% of samples) were included, which left only six of the PPCPs. Of these, caffeine had the highest values. For a detailed discussion of these PPCPs, please see the related article [19] in this issue.

Table 2. Maximum PPCP concentrations found at each sampling site during the dry and wet seasons.

	PPCP Concentrations (µg/L)							
	RL-S1		RL-S2		RL-S3		NL	
PPCP	Dry	Wet	Dry	Wet	Dry	Wet	Dry	Wet
OCT	ND	0.99	ND	ND	ND	0.01	ND	ND
ATV	ND	0.13	ND	ND	ND	ND	ND	ND
FLX	ND	0.02	ND	0.02	ND	0.07	ND	0.01
IBU	ND	10.95	8.63	9.42	8.53	15.17	15.08	14.13
TST	0.19	8.42	31.94	1.21	1.77	0.92	22.33	48.18
PPB	ND	ND	ND	ND	ND	1.11	ND	ND
CBZ	ND	ND	ND	0.27	ND	0.54	ND	ND
DEET	4.54	22.94	4.21	15.32	4.54	17.31	182.48	76.89
MPR	ND	0.28	ND	ND	ND	ND	ND	ND
SMX	ND	0.07	0.06	0.08	ND	0.10	0.05	0.06
APAP	0.17	1.08	0.10	6.99	0.42	0.75	ND	0.67
COT	8.50	8.96	7.96	8.12	8.82	8.14	4.07	1.31
CAF	454.2	374.91	51.22	16.92	29.13	68.30	197.11	249.50
THE	33.18	371.07	53.02	7.29	28.81	298.64	46.97	151.95

ND—not detected.

Table 3. Mean PPCP concentrations found at each sampling site during the dry and wet seasons.

	PPCP Concentrations (μg/L)															
	RL-S1				RL-S2				RL-S3				NL			
PPCP	Dry	SD	Wet	SD	Dry	SD	Wet	SD	Dry	SD	Wet	SD	Dry	SD	Wet	SD
TST	5.63	7.61	2.87	2.98	8.82	15.43	0.43	0.34	0.87	0.61	0.40	0.26	9.14	10.08	6.32	12.20
DEET	3.89	0.47	13.12	4.85	3.80	0.31	8.78	3.99	3.75	0.57	9.72	4.36	53.37	86.64	14.78	18.00
APAP	0.09	0.10	0.18	0.29	0.03	0.05	0.64	2.11	0.25	0.11	0.15	0.20	0.00	0.00	0.08	0.18
COT	8.16	0.37	6.53	1.40	7.85	0.08	6.83	1.12	8.35	0.44	6.51	1.18	2.96	1.28	0.77	0.23
CAF	147.15	209.10	100.71	104.92	24.67	18.15	6.64	3.62	23.58	5.33	11.71	17.57	91.33	85.64	96.79	64.91
THE	20.54	11.98	78.23	94.93	17.37	23.81	2.67	1.96	12.18	11.30	26.82	79.00	29.49	17.07	71.01	39.15

3.1. *Risk Assesment for Maxiumum PPCP Concentrations*

The RQs calculated for maximum PPCP concentrations (RQ_{max}) at each sampling site are presented in Table 4. RQ_{max} values for octocrylene, caffeine, theophylline, and DEET at residential lake (RL-S1, RL-S2, RL-S3) pose the highest environmental risk, with fish and crustaceans showing the most significant risk. High RQ_{max} values (>1) are prevalent in RL, while the risk in NL generally was lower, but driven from the same compounds. Moderate environmental risk contaminants ($0.1 \leq RQ \leq 1$), such as atorvastatin, fluoxetine, ibuprofen, sulfamethoxazole, and acetaminophen, frequently appear in crustaceans across multiple sites. Low-risk contaminants (RQ < 0.1) are mostly found in NL, with methylparaben, sulfamethoxazole, acetaminophen, and ibuprofen consistently showing minimal ecological threats.

Table 4. RQ_{max} at each site during the sampling period.

	Fish				Crustacean				Algae			
PPCP	RL-S1	RL-S2	RL-S3	NL	RL-S1	RL-S2	RL-S3	NL	RL-S1	RL-S2	RL-S3	NL
OCT	115.116	*	1.163	*	12.222	*	0.123	*	2.357	*	0.024	*
ATV	0.231	*	*	*	0.049	*	*	*	0.008	*	*	*
FLX	0.080	0.072	0.280	0.056	0.105	0.095	0.368	0.074	0.061	0.055	0.212	0.042
IBU	0.222	0.191	0.307	0.305	0.254	0.219	0.352	0.350	0.070	0.060	0.097	0.097
TST	0.286	0.547	0.030	0.825	2.625	5.014	0.278	7.564	0.688	1.314	0.073	1.983
PPB	ND	*	0.273	*	*	*	0.056	*	*	*	0.066	*
CBZ	0.025	0.026	0.051	*	0.022	0.023	0.046	*	0.270	0.281	0.563	*
DEET	4.673	3.127	3.533	37.241	0.400	0.268	0.303	3.190	0.713	0.477	0.539	5.685
MPR	0.014	*	*	*	0.028	*	*	*	0.007	*	*	*
SMX	0.001	0.002	0.002	0.001	0.100	0.114	0.140	0.086	0.001	0.001	0.001	0.001
APAP	0.877	5.637	0.605	0.540	0.576	3.698	0.397	0.354	0.309	1.986	0.213	0.190
COT	0.146	0.132	0.144	0.066	0.008	0.007	0.008	0.004	0.045	0.041	0.045	0.021
CAF	49.694	5.604	7.473	27.298	16.221	1.829	2.439	8.911	17.270	1.948	2.597	9.487
THE	24.904	3.558	20.043	10.198	6.686	0.955	5.381	2.738	8.985	1.284	7.231	3.679

*—PPCP not detected; RQ not applicable.

3.2. *Risk Assessment for Mean PPCP Concentrations*

The RQs calculated for mean PPCP concentrations (RQ_{mean}) at each sampling site are presented in Table 5. RQ_{mean} values for caffeine, theophylline, and DEET at residential lake (RL-S1, RL-S2, RL-S3) pose the highest environmental risk, with fish showing the most significant risk. High RQ_{mean} values (>1) are prevalent in RL, while NL generally had lower risk, except for DEET. Moderate environmental risk contaminants ($0.1 \leq RQ \leq 1$), such as caffeine, theophylline, testosterone, and DEET, frequently appear in crustaceans across multiple sites. Low-risk contaminants (RQ < 0.1) are mostly found in NL, with cotinine and acetaminophen consistently showing minimal ecological threats.

Table 5. RQ_{mean} at each site during the sampling period.

	Fish				Crustacean				Algae			
PPCP	RL-S1	RL-S2	RL-S3	NL	RL-S1	RL-S2	RL-S3	NL	RL-S1	RL-S2	RL-S3	NL
TST	0.060	0.046	0.009	0.113	0.546	0.419	0.080	1.035	0.143	0.110	0.021	0.271
DEET	2.259	1.520	1.712	4.765	0.194	0.130	0.147	0.408	0.345	0.232	0.261	0.727
APAP	0.129	0.387	0.137	0.048	0.085	0.254	0.090	0.032	0.045	0.136	0.048	0.017
COT	0.112	0.116	0.113	0.020	0.006	0.007	0.006	0.001	0.035	0.036	0.035	0.006
CAF	12.148	1.253	1.570	10.457	3.965	0.409	0.513	3.414	4.222	0.435	0.546	3.634
THE	4.390	0.442	1.582	4.146	1.179	0.119	0.425	1.113	1.584	0.160	0.571	1.496

3.3. *Risk Assessment for Maximum PPCP Concentrations During the Dry and Wet Seasons*

Table 6 presents $RQ_{max\text{-}dry}$ and $RQ_{max\text{-}wet}$. $RQ_{max\text{-}dry}$ was the highest for caffeine at each trophic level, indicating high environmental risk. The highest $RQ_{max\text{-}dry}$ was observed for RL-S1.

Table 6. $RQ_{max\text{-}dry}$ and $RQ_{max\text{-}wet}$ at each site during dry and wet seasons.

Dry Season												
	Fish				Crustacean				Algae			
PPCP	RL-S1	RL-S2	RL-S3	NL	RL-S1	RL-S2	RL-S3	NL	RL-S1	RL-S2	RL-S3	NL
OCT	*	*	*	*	*	*	*	*	*	*	*	*
ATV	*	*	*	*	*	*	*	*	*	*	*	*
FLX	*	*	*	*	*	*	*	*	*	*	*	*
IBU	0.188	0.175	0.173	0.305	0.215	0.200	0.198	0.350	0.059	0.055	0.055	0.097
TST	0.286	0.547	0.030	0.382	2.625	5.014	0.278	3.505	0.688	1.314	0.073	0.919
PPB	*	*	*	*	*	*	*	*	*	*	*	*
CBZ	*	*	*	*	*	*	*	*	*	*	*	*
DEET	0.927	0.859	0.927	37.241	0.079	0.074	0.079	3.190	0.141	0.131	0.141	5.685
MPR	*	*	*	*	*	*	*	*	*	*	*	*
SMX	*	0.001	*	0.001	*	0.086	*	0.074	*	0.001	*	0.000
APAP	0.137	0.081	0.339	0.004	0.090	0.053	0.222	0.003	0.048	0.028	0.119	0.001
COT	0.139	0.130	0.144	0.066	0.008	0.007	0.008	0.004	0.043	0.040	0.045	0.021
CAF	49.694	5.604	3.187	21.566	16.221	1.829	1.040	7.040	17.270	1.948	1.108	7.495
THE	2.227	3.558	1.934	3.152	0.598	0.955	0.519	0.846	0.803	1.284	0.698	1.137
Wet Season												
	Fish				Crustacean				Algae			
PPCP	RL-S1	RL-S2	RL-S3	NL	RL-S1	RL-S2	RL-S3	NL	RL-S1	RL-S2	RL-S3	NL
OCT	115.116	*	1.163	*	12.222	*	0.123	*	2.357	*	0.024	*
ATV	0.231	*	*	*	0.049	*	*	*	0.008	*	*	*
FLX	0.080	0.072	0.280	0.056	0.105	0.095	0.368	0.074	0.061	0.055	0.212	0.042
IBU	0.222	0.191	0.307	0.286	0.254	0.219	0.352	0.328	0.070	0.060	0.097	0.091
TST	0.144	0.021	0.016	0.825	1.322	0.190	0.144	7.564	0.347	0.050	0.038	1.983
PPB	*	*	0.273	*	*	*	0.056	*	*	*	0.066	*
CBZ	0.025	0.026	0.051	*	0.022	0.023	0.046	*	0.271	0.281	0.563	*
DEET	4.673	3.127	3.533	15.686	0.400	0.268	0.303	1.344	0.713	0.477	0.539	2.394
MPR	0.014	*	*	*	0.028	*	*	*	0.007	*	*	*
SMX	0.001	0.002	0.002	0.001	0.100	0.114	0.140	0.083	0.001	0.001	0.001	0.001
APAP	0.871	5.637	0.605	0.540	0.571	3.698	0.397	0.354	0.307	1.986	0.213	0.190
COT	0.146	0.132	0.132	0.021	0.008	0.007	0.007	0.001	0.045	0.041	0.041	0.007
CAF	41.019	1.851	7.473	27.298	13.390	0.604	2.439	8.911	14.255	0.643	2.597	9.487
THE	24.904	0.489	20.043	10.198	6.686	0.131	5.381	2.738	8.985	0.177	7.231	3.679

*—PPCP not detected; RQ not applicable.

During the wet season, a greater number of PPCPs were detected in the lake waters, and more PPCPs exceeded high-risk thresholds, indicating an increase in environmental risk. During the wet season, $RQ_{max\text{-}wet}$ was high in octocrylene, caffeine, theophylline, and DEET, especially at sites in RL. Moderate-risk contaminants ($0.1 \leq RQ \leq 1$) were more widespread in the wet season, affecting a larger number of species, especially crustaceans and fish.

NL generally had lower contamination levels, but DEET, caffeine, theophylline, and testosterone posed high or moderate risk across both seasons. The overall trend across seasons suggests that residential lakes remain more polluted than the natural lake, with fish being the most affected organism, followed by crustaceans and algae.

The RQ values across the four study sites demonstrate significant differences in contamination levels, with RL generally exhibiting higher risks. High environmental risk for octocrylene, caffeine, theophylline, and DEET was predominantly found in RL-S1, followed by RL-S2 and RL-S3. Moderate-risk contaminants were present across all residential lakes but were more frequent in RL-S3 and RL-S2. Low-risk contaminants were found mostly in NL.

3.4. Risk Assessment for Mean PPCP Concentrations During the Dry and Wet Seasons

The environmental risk assessed by RQ for mean PPCP concentrations varies across the four sampling sites (Table 7).

Table 7. $RQ_{mean\text{-}dry}$ and $RQ_{mean\text{-}wet}$ at each site during dry and wet seasons.

Dry Season												
	Fish				**Crustacean**				**Algae**			
PPCP	RL-S1	RL-S2	RL-S3	NL	RL-S1	RL-S2	RL-S3	NL	RL-S1	RL-S2	RL-S3	NL
TST	0.096	0.151	0.015	0.157	0.884	1.385	0.137	1.435	0.232	0.363	0.036	0.376
DEET	0.793	0.776	0.765	10.892	0.068	0.066	0.066	0.933	0.121	0.118	0.117	1.663
APAP	0.073	0.024	0.202	0.000	0.048	0.016	0.132	0.000	0.026	0.009	0.071	0.001
COT	0.133	0.128	0.136	0.048	0.007	0.007	0.008	0.003	0.041	0.040	0.042	0.015
CAF	16.100	2.699	2.582	9.992	5.255	0.881	0.842	3.262	5.595	0.938	0.897	3.473
THE	1.379	1.162	0.817	3.979	0.370	0.312	0.219	0.531	0.497	0.419	0.295	0.714
Wet Season												
	Fish				**Crustacean**				**Algae**			
PPCP	RL-S1	RL-S2	RL-S3	NL	RL-S1	RL-S2	RL-S3	NL	RL-S1	RL-S2	RL-S3	NL
TST	0.049	0.007	0.007	0.108	0.451	0.068	0.063	0.992	0.118	0.018	0.016	0.260
DEET	2.678	1.792	1.984	3.016	0.229	0.153	0.170	0.258	0.409	0.274	0.303	0.460
APAP	0.145	0.516	0.121	0.065	0.095	0.339	0.079	0.042	0.051	0.182	0.043	0.023
COT	0.108	0.111	0.106	0.013	0.006	0.006	0.006	0.001	0.033	0.034	0.033	0.004
CAF	11.019	0.726	1.281	10.590	3.597	0.237	0.418	3.457	3.829	0.252	0.445	3.680
THE	5.250	0.179	1.800	4.766	1.410	0.048	0.483	1.279	1.894	0.065	0.649	1.719

RL-S1 showed severe pollution in both dry and wet seasons. During the dry season, caffeine (RQ = 16.100) and theophylline (RQ = 1.379) presented high environmental risk for fish, while crustaceans and algae experienced significant exposure to caffeine (RQ = 5.255 and 5.595, respectively). DEET, acetaminophen, and cotinine posed a moderate environmental risk in fish. Moderate risk was posed for crustaceans and algae by testosterone and theophylline. In the wet season, RQ for caffeine decreased to 13.212, and theophylline increased to 6.007 in fish, while DEET increased to 2.622. For crustaceans and algae, theophylline increased in the wet season.

RL-S2 shows moderate contamination levels compared to RL-S1. In the dry season, caffeine and theophylline posed a high environmental risk for fish, and a moderate environmental risk in fish was posed by testosterone, DEET, and cotinine. Otherwise, the remaining contaminants presented low-risk range. The wet season increases the risk of acetaminophen and DEET in fish, moving it into the high-risk category. However, the overall increase in contamination is less pronounced than in RL-S1.

RL-S3 appears to be the least contaminated among the residential lakes. In the dry season, only caffeine poses high risk for fish ($RQ_{mean\text{-}dry}$ = 2.582). Moderate risk was posed by DEET, theophylline, acetaminophen, and cotinine. During the wet season, the highest environmental risk in fish was from DEET.

For the NL, the dry season held high environmental risk to fish from DEET, caffeine, and theophylline. Daphnia's high risk arose from caffeine, testosterone, and DEET, while green algae's risk came from caffeine, DEET, and theophylline. Moderate risk was posed by testosterone for fish and algae. Crustacean's moderate risk came from theophylline. In the wet season, high RQs for fish were found for caffeine, DEET, and theophylline. High risk for crustacean and algae was posed by caffeine, testosterone, theophylline, and DEET. A moderate environmental risk to fish was assessed to be testosterone and acetaminophen. Acetaminophen also posed a moderate risk for daphnia and algae.

A one-way ANOVA was conducted to assess whether there were statistically significant differences in risk quotients values. The results indicated that none of the compounds showed statistically significant variation in dry and wet seasons (Tables S7–S10).

4. Discussion

The results demonstrate that all detected PPCPs pose a high or moderate environmental risk to aquatic organisms for at least one of the trophic levels. As such, our supposition that the PPCPs pose little risk is incorrect. Our findings agree with other studies assessing environmental risk of aquatic organisms particularly for caffeine, carbamazepine, acetaminophen, and DEET [7,21,25–27].

4.1. Aquatic Trophic Levels and Environmental Risk Assessment

An understanding of the ecological risk that PPCPs pose to species at various trophic levels in small urban water bodies is essential for evaluating their potential impact on the ecosystem. Our results show that PPCPs present the highest risk for the uppermost trophic level (fish) in each sampling site. Predation by fish of lower trophic species provides the avenue for PPCP bioaccumulation [28]. Mojiri et al. [29] found PPCPs from various classes in tissues, blood plasma, muscle, liver, gill, and the homogenized bodies of fish in waters around the world. For our study, the apex predator is the American alligator, *Alligator mississippiensis*, which occupied the residential lake. This animal would be exposed to the detected PPCPs for the longest period of any species in the lake. A study of alligators hatched in a polluted Lake Apopka in Florida showed altered gonadal development, plasma hormone concentrations, and growth [30].

Planktonic crustaceans (Daphnia) have the second highest risk, followed by algae. Only for caffeine is the risk for algae slightly higher than for daphnia organisms. Daphnia's susceptibility to PPCPs was shown by Luna et al. [31], who exposed *Daphnia magna* to fluoxetine over a 40-day period during which population growth rates decreased. Flaherty and Dodson [32] subjected Daphnia to sulfamethoxazole and determined a significant increase in the male/female ratio for offspring. Ecological impacts can be intensified when aquatic organisms are simultaneously exposed to multiple PPCPs [33]. The combination of individual PPCPs may result in harmful effects even when each compound is present at its no-observed-effect concentration (NOEC) [34]. Backhaus et al. [35] found that a mixture of five PPCPs caused toxicity in marine microalgal communities, despite each being at its individual NOEC.

The risk to algae posed by PPCPs comes through various processes such as chelation adsorption, ion exchange or complexation, electrostatic interaction, and microprecipitation [36,37]. Concentrations of PPCPs below two mg/L do not significantly affect algae [38]. However, some compounds, such as antibiotics and antibacterial agents, can decrease algal growth at concentrations below one mg/L, but in most cases algae can survive low concentrations of PPCPs [29]. These authors found that *Nannochloris* sp. green algae were very effective at removing triclosan from wastewater. Algae's ability to effectively remove PPCPs from wastewater highlights their potential for bioremediation [39].

4.2. Seasonal Risk to Aquatic Organisms

RQ values vary between the seasons, with the larger number of PPCPs having a higher RQ in the wet season. Only testosterone had a higher RQ_{max} in the dry season. The variable life cycles of aquatic organisms can influence the degree of negative impact of PPCPs [40]. For example, mosquitofish (*Gambusia holbrooki*), which are found in both lakes, are reproductively active from spring to fall. During the winter, the female fish store the sperm, and the offspring fertilizes in the spring [40]. Elevated testosterone and cotinine can impact these aquatic organisms during hatching and in the wet season influence the organism's adult life cycle, thereby adversely affecting survival rates. Seasonal spikes in the contamination we found could affect population dynamics and ecosystem stability, highlighting the critical need for monitoring and managing PPCP levels during this vulnerable time for aquatic life.

4.3. Concentration Level: Maximum or Mean?

Maximum RQs highlight the worst-case ecological scenario, especially if the lake might have a potentially sensitive species to elevated PPCP levels. This study provides a precautionary approach to environmental management. As many PPCPs have not been tested for their toxicity to aquatic organisms, calculating the maximum RQ value provides some mechanism for testing the state of risk [41]. An inherent limitation of concentrating on only maximum values is that having one short-term spike in PPCPs, while otherwise values are relatively low, could suggest a more dire situation than is actually the case. Conversely, using mean RQ concentrations removes the influence of both the highest and lowest values, which could create a situation of undue alarm or a sense of complacency.

From an environmental management perspective, regular monitoring of PPCP concentrations in lake waters using RQs is needed to reduce the likelihood of unexpected environmental degradation of our lakes [42]. As shown in our study, sampling only several times a year will create an inaccurate view of the state of the risk posed to aquatic organisms.

4.4. Assumptions and Limitations

We calculated the RQ values using predicted chronic toxicity values (ChV) derived from the ECOSAR model. Hence, it should be recognized that this prediction results in the RQs being estimates. For consistency and to streamline the assessment process, we used exclusively the ECOSAR-derived ChVs as the source of toxicity information for this analysis [21]. This model/program is commonly used by the US EPA, from which we infer our approach is justified; the agency updated its ECOSAR software in 2024. This method is also recommended in the Guidelines on Environmental Risk Assessment of the European Medicines Agency. We recognize there are other approaches that are better suited to chronic exposures. For example, Raimondo and Forbes [43] suggest the use of other comprehensive modeling frameworks, such as the Population Modeling Guidance for Use, Interpretation, and Development in Ecological Risk Assessment (Pop-GUIDE), which offers better leverage of existing data for evaluating potential ecological risks.

There are several limitations to using RQs for determining environmental hazards to any species. First, many of the studies on fish are conducted in laboratory settings, which have certain drawbacks [44]. These studies only capture a narrow view of the complexity of natural environments and the dynamic processes involved. Interactions and conditions in the natural environment are far more variable and intricate than those in a laboratory. Therefore, many of the processes related to PPCPs and aquatic life are not yet fully understood. Additionally, laboratory experiments often have short durations and cannot measure chronic effects that may take many months to become apparent. However, these studies are important first steps for determining the effects of PPCPs due in part to the vast numbers of these contaminants. Second, risk quotients do not describe what the actual risk is to the organism being exposed to contaminants [45]. Third, Raimondo and Forbes [43] strongly suggest that RQs are too simplistic for assessing chronic risks and there is a need to use population models for more accurate assessments that can help inform better regulatory decision-making. The USEPA acknowledges that RQs are not necessarily correlated with the magnitude or risk [46].

We recognize that using data from only two lakes restricts our ability to make definitive conclusions and incorporating more lakes would be optimal. As such, we consider this research as a pilot study to guide future investigations. Additionally, an entire year's worth of data could provide a better understanding of the dry season; however, the focus of the study was the onset of the wet season. Also, a greater suite of PPCPs could have been analyzed, but once again, time and cost were limiting factors. Finally, investigating the impact of individual precipitation events may provide insight into extreme events, but their short time span might not be significant for the chronic impact on the health of the lake species.

5. Conclusions

Our results found that the concentrations of PPCPs in the residential lake, as defined by the risk quotient, pose an environmental risk to the aquatic organisms. The results demonstrated that all the detected PPCPs in the lakes present a high or moderate environmental risk to aquatic organisms, which include one species from each trophic level: algae, daphnia, and fish. We should stress that this research should be considered a pilot study. To address the limitations listed above, we recommend the following: (1) increase the sampling period to a full year to provide a better understanding of how PPCP concentrations change after the end of the wet season; (2) incorporate more lakes to produce more definitive conclusions; (3) increase the number of PPCPs analyzed; and (4) incorporate population models in determining risk.

Supplementary Materials: The supporting information can be downloaded at https://www.mdpi.com/article/10.3390/environments12070231/s1.

Author Contributions: Conceptualization, E.B.-J. and P.v.B.; methodology, E.B.-J., L.C. and P.v.B.; software, E.B.-J. and L.C.; validation, E.B.-J., P.v.B. and L.C.; formal analysis, E.B.-J., L.C. and P.v.B.; investigation, E.B.-J. and P.v.B.; resources, P.v.B., L.C. and E.B.-J.; data curation, E.B.-J.; writing—original draft preparation, E.B.-J. and P.v.B.; writing—review and editing, E.B.-J. and P.v.B.; visualization, E.B.-J.; supervision, P.v.B.; project administration, P.v.B. All authors have read and agreed to the published version of the manuscript.

Funding: This research received no external funding.

Data Availability Statement: The original contributions presented in this study are included in the article/Supplementary Materials. Further reasonable inquiries can be directed to the corresponding author.

Acknowledgments: We thank the homeowners for granting access to the residential lake and Rae and Jenna for helping with sample collection.

Conflicts of Interest: The authors declare no conflicts of interest.

References

1. Gao, Q.; Blum, K.M.; Gago-Ferrero, P.; Wiberg, K.; Ahrens, L.; Andersson, P.L. Impact of On-Site Wastewater Infiltration Systems on Organic Contaminants in Groundwater and Recipient Waters. *Sci. Total Environ.* **2019**, *651*, 1670–1679. [CrossRef] [PubMed]
2. Junaid, M.; Wang, Y.; Hamid, N.; Deng, S.; Li, W. Prioritizing Selected PPCPs on the Basis of Environmental and Toxicogenetic Concerns: A Toxicity Estimation to Confirmation Approach. *J. Hazard. Mater.* **2019**, *380*, 120828. [CrossRef]
3. Brausch, J.M.; Connors, K.; Brooks, B.W.; Rand, G.M. Human Pharmaceuticals in the Aquatic Environment: A Review of Recent Toxicological Studies and Considerations for Toxicity Testing. In *Reviews of Environmental Contamination and Toxicology*; Whitacre, D.M., Ed.; Springer: Boston, MA, USA, 2012; pp. 1–99.
4. Keerthanan, S.; Jayasinghe, C.; Biswas, J.K.; Vithanage, M. Pharmaceutical and Personal Care Products (PPCPs) in the Environment: Plant Uptake, Translocation, Bioaccumulation, and Human Health Risks. *Crit. Rev. Environ. Sci. Technol.* **2021**, *51*, 1221–1258. [CrossRef]
5. Liu, N.; Jin, X.; Feng, C.; Wang, Z.; Wu, F.; Johnson, A.C.; Xiao, H.; Hollert, H.; Giesy, J.P. Ecological Risk Assessment of Fifty Pharmaceuticals and Personal Care Products (PPCPs) in Chinese Surface Waters: A Proposed Multiple-Level System. *Environ. Int.* **2020**, *136*, 105454. [CrossRef] [PubMed]
6. Hernando, M.D.; Mezcua, M.; Fern, A.R.; Barcel, D. Environmental Risk Assessment of Pharmaceutical Residues in Wastewater Effluents, Surface Waters and Sediments. *Talanta* **2006**, *69*, 334–342. [CrossRef]
7. Blair, B.D.; Crago, J.P.; Hedman, C.J.; Klaper, R.D. Pharmaceuticals and Personal Care Products Found in the Great Lakes above Concentrations of Environmental Concern. *Chemosphere* **2013**, *93*, 2116–2123. [CrossRef]
8. Tewari, S.; Jindal, R.; Kho, Y.L.; Eo, S.; Choi, K. Chemosphere Major Pharmaceutical Residues in Wastewater Treatment Plants and Receiving Waters in Bangkok, Thailand, and Associated Ecological Risks. *Chemosphere* **2013**, *91*, 697–704. [CrossRef]
9. Beiras, R. Environmental Risk Assessment of Pharmaceutical and Personal Care Products in Estuarine and Coastal Waters. In *Pharmaceuticals in Marine and Coastal Environments*; Duran-Alvarez, J.C., Jiménez-Cisneros, B., Eds.; Elsevier: Amsterdam, The Netherlands, 2021; pp. 195–252.
10. Li, Z.; Xiang, X.; Li, M.; Ma, Y.; Wang, J.; Liu, X. Occurrence and Risk Assessment of Pharmaceuticals and Personal Care Products and Endocrine Disrupting Chemicals in Reclaimed Water and Receiving Groundwater in China. *Ecotoxicol. Environ. Saf.* **2015**, *119*, 74–80. [CrossRef]
11. Gilart, N.; Marcé, R.M.; Borrull, F.; Fontanals, N. Determination of Pharmaceuticals in Wastewaters Using Solid-Phase Extraction-Liquid Chromatography-Tandem Mass Spectrometry. *J. Sep. Sci.* **2012**, *35*, 875–882. [CrossRef]
12. Petrie, B.; Youdan, J.; Barden, R.; Kasprzyk-Hordern, B. Multi-Residue Analysis of 90 Emerging Contaminants in Liquid and Solid Environmental Matrices by Ultra-High-Performance Liquid Chromatography Tandem Mass Spectrometry. *J. Chromatogr. A* **2016**, *1431*, 64–78. [CrossRef]
13. Althakafy, J.T.; Kulsing, C.; Grace, M.R.; Marriott, P.J. Liquid Chromatography–Quadrupole Orbitrap Mass Spectrometry Method for Selected Pharmaceuticals in Water Samples. *J. Chromatogr. A* **2017**, *1515*, 164–171. [CrossRef] [PubMed]
14. Archer, E.; Petrie, B.; Kasprzyk-Hordern, B.; Wolfaardt, G.M. The Fate of Pharmaceuticals and Personal Care Products (PPCPs), Endocrine Disrupting Contaminants (EDCs), Metabolites and Illicit Drugs in a WWTW and Environmental Waters. *Chemosphere* **2017**, *174*, 437–446. [CrossRef] [PubMed]
15. Styszko, K.; Proctor, K.; Castrignanò, E.; Kasprzyk-Hordern, B. Occurrence of Pharmaceutical Residues, Personal Care Products, Lifestyle Chemicals, Illicit Drugs and Metabolites in Wastewater and Receiving Surface Waters of Krakow Agglomeration in South Poland. *Sci. Total Environ.* **2021**, *768*, 144360. [CrossRef] [PubMed]
16. Center for Drug Evaluation and Research. *Validation of Analytical Procedures*; Center for Drug Evaluation and Research: Silver Spring, MD, USA, 2024.
17. Battaglin, W.A.; Bradley, P.M.; Iwanowicz, L.; Journey, C.A.; Walsh, H.L.; Blazer, V.S. Pharmaceuticals, Hormones, Pesticides, and Other Bioactive Contaminants in Water, Sediment, and Tissue from Rocky Mountain National Park, 2012–2013. *Sci. Total Environ.* **2018**, *643*, 651–673. [CrossRef]
18. Anumol, T.; Merel, S.; Clarke, B.O.; Snyder, S.A. Ultra High Performance Liquid Chromatography Tandem Mass Spectrometry for Rapid Analysis of Trace Organic Contaminants in Water. *Chem. Cent. J.* **2013**, *7*, 1–14. [CrossRef]
19. Bialkowska-Jelinska, E.; van Beynen, P.; Calcul, L. Seasonality of Pharmaceuticals and Personal Care Products in Shallow Lakes, Florida, USA—Part A. *Environments* **2025**, *12*, 219. [CrossRef]

20. Lin, K.; Wang, R.; Han, T.; Tan, L.; Yang, X.; Wan, M.; Chen, Y.; Zhao, T.; Jiang, S.; Wang, J. Seasonal Variation and Ecological Risk Assessment of Pharmaceuticals and Personal Care Products (PPCPs) in a Typical Semi-Enclosed Bay—The Bohai Bay in Northern China. *Sci. Total Environ.* **2023**, *857*, 159682. [CrossRef]
21. Sengar, A.; Vijayanandan, A. Human Health and Ecological Risk Assessment of 98 Pharmaceuticals and Personal Care Products (PPCPs) Detected in Indian Surface and Wastewaters. *Sci. Total Environ.* **2022**, *807*, 150677. [CrossRef]
22. Ong, T.T.X.; Blanch, E.W.; Jones, O.A.H. Predicted Environmental Concentration and Fate of the Top 10 Most Dispensed Australian Prescription Pharmaceuticals. *Environ. Sci. Pollut. Res.* **2018**, *25*, 10966–10976. [CrossRef]
23. United States Environmental Protection Agency (EPA) Ecological Structure Activity Relationships (ECOSAR) Predictive Model. Available online: https://www.epa.gov/tsca-screening-tools/ecological-structure-activity-relationships-ecosar-predictive-model (accessed on 12 November 2023).
24. Li, Y.; Zhang, L.; Ding, J.; Liu, X. Prioritization of Pharmaceuticals in Water Environment in China Based on Environmental Criteria and Risk Analysis of Top-Priority Pharmaceuticals. *J. Environ. Manag.* **2020**, *253*, 109732. [CrossRef]
25. Gao, X.; Wang, X.; Li, J.; Ai, S.; Fu, X.; Fan, B.; Li, W.; Liu, Z. Aquatic Life Criteria Derivation and Ecological Risk Assessment of DEET in China. *Ecotoxicol. Environ. Saf.* **2020**, *188*, 109881. [CrossRef] [PubMed]
26. Anagnostopoulou, K.; Nannou, C.; Aschonitis, V.G.; Lambropoulou, D.A. Screening of Pesticides and Emerging Contaminants in Eighteen Greek Lakes by Using Target and Non-Target HRMS Approaches: Occurrence and Ecological Risk Assessment. *Sci. Total Environ.* **2022**, *849*, 157887. [CrossRef] [PubMed]
27. Xu, M.; Huang, H.; Li, N.; Li, F.; Wang, D.; Luo, Q. Occurrence and Ecological Risk of Pharmaceuticals and Personal Care Products (PPCPs) and Pesticides in Typical Surface Watersheds, China. *Ecotoxicol. Environ. Saf.* **2019**, *175*, 289–298. [CrossRef] [PubMed]
28. Yang, H.; Lu, G.; Yan, Z.; Liu, J.; Dong, H.; Bao, X.; Zhang, X.; Sun, Y. Residues, Bioaccumulation, and Trophic Transfer of Pharmaceuticals and Personal Care Products in Highly Urbanized Rivers Affected by Water Diversion. *J. Hazard. Mater.* **2020**, *391*, 122245. [CrossRef]
29. Mojiri, A.; Zhou, J.L.; Ratnaweera, H.; Rezania, S.; Nazari, V.M. Pharmaceuticals and Personal Care Products in Aquatic Environments and Their Removal by Algae-Based Systems. *Chemosphere* **2022**, *288*, 132580. [CrossRef]
30. Moore, B.C.; Roark, A.M.; Kohno, S.; Hamlin, H.J.; Guillette, L.J. Gene-Environment Interactions: The Potential Role of Contaminants in Somatic Growth and the Development of the Reproductive System of the American Alligator. *Mol. Cell. Endocrinol.* **2012**, *354*, 111–120. [CrossRef]
31. Luna, T.O.; Plautz, S.C.; Salice, C.J. Chronic Effects of 17α-Ethinylestradiol, Fluoxetine, and the Mixture on Individual and Population-Level End Points in *Daphnia magna*. *Arch. Environ. Contam. Toxicol.* **2015**, *68*, 603–611. [CrossRef]
32. Flaherty, C.M.; Dodson, S.I. Effects of Pharmaceuticals on Daphnia Survival, Growth, and Reproduction. *Chemosphere* **2005**, *61*, 200–207. [CrossRef]
33. Bradley, P.M.; Journey, C.A.; Romanok, K.M.; Barber, L.B.; Buxton, H.T.; Foreman, W.T.; Furlong, E.T.; Glassmeyer, S.T.; Hladik, M.L.; Iwanowicz, L.R.; et al. Expanded Target-Chemical Analysis Reveals Extensive Mixed-Organic-Contaminant Exposure in U.S. Streams. *Environ. Sci. Technol.* **2017**, *51*, 4792–4802. [CrossRef]
34. Fent, K.; Weston, A.A.; Caminada, D. Ecotoxicology of Human Pharmaceuticals. *Aquat. Toxicol.* **2006**, *76*, 122–159. [CrossRef]
35. Backhaus, T.; Porsbring, T.; Arrhenius, Å.; Brosche, S.; Johansson, P.; Blanck, H. Single-Substance and Mixture Toxicity of Five Pharmaceuticals and Personal Care Products to Marine Periphyton Communities. *Environ. Toxicol. Chem.* **2011**, *30*, 2030–2040. [CrossRef] [PubMed]
36. Bulgariu, L.; Gavrilescu, M. Bioremediation of Heavy Metals by Microalgae. In *Handbook of Marine Microalgae: Biotechnology Advances*; Academic Press: Cambridge, MA, USA, 2015; pp. 457–469. ISBN 978-0-12-800776-1.
37. Zeraatkar, A.K.; Ahmadzadeh, H.; Talebi, A.F.; Moheimani, N.R.; McHenry, M.P. Potential Use of Algae for Heavy Metal Bioremediation, a Critical Review. *J. Environ. Manag.* **2016**, *181*, 817–831. [CrossRef] [PubMed]
38. Xiong, J.Q.; Cui, P.; Ru, S. Biodegradation of Doxylamine From Wastewater by a Green Microalga, Scenedesmus Obliquus. *Front. Microbiol.* **2020**, *11*, 584020. [CrossRef] [PubMed]
39. Couto, E.; Assemany, P.P.; Assis Carneiro, G.C.; Ferreira Soares, D.C. The Potential of Algae and Aquatic Macrophytes in the Pharmaceutical and Personal Care Products (PPCPs) Environmental Removal: A Review. *Chemosphere* **2022**, *302*, 134808. [CrossRef]
40. Edwards, T.M.; Miller, H.D.; Toft, G.; Guillette, L.J. Seasonal Reproduction of Male *Gambusia holbrooki* (Eastern Mosquitofish) from Two Florida Lakes. *Fish Physiol. Biochem.* **2013**, *39*, 1165–1180. [CrossRef]
41. Zillien, C.; van Loon, C.; Gülpen, M.; Tipatet, K.; Hanssen, B.; Beeltje, H.; Roex, E.; Oldenkamp, R.; Posthuma, L.; Ragas, A.M.J. Risk-Management Tool for Environmental Prioritization of Pharmaceuticals Based on Emissions from Hospitals. *Sci. Total Environ.* **2019**, *694*, 133733. [CrossRef]
42. Nkoom, M.; Lu, G.; Liu, J. Occurrence and Ecological Risk Assessment of Pharmaceuticals and Personal Care Products in Taihu Lake, China: A Review. *Environ. Sci. Process. Impacts* **2018**, *20*, 1640–1648. [CrossRef]

43. Raimondo, S.; Forbes, V.E. Moving beyond Risk Quotients: Advancing Ecological Risk Assessment to Reflect Better, More Robust and Relevant Methods. *Ecologies* **2022**, *3*, 145–160. [CrossRef]
44. Amiard, J.-C.; Amiard-Triquet, C. Conventional Risk Assessment of Environmental Contaminants. In *Aquatic Ecotoxicology: Advancing Tools for Dealing with Emerging Risks*; Elsevier: Amsterdam, The Netherlands, 2015; pp. 25–49.
45. Garber, K.; Etterson, M.; Odenkirchen, E.; Anderson, B. Use of Risk Quotient and Probabilistic Approaches to Assess Risks of Pesticides to Birds. In Proceedings of the Assessing Risks of Pesticides to Federally Listed (Threatened and Endangered) Species at a National Level, SETAC, Vancouver, BC, Canada, 9–13 November 2014.
46. United States Environmental Protection Agency (USEPA). *Guidelines for Ecological Risk Assessment, EPA/630/R-95/002F*; United States Environmental Protection Agency (USEPA): Washington, DC, USA, 1998.

environments

MDPI

Article

Groundwater Quality in a Rural and Urbanized Region in Limpopo Province, South Africa

Ebrahim Shokoohi [1,*] and Ngoni Moyo [2]

1 Department of Biochemistry, Microbiology and Biotechnology, University of Limpopo, Private Bag X1106, Sovenga 0727, South Africa
2 Aquaculture Research Unit, School of Agricultural and Environmental Sciences, Faculty of Science and Agriculture, University of Limpopo, Turfloop Campus, Private Bag X1106, Sovenga 0727, South Africa
* Correspondence: ebrahim.shokoohi@ul.ac.za

Abstract: The Limpopo Province, situated in the northern part of South Africa, is mainly comprising rural areas that lack adequate facilities for drinking water. Boreholes are the main source of drinking water in rural and urbanized areas of Limpopo Province. Sixty-three water samples, from three locations in Limpopo Province, namely Mankweng, Dalmada, and Polokwane, plus two samples from a river in Magoebaskloof and still water as controls, were collected and subjected to analysis. The Sodium Absorption Ratio (SAR) analysis ranged from 1.4 to 35.6, revealing that 25% of the samples from Mankweng bear low quality with a high amount of sodium. Piper plot showed that two major water types exist in the samples, 33% and 67% of the water samples were of Na-Cl and Ca-Cl types, respectively. To identify the leading natural and anthropogenic processes causing variation in groundwater chemistry, principal component analysis (PCA) was used. The most detected heavy metal was V (vanadium) with 0.00 to 0.59 (mg/mL). The PCA results grouped all water samples from Dalmada together. However, the water samples from Mankweng were divided into three groups by PCA, with borehole samples showing a correlation with heavy metals. In conclusion, the study revealed that natural and anthropogenic activities cause groundwater variation in the Limpopo Province. All the boreholes sampled showed the presence of total coliform, but no *E. coli* was detected. In addition, regarding microbial contamination, water samples were suitable for drinking and irrigation purposes.

Keywords: bacteria; borehole water; drinking water; multivariant analysis; pollution

1. Introduction

The World Health Organization (WHO) [1] has highlighted a concerning issue regarding limited access to drinking water in the Sub-Saharan region. However, the lack of proper water quality management has resulted in several health problems and environmental issues. Water quality is a key factor in ensuring that it is suitable for meeting various needs, including human consumption. As pointed out by Jakeman et al. [2], it is just as important as the availability and quantity of water. The quality of groundwater can be influenced by various factors, including natural ones such as the type of rock, groundwater velocity, geochemical reactions, and soluble salts. Additionally, human activities such as agriculture and industrialization can also have an impact on the quality of water that is recharged into aquifers. This can be seen in the example of WHO [1], where human intervention has resulted in changes to the chemical and physical properties of the groundwater. In order to implement appropriate treatment methods, it is important to identify and understand the sources of pollution that may exist within the groundwater system. This knowledge is essential for effective groundwater management and ensuring its safe use [3]. To ensure that water is safe to drink and use for irrigation, it is necessary to analyze its physicochemical properties. This analysis provides crucial information about the interactions between water and water-bearing rocks, which can have a significant impact on the quality and availability of water. By understanding the mechanics and nature of these chemical reactions, we can develop effective strategies to protect and manage our water resources. This, in turn, can help ensure that everyone has access to clean and safe drinking water while also supporting sustainable agricultural practices [4]. Water quality plays a crucial role in determining the suitability of soil for crop cultivation and population centers [5]. Previous studies have demonstrated that human activity can lead to microbial contamination of groundwater [6,7].

In South Africa, the degradation of improved water sources often occurs due to insufficient operational procedures and inadequate maintenance practices. As a result, these issues lead to frequent system failures and subsequent contamination of the water supply [8]. The Limpopo Province serves as a quintessential example of a predominantly rural area and stands out as one of the poorest provinces in South Africa, characterized by its limited water resources. Nearly all of the available water sources within the province are currently fully exploited and allocated, leaving minimal room for further resource development. This is primarily due to the region's arid climate, challenging topography, presence of sandy rivers, and constraints on increased groundwater [9]. In addition, in the Limpopo Province of South Africa, research studies have revealed concerning findings about the quality of borehole water used by local communities for drinking purposes. The studies detected contamination and identified risks associated with the presence of *E. coli* in this water source. These findings raise important concerns about the safety and suitability of the water supply for the affected communities [10].

In Limpopo Province, there are two main settlements, which include rural and urban areas. In both regions, the primary source of drinking water and irrigation is boreholes [9]. However, it is difficult to distinguish the source of any contamination. Additionally, both areas have the same underground geological plateau. The studies indicate that Limpopo Province, particularly the area around Polokwane, is characterized by supracrustal rocks. Beneath the surface, gneisses and granites from the Swazian to Randian age form the foundation of the region. The supracrustal rocks consist of various types, including schists, quartzite, magnetic quartzite, shale, metavolcanics, serpentinite, and metapyroxenite. Additionally, in the northern part of the region, highly deformed keels made up of marble, calc-silicate rocks, metaquartzite, metapelite, and amphibolite can be found within the gneisses [11]. Therefore, it is less likely that contamination is due to dissolution from the underground plateau. On the other hand, agriculture and mining activities could potentially be the source of contamination for the underground water system. Therefore, it is crucial to analyze the water systems in these areas and identify the sources of pollution using various approaches, such as multivariate analysis.

However, the few studies on groundwater quality in Southern Africa have been highly localized [4,8–10,12]. Studies on groundwater in South Africa showed that arsenic, lead, and chlorine were the chemicals most commonly detected [4]. In addition, through a tool development for using groundwater in the Vhembe district of Limpopo Province [8], results showed that the quantitative availability of drinking water was a problem due to poorly constructed boreholes. Another study conducted in Siloam Village, located in the Vhembe district of Limpopo Province, showed a negative impact of high fluoride on human health in the mentioned location [9]. The microbial analysis of the groundwater in the Vhembe district of Limpopo Province showed that remarkable levels of total coliforms, including fecal coliforms, fecal enterococci, *Clostridium perfringens*, and somatic coliphages, were detected for community and privately owned borehole water [10]. Additionally, water source information in South Africa is required to register the available data to the national Blue Drop System (BDS) of South Africa [12], which, due to the insufficient information, creates a challenge in monitoring the country's water system.

A comprehensive assessment of global groundwater quality is crucial in light of the increasing threats posed by anthropogenic (human-made) and geogenic (naturally occurring) contaminants. This evaluation should encompass a thorough analysis of various pollutants, including heavy metals, agricultural runoff, and industrial chemicals, as well as naturally occurring substances like arsenic and fluoride. Understanding the extent and impact of these contaminants on groundwater sources is vital for safeguarding public health and ensuring the sustainability of freshwater resources [13].

Limpopo Province is characterized predominantly by its vast rural areas, with Mankweng being one of the major locations. This region is notable for its agricultural activities, which are essential to the local community's livelihood. However, it is important to highlight that nearby urban areas, such as Dalmada, are facing significant challenges related to water quality. Dalmada, while more urbanized, grapples with pollution and contamination issues that affect the health and well-being of its residents. Addressing these water quality concerns is crucial for ensuring safe access to clean water for all inhabitants in both Mankweng and Dalmada. The water quality in South Africa has been the subject of numerous studies [4,8–10,12]; however, the specific regions of Mankweng and Dalmada have not been thoroughly examined. This gap in research presents a unique opportunity for the current study, which aims to assess and provide comprehensive insights into the water quality of these areas. Understanding the water quality in Mankweng and Dalmada is crucial, as it can offer valuable information that supports various applications, including agricultural practices, public health initiatives, and local economic development. By addressing this underexplored topic, the study seeks to contribute significantly to the existing body of knowledge and promote sustainable water management in these communities.

Therefore, the study aims to (1) identify the physicochemical linkages with groundwater resources in Limpopo Province, South Africa, and (2) investigate microbial contamination of the water samples from Limpopo Province.

2. Materials and Methods

2.1. Study Sites

Limpopo Province is located in the northern region of South Africa, and it has been divided into five districts, as shown in Figure S1. The study sites were in the Capricorn district, namely Mankweng, Dalmada, and Polokwane regions of the Limpopo Province, South Africa. Mankweng is a rural area, and Dalmada is an urbanized area close to Polokwane, the capital of Limpopo Province. Both sites relied on boreholes for drinking purposes. From March 2022 to December 2022, twenty-one water samples were randomly selected to capture the main dominant natural and anthropogenic factors of the underground water in Limpopo Province. The water samples were included (1) seven boreholes from Dalmada, (2) eleven boreholes from Mankweng in the Capricorn district, (3) one still water sample from Turfloop campus, which was prepared by water facilities of the University of Limpopo, South Africa) as a control, (4) one samples from a borehole in Polokwane, and (5) one sample from a river in Magoebaskloof mountains. The maps in Figure 1A,B were generated using open-source QGIS (http://www.qgis.org (accessed on 1 April 2023)). Water samples were collected from these boreholes and analyzed for physico-chemical parameters. For quality analysis, a sterilized plastic bottle was utilized, while sterilized glass sampling bottles were employed for bacterial analysis. The GPS coordinates of the samples were recorded and are listed in Figure 1 (Table S1). The water samples were collected in 1 L sample bottles and were transported to the laboratory on ice for analysis.

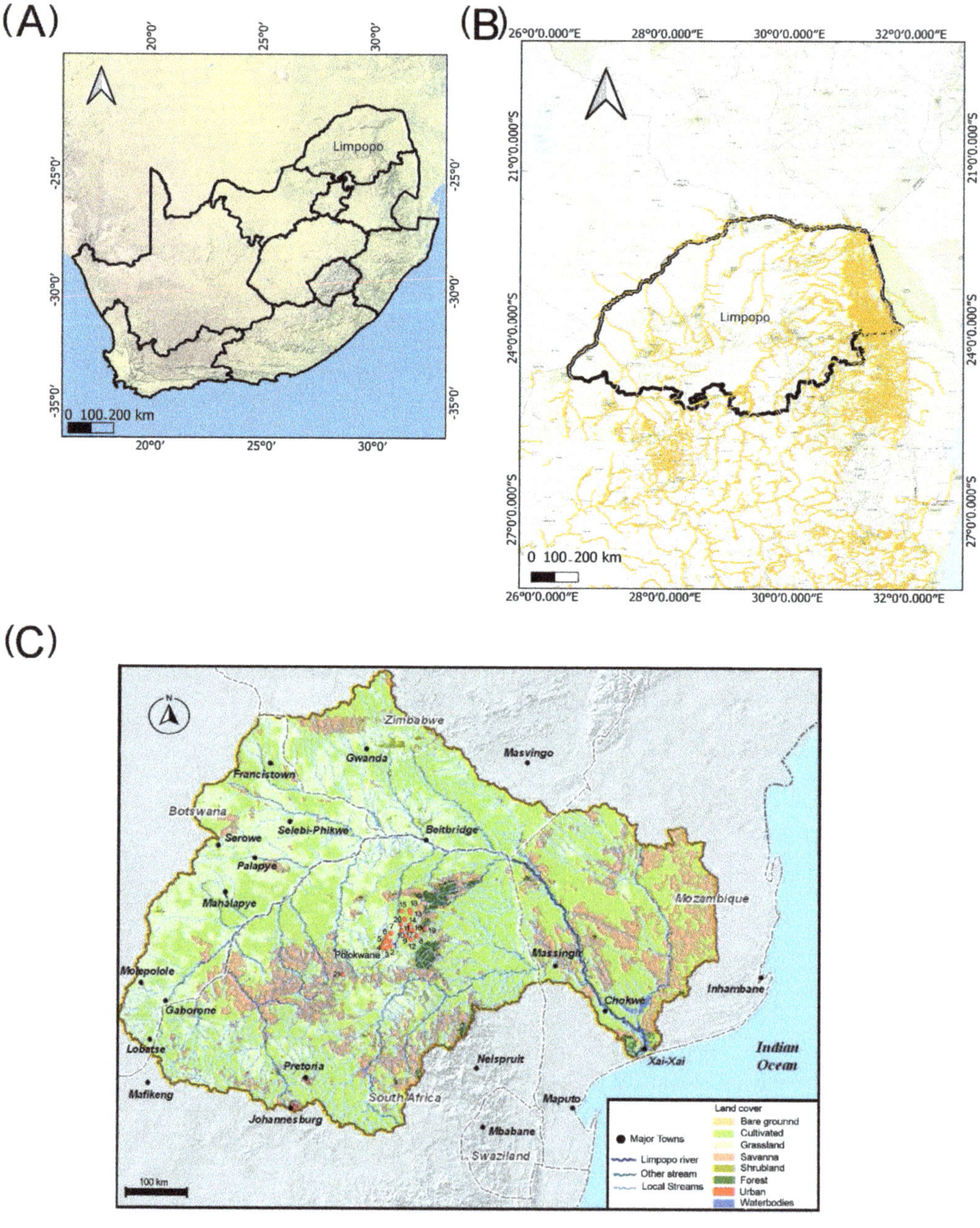

Figure 1. Sampling sites for the water analysis in Limpopo Province, South Africa. (**A**): South Africa map; (**B**): Limpopo map; (**C**): sampling sites (Table S1: site number: 1–21; red dots).

2.2. Physico-Chemical Analysis

Water samples were analyzed in the water analysis laboratory of the University of Limpopo. A volume of 100 mL of each water sample was passed through 0.45 μm pore size, 47 mm diameter sterile filter membranes. The membranes were placed on the appropriate substrate medium and incubated. The water samples were analyzed using standard APHA [14] methods. The metals were analyzed according to USEPA PhosVer 3, and the results were obtained in mg/L [15].

pH, electrical conductivity (measured in mS/cm), salinity (measured in ppt), and turbidity were measured on-site using a handheld multiparameter meter (Professional plus YSI 605000) and a WTW turbidity meter (Turb 430 IR; Xylem, Weilheim, Germany), respectively. Alkalinity as bicarbonate (measured in mg/L) and carbonate (measured in mg/L; Method 2320 B), potassium as K (measured in mg/L; Method 3120 B and EPA method 200.7) were analyzed [14]. Total nitrogen as N (measured in mg/L) was analyzed using spectroquant nitrogen test number 1.14537.0001, ammonia and ammonium (measured in mg/L; spectroquant ammonium-test 1.14752.0001/1.14752.0002/1.00683.0001 method), total phosphate as P (measured in mg/L; spectroquant test 1.14848.0001), and chemical oxygen demand (measured in mg/L; spectroquant test 1.14848.0001 and spectroquant 1.14541.0001 closed reflux method).

The level of hardness was measured using calcium and magnesium; Calmagite Colorimetric Method 8030 of DOC316.53.01043. Nitrate was measured using the Cadmium reduction method no. 8171 of DOC316.53.01069. Phosphate, ammonia, and copper were measured using USEPA PhosVer 3 as per the guidelines of HACH [15]. The results were reported in mg/L. The analysis of all samples was conducted using analytical reagent grade chemicals, and the same quality assurance/quality control procedures were applied to all samples.

For metal analysis, samples were filtered using a pre-washed 0.45 μm pore membrane filter to remove all solid materials before analysis. The trace metals were determined using an Inductively Coupled Plasma-Mass Spectrometer (ICP-MS: NexION 300D, Perkin Elmer, Waltham, MA, USA).

To assess water quality for irrigation, sodium adsorption ratio (SAR) [16] using Equation (1), soluble sodium percentage (SSP) [17] using Equation (2), Magnesium adsorption ratio (MAR) [18] using Equation (3), and Kelley's ratio (KR) [19] using Equation (4), were measured. The equations for the ratios are indicated below:

$$SAR = \frac{Na^{+}}{\sqrt{\left(Ca^{2+} + Mg^{2+}\right)/2}} \tag{1}$$

$$SSP = \frac{Na^{+} + K^{+}}{Na^{+} + K^{+} + Ca^{2+} + Mg^{2+}} \times 100 \tag{2}$$

$$MAR = \frac{Mg^{2+}}{Ca^{2+} + Mg^{2+}} \times 100 \tag{3}$$

$$KR = \frac{Na^{+}}{Ca^{2+} + Mg^{2+}} \tag{4}$$

2.3. Water Character Analysis

The Aquachem software V. 10 [20] was used to submit the sample sites and variables, and the Piper plot was extracted. To ensure quality assurance (QA) and quality control (QC), standard reference materials were utilized. Each batch of samples included a blank and a standard for quality control. The accuracy of the analysis was confirmed using certified standards from De-Bruyn spectropic solutions (500 MUL20-50STD2), and the recoveries were within 10% of the certified values.

2.4. Statistical Analysis

The study examined the correlation between the samples by using a Pearson correlation through XLSTAT software [21]. To evaluate the relationship between the water physicochemical properties and sampling sites, a principal component analysis (PCA) was conducted. Measures were normalized using XLSTAT software. Score values were calculated for each variable based on principal components. The first two components were used to create a two-dimensional plot (F1 and F2) based on eigenvalues provided by XLSTAT.

2.5. Bacterial Contamination

After filtering the water samples using a 0.45 µm pore size cellulose ester membrane, 100 mL of the sample was filtered. Next, incubate total coliform plates at 37 °C for 24 h, fecal coliform plates at 44.5 °C for 24 h, fecal enterococci plates at 37 °C for 48 h, and Clostridium perfringens plates at 37 °C for 24 h under micro-aerophilic conditions using Anaerogen sachets. Once the incubation period is over, count the resulting colonies to identify the types of bacteria present.

For counting *E. coli*, place membranes from fecal coliform agar plates onto Nutrient MUG agar plates and incubate them in an aerobic environment at 37 °C for 24 h. Count fluorescent colonies as presumptive *E. coli* bacteria and confirm them using Gram staining and indole tests with Kovac's reagent. Any Gram-negative, indole-positive colonies were recorded as *E. coli* isolates according to the techniques described by MacFaddin [22].

3. Results

3.1. Water Quality Parameters

Typical properties of groundwater in the examined regions of Limpopo Province are shown in Table 1. Groundwater pH tends to be slightly alkaline. According to the Piper diagram, chloride is the primary anion in the majority of groundwater samples (about 86%), as illustrated in Figure 2G.

In addition, 14% of the groundwater samples contain sulfate (Figure 2F). However, in the case of cations, 29% of samples had more sodium (Figure 2D), 19% had more calcium (Figure 2A), and 29% had more magnesium. In contrast, 23% of samples contain no dominant cations. The study area showed that the average trends for cations and anions were $Na^+ > Mg^{2+} > Ca^{2+} > K^+ > Fe^{2+}$ and $Cl > SO_4$, respectively. The collected water samples were broadly classified into two types, NaCl type (Figure 2; green arrow) and CaCl type (Figure 2; yellow arrow), with 29% and 71% of the total samples, respectively. The Piper diagram (Figure 2) showed that the Mankweng water samples represented NaCl type, whereas the Dalmada water samples represented CaCl type. The results showed that 25% of the water samples collected from the Mankweng area had a Na:Ca ratio over 3, indicating poor soil infiltration. However, the rest of the samples showed a Na:Ca ratio below 3, which means no infiltration might occur in the soil where the water samples were collected.

Table 1. General characteristics of groundwater studied from Mankweng and Dalmada, Limpopo Province, South Africa. The amount of the measures is mg/mL, except for EC, which is mS/m [- = zero or not detected].

Location	Mankweng	Mankweng	Mankweng	Mankweng	Mankweng	Mankweng	Mankweng	Mankweng	Mankweng	Mankweng	Mankweng
Sample code	Mankanye-T	Frans-MBH-3	Bankoeng BH-1	Mankoeng BH5	Makanye-BH	Klas-MBH	Mankoeng-T	Klas-M-TW3	Mankoeng-TW3	Sogoreng-BH-1	Klass BH5
pH	8.40	8.06	8.15	7.93	7.58	7.04	8.33	7.66	7.29	7.88	7.38
EC	63.30	104.20	115.50	114.80	80.80	61.00	63.10	4.00	6.63	112.90	64.50
TDS	411.45	677.30	750.80	746.20	525.20	396.50	410.15	26.00	43.10	733.85	419.25
TH	119.27	146.04	284.14	237.78	238.33	168.64	116.65	16.47	21.28	321.75	190.88
F	0.40	1.73	0.68	0.78	0.65	0.42	0.39	0.00	0.00	1.39	0.51
Cl	22.21	67.20	53.28	64.10	60.93	44.25	22.04	4.89	5.73	111.49	58.06
SO_4	7.22	17.31	54.34	58.07	40.84	24.21	7.57	0.31	1.59	43.26	28.72
K	1.78	5.89	1.94	1.87	15.09	8.02	1.75	0.68	0.91	4.87	7.86
Ca	17.22	25.16	49.22	34.92	46.75	29.80	16.83	5.45	5.30	66.47	35.53
Mg	18.50	20.18	39.10	36.52	29.48	22.85	18.10	0.69	1.95	37.76	24.77
Na	94.22	169.69	151.86	140.53	64.42	53.52	93.67	3.32	6.31	121.30	51.85
Zn	-	0.01	0.03	-	0.10	0.58	-	-	-	0.14	0.27
Pb	0.01	0.01	0.01	-	0.01	0.01	-	-	-	0.01	0.01
Mn	-	0.01	-	-	-	-	-	-	-	0.02	-
V	0.06	0.07	0.11	0.14	0.08	0.06	0.06	0.00	0.00	0.10	0.09
Cu	-	-	-	0.01	-	-	-	-	-	0.01	0.01
Fe	-	-	-	0.00	-	-	-	-	-	-	-
Ni	-	-	-	0.45	-	-	-	-	-	-	0.01
B	-	0.14	0.03	0.05	-	-	-	-	-	0.03	0.03
Se	0.16	0.55	0.63	2.05	-	0.18	0.19	0.20	0.23	0.23	0.06
Ba	0.02	0.05	0.03	0.01	0.06	0.03	0.02	0.02	0.02	0.04	0.03
Si	6.51	14.11	9.92	9.60	12.00	13.17	6.36	1.60	1.99	7.52	12.53
Location	**Dalmada**	**Dalmada**	**Dalmada**	**Dalmada**	**Dalmada**	**Dalmada**	**Dalmada**	**Control**	**Polokwane**	**Magobeskloof**	
Sample code	Baskop	DG-1 26	D22	D34	D56	B51	D136	DW	Filip-N12	River-M	
pH	7.32	7.40	7.31	7.62	7.54	7.83	7.35	7.63	7.11	7.82	
EC	116.70	4.92	123.40	104.40	99.50	97.10	129.40	0.90	1.30	5.73	
TDS	758.55	31.98	802.10	678.60	646.75	631.15	841.10	5.85	8.45	37.24	
TH	519.35	16.17	462.64	431.41	404.94	423.10	583.04	-	-	23.95	
F	0.17	-	0.10	0.08	0.10	0.11	-	-	-	-	

Table 1. *Cont.*

Location	Dalmada	Dalmada	Dalmada	Dalmada	Dalmada	Dalmada	Dalmada	Control	Polokwane	Magobeskloof
Cl	84.09	4.18	16.67	29.40	57.94	57.53	78.62	0.07	2.13	2.98
SO_4	18.42	0.59	228.38	73.96	45.95	43.50	45.06	0.14	0.60	0.53
K	7.04	0.96	10.98	7.34	12.30	6.16	6.16	0.15	0.10	0.77
Ca	14.59	4.90	40.94	46.41	51.76	42.40	39.61	-	2.30	8.50
Mg	117.20	0.95	87.45	76.55	66.88	76.97	117.48	-	0.00	0.66
Na	21.15	2.67	65.55	36.68	45.79	33.49	21.78	-	1.98	2.91
Zn	0.01	0.25	0.02	0.01	0.02	0.01	0.02	-	-	0.01
Pb	0.01	-	0.01	0.01	0.01	-	0.01	-	-	-
Mn	0.00	0.01	-	-	-	-	-	-	-	-
V	0.56	0.01	0.43	0.37	0.26	0.31	0.59	-	-	-
Al	-	-	-	-	-	-	-	-	-	0.04
Cr	-	-	0.03	0.01	0.01	-	-	-	-	-
Cu	-	0.01	0.00	0.01	0.03	0.03	0.02	-	-	0.01
Fe	-	0.08	-	-	-	-	-	-	-	0.12
Ni	0.01	-	0.01	-	0.01	0.01	0.02	-	-	0.10
B	0.02	-	0.22	0.10	0.08	0.04	0.04	-	-	0.01
Se	2.16	1.21	1.88	1.75	0.73	0.44	1.26	0.46	-	0.42
Ba	0.17	0.01	0.09	0.04	0.09	0.08	0.08	-	-	0.04
Si	10.28	1.70	12.23	10.66	13.86	12.85	13.11	-	0.18	2.38

Electrical conductivity as EC; total dissolved solids as TDS; total hardness as TH; fluoride as F; chloride as Cl; sulfate as SO_4; potassium as K; calcium as Ca; magnesium as Mg; sodium as Na; zinc as Zn; lead as Pb; manganese as Mn; vanadium as V; aluminum as Al; chromium as Cr; copper as Cu; iron as Fe; nickel as Ni; boron as B; selenium as Se; barium as Ba; silica as Si.

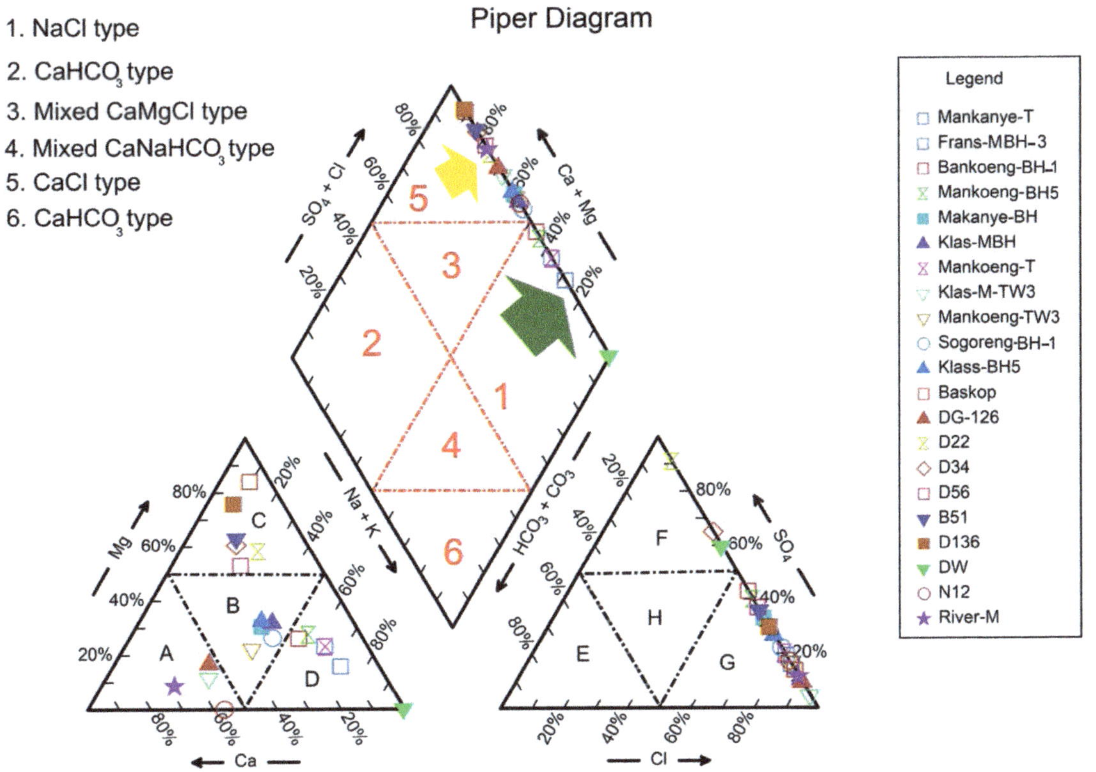

Figure 2. Piper diagram of the borehole water samples from Mankweng and Dalmada, Limpopo Province, South Africa. [A: calcium type; B, H: no dominant type; C: magnesium type; D: sodium type; E: bicarbonate type; F: sulfate type; G: chloride type].

According to the Pearson correlation shown in Figure 3 (Table S2), there was a strong positive correlation between EC and TDS (r = 1.00), total hardness (r = 0.881), chloride (r = 0.794), calcium (r = 0.810), magnesium (r = 0.801), vanadium (r = 0.728), and silica (r = 0.852). In addition the result indicated a positive significant correlation of the following properties, including fluride with sodium (r = 0.862), sulfate with EC and TDS (r = 0.591), chromium and sulfate (r = 0.894) and boron (r = 0.803), magnesium with total hardness (r = 0.975), vanadium with barium (r = 0.813) and total hardness (r = 0.935), sodium with fluoride (r = 0.862), aluminum with iron (r = 0.836), and barium with magnesium (r = 0.829). However, pH does not correlate with the anions and cations of the groundwater samples, except with sodium (r = 0.635) and zinc (r = −0.425).

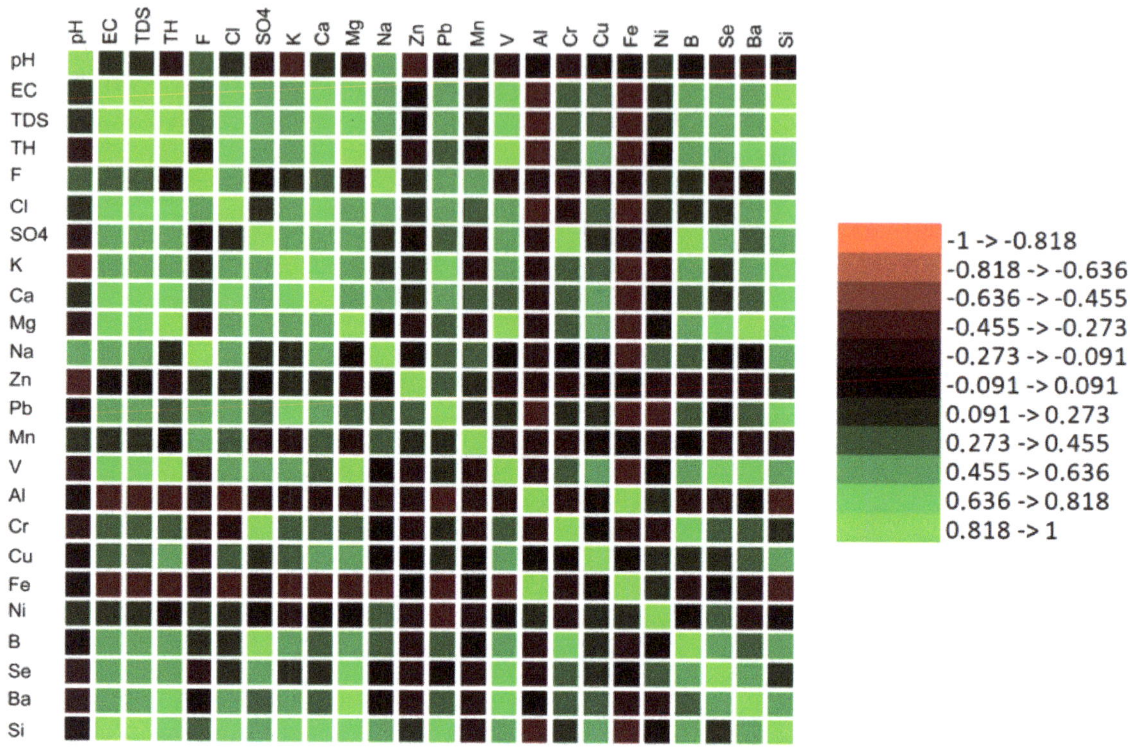

Figure 3. Correlation of the physicochemical properties of the water samples from Mankweng and Dalmada, Limpopo Province, South Africa. All measurements were expressed in mg/L, except for electrical conductivity (EC), which was measured in mS/m.

3.2. Water Quality Indices

Regarding the water quality indices (Table 2), the highest SAR value was 35.6 found for Frans-MBH-3 located in Mankweng, where the concentration of Na^+ (169.7 mg/L) was also the highest among all locations. In contrast, the lowest SAR value was 1.6, which was found for a borehole (DG-126) in Dalmada with a concentration of Na^+ (2.67 mg/L), which was within the standard level and very good for irrigation. Overall, the samples from Mankweng had higher SAR values than Dalmada water samples, indicating that there was a higher amount of sodium and EC in the Mankweng water samples. Regarding the soluble sodium percentage (SSP) values, suitability ranged from 79.5 (Frans-MBH-3 in Mankweng) to 15.1 (D136 in Dalmada). The result indicated a higher SSP of Mankweng water samples than Dalamad, showing a higher amount of sodium in the groundwater samples.

Regarding the MAR, the result ranged from 0.0 (N12 sample in Polokwane) to 88.9 (Baskop sample, Dalmada). On the other hand, the result indicated a higher amount of Mg in Dalamada groundwater samples than in Mankweng and Polokwane. A very low amount of Mg was detected in a sample taken from Polokwane. In 52% of the water samples, MAR was higher than 50, indicating an infiltration problem in the area studied.

KR indicated balance among Na^+, Ca^{2+}, and Mg^{2+} ions in water; the result ranged from 0.9 (N12 sample in Polokwane) to 118.6 (Baskop sample, Dalmada). In 90% of the water samples, the KR value measured was over 1.0, indicating an excessive amount of sodium in the water of the locations studied (Table 2).

Table 2. Water quality indices for the groundwater samples from Mankweng and Dalmada, Limpopo Province, South Africa. [sodium adsorption ratio (SAR); soluble sodium percentage (SSP); Magnesium adsorption ratio (MAR); Kelley's ratio (KR)].

Sample Location	Sample Code	Sample Depth (m)	SAR	SSP	MAR	KR	Na/Ca
Mankweng	Mankanye-T	70	22.3	72.9	51.8	24.0	5.5
Mankweng	Frans-MBH-3	70	35.6	79.5	44.5	26.9	6.7
Mankweng	Bankoeng-BH-1	70	22.9	63.5	44.3	42.2	3.1
Mankweng	Mankoeng BH5	70	23.5	66.6	51.1	40.5	4.0
Mankweng	Makanye-BH	70	10.4	51.1	38.7	30.9	1.4
Mankweng	Klas-MBH	70	10.4	53.9	43.4	24.6	1.8
Mankweng	Mankoeng-T	70	22.4	73.2	51.8	23.7	5.6
Mankweng	Klas-M-TW3	70	1.9	39.4	11.2	1.3	0.6
Mankweng	Mankoeng-TW3	70	3.3	49.9	26.9	3.1	1.2
Mankweng	Sogoreng-BH-1	70	16.8	54.8	36.2	39.6	1.8
Mankweng	Klass BH5	70	9.4	49.8	41.1	26.2	1.5
Dalmada	Baskop	100	2.6	17.6	88.9	118.6	1.4
Dalmada	DG-126	50	1.6	38.3	16.2	1.5	0.5
Dalmada	D22	60	8.2	37.3	68.1	89.1	1.6
Dalmada	D34	120	4.7	26.4	62.3	77.3	0.8
Dalmada	D56	180	5.9	32.9	56.4	67.8	0.9
Dalmada	B51	55	4.3	24.9	64.5	77.8	0.8
Dalmada	D136	70	2.5	15.1	74.8	118.0	0.5
Polokwane	Filip-N12	70	1.8	47.5	0.0	0.9	0.9
Magoebaskloof	River-M	1	1.4	28.7	7.2	1.0	0.3

3.3. Multivariate Analysis

The analysis revealed that the Dalmada samples were linked to total hardness and sulfate, while the Mankweng samples were associated with sodium, pH, chloride, fluoride, total dissolved solids, EC, and calcium. Furthermore, the PCA indicated that two samples from Mankweng (klas-M-TW3 and Mankoeng-TW3), one sample from Dalmada (DG-1), a sample from the Magoebaskloof river, a sample from Polokwane (N12), and a still water sample, which was used as a control, only showed correlations with aluminum and iron (Figure 4).

The PCA accounted for 56.45% of the variability in the water quality parameters in the Mankweng, Dalmada, and Polokwane regions of Limpopo Province (Figure 4). The result indicated that 39.12% was explained by F1 and 17.32% was explained by F2, of which the water quality parameters had a significant effect on the tested water in the Mankweng and Dalmada regions of Limpopo Province. The highest contribution of the variability in F1 was measured for EC and TDS with 9.4%. The lowest contribution to the variability in F1 was measured for Zinc (0.009%) (Table 3). In contrast, the highest variability contribution in F2 was measured for fluoride and sodium, with 17.8% and 13.6%, respectively. The lowest contribution to the variability in F2 was measured for Nickel (0.0%) (Table 3).

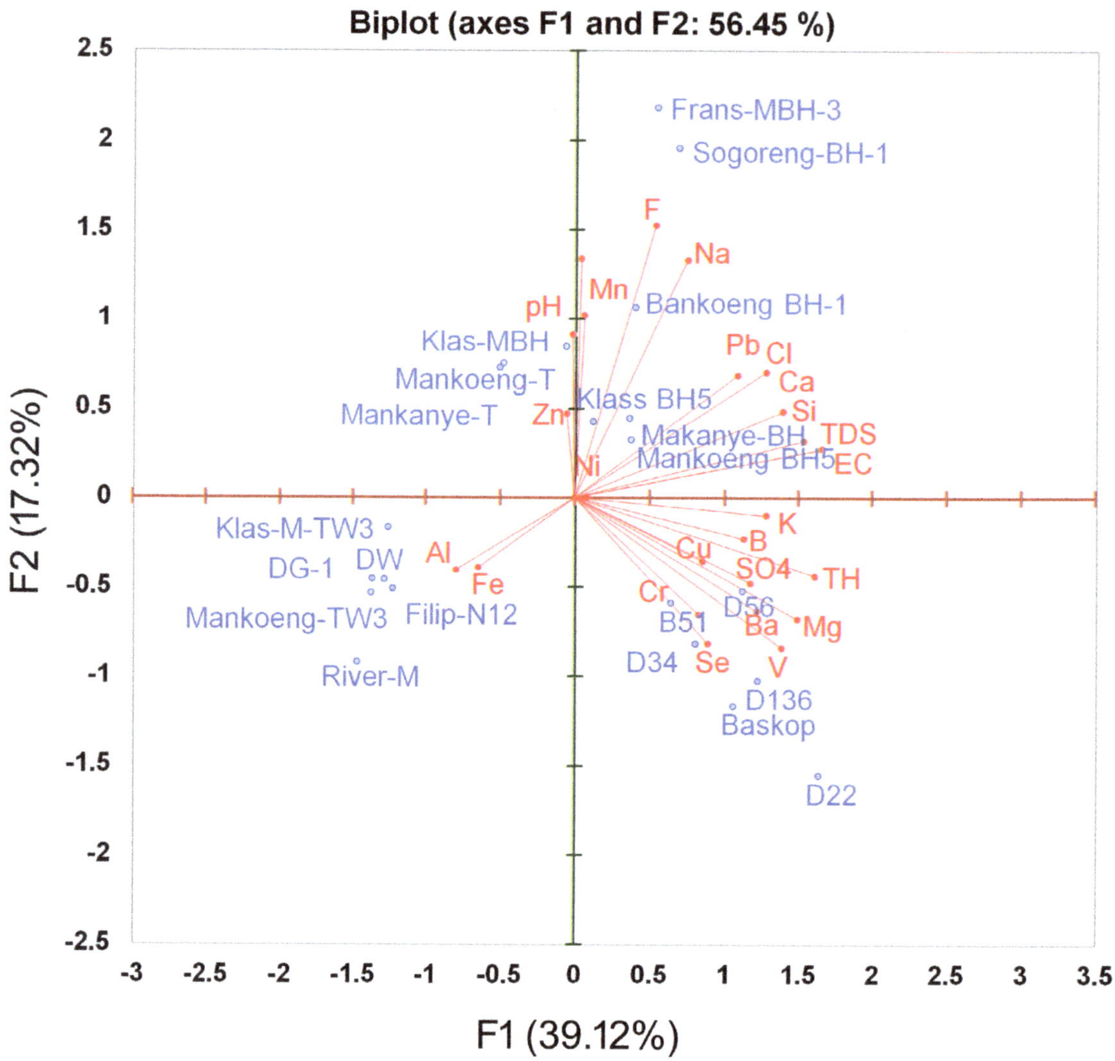

Figure 4. Principal component analysis (PCA) of the relationship between the physicochemical and the sampling location from Mankweng and Dalmada, Limpopo Province, South Africa.

The contribution of inorganic, physical, and metal components of the tested water and sampling locations indicated that Dalmada samples were correlated with heavy metals such as chromium, selenium, and barium (Figure 4; Table 2). The factor score, representing the variability in F1, was highest for D22 (5.128) in Dalmada and lowest for River-M (−4.596) in Magoebaskloof. In contrast, the factor score of the F2 was measured as the highest for Frans-MBH-3 (4.543) in Mankweng and D22 (−3.238) in Dalmada (Figure 4; Table 4).

Table 3. Contribution of the variables (%) of the water samples from Mankweng and Dalmada, Limpopo Province, South Africa.

Properties	F1	F2
pH	0.001	6.379
EC	9.403	0.570
TDS	9.403	0.570
Total Hardness	8.865	1.478
F	0.982	17.844
Cl	5.596	3.801
SO_4	4.721	1.752
K	5.611	0.076
Ca	6.644	1.795
Mg	7.603	3.536
Na	1.923	13.603
Zn	0.009	1.700
Pb	4.025	3.613
Mn	0.013	7.973
V	6.559	5.415
Al	1.419	1.178
Cr	2.338	3.298
Cu	2.494	0.975
Fe	2.162	1.256
Ni	0.021	0.000
B	4.372	0.404
Se	2.706	5.130
Ba	5.084	3.089
Si	8.039	0.767

Table 4. Factor score of the water samples from Mankweng and Dalmada, Limpopo Province, South Africa.

Location	Sample Code	F1	F2
Mankweng	Mankanye-T	−1.487	1.569
Mankweng	Frans-MBH-3	1.717	4.543
Mankweng	Bankoeng-BH-1	1.260	2.214
Mankweng	Mankoeng-BH5	1.172	0.675
Mankweng	Makanye-BH	1.138	0.928
Mankweng	Klas-MBH	−0.176	1.765
Mankweng	Mankoeng-T	−1.567	1.518
Mankweng	Klas-M-TW3	−3.933	−0.349
Mankweng	Mankoeng-TW3	−3.825	−1.062

Table 4. *Cont.*

Location	Sample Code	F1	F2
Mankweng	Sogoreng-BH-1	2.157	4.071
Mankweng	Klass-BH5	0.378	0.890
Dalmada	Baskop	3.320	−2.428
Dalmada	DG-126	−4.007	−0.956
Dalmada	D22	5.128	−3.238
Dalmada	D34	2.524	−1.703
Dalmada	D56	3.512	−1.089
Dalmada	B51	2.018	−1.230
Dalmada	D136	3.841	−2.135
Turfloop	DW	−4.276	−0.950
Polokwane	N12	−4.300	−1.114
Magoebaskloof	River-M	−4.596	−1.919

3.4. Microbiology Analysis

The water from all the boreholes in different locations was found to be of acceptable microbiological quality for irrigation, as shown in Table 5. However, some of the water samples, such as Bankoeng-BH-1, Mankoeng-BH5, and Klas-MBH in Mankweng, and Baskop, D22, and D136 in Dalmada TC, were detected to have over 201 cfu per 100 mL, which is not safe for drinking purposes. Similarly, in River-M in Magoebaskloof, a high number of TC (>201) per 100 mL was detected. It is noteworthy that no high cfu per 100 mL of *E. coli* was found in any of the water samples (Table 5).

Table 5. Total coliform (TC) and *E. coli* in the water samples from Mankweng and Dalmada, Limpopo Province, South Africa.

Location.	Sample Code	TC (cfu/100 mL)	*E. coli* (cfu/100 mL)
Mankweng	Mankanye-T	<1	<1
Mankweng	Frans-MBH-3	<1	<1
Mankweng	Bankoeng-BH-1	>201	<1
Mankweng	Mankoeng-BH5	>201	<1
Mankweng	Makanye-BH	<1	<1
Mankweng	Klas-MBH	>201	<1
Mankweng	Mankoeng-T	<1	<1
Mankweng	Klas-M-TW3	<1	<1
Mankweng	Mankoeng-TW3	<1	<1
Mankweng	Sogoreng-BH-1	<1	<1
Mankweng	Klass-BH5	<1	<1
Dalmada	Baskop	>201	1
Dalmada	DG-126	<1	<1
Dalmada	D22	>201	<1
Dalmada	D34	36	<1

Table 5. *Cont.*

Location.	Sample Code	TC (cfu/100 mL)	*E. coli* (cfu/100 mL)
Dalmada	D56	95	<1
Dalmada	B51	118	<1
Dalmada	D136	>201	<1
Turfloop	DW	<1	<1
Polokwane	N12	4	<1
Magoebaskloof	River-M	>201	<1

4. Discussion

This research study provides detailed insights into the characteristics of borehole water quality in specific rural and urbanized areas within South Africa's Limpopo Province. The results of the study were based on the physicochemical, microbiological, and trace metal properties of the water samples collected from Mankweng, Dalmada, and Polokwane.

The inhabitants of rural and economically disadvantaged communities often rely on water sources that have been found to possess substandard quality. This issue has gained international recognition as a pressing concern for public health [23]. The effective operation and maintenance of these water supplies may be hindered by insufficient resources and a lack of understanding of the various factors that can influence water quality [8]. Mankweng is a rural area located in the Limpopo Province, known for its ongoing development initiatives. The inhabitants of this area rely primarily on boreholes as their main source of water supply. Dalmada is an urbanized location with boreholes as the main source of water for the majority of inhabitants. Therefore, water quality is crucial for both of these areas.

4.1. Water Quality Indices and Parameters

SAR is an indicator that shows the level of sodium hazard in a water sample. High levels of sodium ions in water can impact the soil's permeability and cause infiltration problems. The SAR value of irrigation water is a crucial factor that affects the amount of sodium absorbed by the soil. Higher SAR values in irrigation water can lead to an increase in the amount of sodium absorbed by the soil, which can impact soil quality and crop growth. Therefore, it is important to monitor and manage SAR values in irrigation water to ensure optimal soil health [24]. The SAR index classifies water into four categories: S1 (<10; excellent), S2 (10–18; good), S3 (18–26; doubtful), and S4 (>26; unsuitable). The results of a recent study indicated that in the Mankweng area, 20% of the rural area of Mankweng had a SAR value of over 20. This means that the water is doubtful for use, but still suitable for maize production, which is the dominant crop in this region.

When water with high levels of sodium (Na^+) and low levels of calcium (Ca^{2+}) is used for irrigation, the cation exchange complex in the soil may become overwhelmed with sodium. This can destroy the soil structure due to the dispersion of clay particles, leading to soil erosion and decreased fertility. The cation exchange complex is responsible for holding onto important nutrients like calcium and magnesium for plant uptake. However, when sodium levels are high, they can displace these essential nutrients and wreak havoc on the soil's physical and chemical properties. Therefore, it is essential to monitor the quality of water used for irrigation to prevent soil degradation and ensure long-term productivity [17]. Additionally, sodium reacts with the soil, thus altering the soil's permeability. Furthermore, high levels of sodium in the soil make it alkaline and also cause a deficiency of calcium. In rainy seasons, excess amounts of sodium make the soil sticky, which is often observed in rural areas like Mankweng, where the SAR value was over 20. However, in Dalmada, the water SAR value was below 10, which is excellent for irrigation purposes. A study of the SAR values of water samples in Limpopo Province [25] indicated that most sampling sites were suitable for irrigating various crops. This finding aligns with the current study, which identified only five sites in Mankweng that had a high SAR value, showing that most of the sample sites were suitable for the irrigation of various crops. This is in agreement with the present study, in which only five sites of Mankweng had a high value of SAR.

Measuring the concentrations of calcium (Ca^{+2}), magnesium (Mg^{+2}), and sodium (Na^{+}) is essential for producing accurate predictions of the SAR value, a key parameter in assessing water quality for irrigation [26]. Understanding SAR is particularly important in the context of irrigation practices as it influences soil structure, fertility, and plant health. This approach is directly aligned with the findings of the current study, which provides critical insights into the optimal irrigation strategies for a diverse range of crops cultivated in the Limpopo Province of South Africa. Such information is vital for enhancing agricultural productivity and sustainability in the region, where effective water management practices can significantly impact crop yield and overall farm viability.

When the concentration of sodium solution (SSP) rises in water, it can adversely affect both plant growth and the soil's capacity to absorb and distribute water. This means that the plants may not be able to grow to their full potential, and the soil may become more compacted, making it harder for water to penetrate the surface. Therefore, it is important to monitor the level of SSP value in water to ensure healthy plant growth and maintain proper soil permeability [27,28].

Groundwater is the water that fills the spaces between soil and rock particles underground. It is common for groundwater to contain sodium, as most rocks and soils have sodium compounds that easily dissolve in it. Despite this, the concentration of sodium in groundwater is typically lower than that of calcium and magnesium in freshwater. Calcium and magnesium are essential minerals for human health, whereas high levels of sodium in drinking water can be harmful to health [29]. The general consensus is that consuming higher amounts of dietary sodium is connected to an increased risk of developing hypertension within the general population. Furthermore, elevated blood pressure has been found to be associated with greater salinity levels in drinking water [29]. Therefore, it is important to monitor the levels of these minerals in our drinking water [29]. Elevated levels of sodium in groundwater primarily originate from three main sources. The first source is the weathering of rocks that contain sodium, which can leach into the groundwater. The second source is irrigation returns, where the excess water used in irrigation may contain high levels of sodium that can infiltrate into the groundwater. Lastly, pollution from sewage effluent and seawater intrusion can also contribute to elevated levels of sodium in groundwater. These sources can significantly impact the quality and safety of groundwater for human consumption [30]. In contrast, in Dalmada, the urbanized area, the primary source of sodium is likely rocks or irrigation return [31]. Furthermore, Limpopo Province is situated within a Granulite-facies rock belt, which is a source of both sodium and calcium [11,31].

The MAR causes a harmful effect on the soil when it exceeds 50 [32]. The result indicated that in 52% of the water samples, the MAR value was over 50 [33]. In areas with high amounts of Mg^{2+}, the soil faces an infiltration problem [34]. Furthermore, the MAR value plays a significant role in influencing the alkalinity of the soil, which can negatively impact soil quality and ultimately diminish crop yields [26,35]. A comprehensive study conducted in the Limpopo Province of South Africa revealed that a majority of the water samples analyzed exhibited a MAR value of less than 50. This indicates a favorable condition for irrigation practices [25]. However, these findings are in contrast to those observed in the current study, highlighting potential discrepancies in water quality and usability for agricultural purposes.

This suggests that, while the groundwater resources might generally be deemed suitable for irrigation based on MAR assessments, there are underlying challenges that must be addressed to optimize their use. In particular, the potential effects of rising alkalinity on soil health warrant careful consideration. Therefore, it is imperative to implement continuous monitoring of calcium, magnesium, water quality, and soil conditions, coupled with effective management strategies, to ensure sustainable agricultural practices and safeguard crop productivity in the face of varying environmental factors. A comprehensive analysis of groundwater quality in Yemen has disclosed that 55.17% of the sampled sites are deemed unsuitable for irrigation purposes, primarily due to elevated levels of alkalinity [36]. This significant finding underscores persistent concerns regarding soil and water management in arid regions. Moreover, it aligns with ongoing research emphasizing the importance of global monitoring and assessment of key mineral levels, specifically calcium (Ca^{+2}) and magnesium (Mg^{+2}). Monitoring these elements is crucial as they play vital roles in soil health and plant growth, impacting agricultural productivity and food security in vulnerable regions. The results call for the implementation of effective strategies to address the challenges associated with water quality and sustainable irrigation practices.

Calcium and magnesium are essential minerals that play a vital role in numerous biological processes, including bone formation, muscle contraction, and nerve function [32]. They are widely available in nature, and their presence in all-natural water sources makes them easily accessible to living organisms. Groundwater is a significant source of these minerals, including calcium and magnesium [37]. These minerals primarily originate from the weathering of rocks, particularly limestone and dolomite. The presence of these minerals in groundwater can have significant implications for water quality, as high concentrations can have adverse effects on human health and the environment. Several epidemiological studies have provided evidence of a correlation between the hardness of drinking water or its magnesium and calcium content, and an increased risk for cardiovascular disease, growth retardation, reproductive failure, and other health issues [38].

Magnesium was detected in higher amounts in Dalmada than in Mankweng. The previous result indicated more limestone distribution in the Capricorn district than in other locations in Limpopo Province [39]. Therefore, more magnesium was detected in Dalmada than in Mankweng.

Based on the KR value, which balances Na^+, Ca^{2+}, and Mg^{2+}, a KR value over 1 indicates an excess amount of Na^+ [19]. The result of the present study indicated that Polokwane and Magoebaskloof water samples had excellent quality for irrigation purposes.

Calcium is a vital mineral that is essential for healthy bones and teeth. The primary sources of this important nutrient are carbonate rocks, such as limestones and dolomites [40]. These rocks dissolve in groundwater that contains carbonic acid, which releases calcium ions into the water. The dissolved calcium is then available for uptake by plants and animals, and ultimately ends up in the food chain. Without these sources of calcium, life as we know it would not be possible [41]. The geological map of Limpopo Province revealed a high distribution of limestone in Dalmada, a reason why this area has more calcium than other localities [42]. As groundwater infiltrates or flows through rocks, it has the potential to dissolve naturally occurring minerals such as calcium carbonate ($CaCO_3$) or calcium magnesium carbonate ($CaMg\ (CO_3)_2$). This process can result in an increase in the concentration of Ca^{2+} and Mg^{2+} ions present in the water. These ions, commonly referred to as hardness ions, can have an impact on water quality and may lead to problems such as scale buildup in plumbing systems [43].

4.2. Correlation and Principal Component Analysis

The results of the present study indicated a positive correlation between pH and sodium, while a negative correlation was found between pH and zinc. Similar findings were reported by Zou et al. [44] and Zhao et al. [45], who showed that pH was negatively correlated with zinc levels in China. Additionally, electrical conductivity (EC) exhibited a positive correlation with TDS and zinc, consistent with previous studies.

The accumulation of heavy metals in aquatic environments poses significant risks to aquatic ecosystems [46]. In this study, neither pH nor EC showed a significant correlation with nickel (Ni). Zinc is an essential trace metal; it is a necessary nutrient for many organisms and is found in the tissues and organs of humans [47]. South Africa recommends a maximum of 5 mg/L for zinc and 0.07 mg/L for nickel in drinking water, and 2 mg/L in effluent water samples [48]. It is important to note that the toxicity of zinc in water samples is minimal and is unlikely to cause secondary pollution in the receiving surface water used for irrigation or drinking purposes.

In regard to anthropogenic pollution in the groundwater of the Limpopo region, previous research [49] has revealed concerning findings regarding water quality. Specifically, the study identified the presence of several contaminants in river water samples collected from the area, including elevated levels of salt, phosphate, vanadium, lead, and silicon (Figure S3). These contaminants can significantly alter the chemical composition of the tested water samples, potentially posing risks to both environmental health and human safety. The detection of these substances highlights the ongoing challenges of managing water quality in this region, emphasizing the need for effective monitoring and remediation strategies to mitigate the impact of pollution on the groundwater resources. The PCA suggested that groundwater quality is mainly controlled by TDS, EC, and sodium in the areas studied. Water hardness was high in the samples collected from Dalmada, reflecting the high concentrations of calcium and magnesium.

When assessing the quality of water for irrigation purposes, electrical conductivity is a crucial parameter to consider. It directly relates to salinity issues and reflects the TDS present in groundwater. As it indicates the salinity hazard to crops, measuring electrical conductivity is an excellent way to evaluate water quality for irrigation [50]. Groundwater TDS concentrations of the present study ranged from 5.8 to 841.0 mg/L (avg. 456.3 mg/L), all within desirable limits according to South African Water Quality Guidelines (SAWQG) [51] and WHO [52] standards. Safe EC range for health is 0–150 mS/m and acceptable TDS range is 0–1000. Therefore, EC and TDS of the water samples in Dalmada and Mankweng were acceptable for irrigation and drinking. PCA in hydrogeochemical characteristics of groundwater quality in the Gwayi area of Zimbabwe showed a high concentration of calcium and magnesium, contributing to the hardness of the water [7].

Hard water is high in calcium and magnesium, common in dolomitic aquifers. It causes mineral buildup in kettles and wastes soap and synthetic detergents. This character was observed in Dalmada, where the total hardness of the water ranges from 404.9 to 583.0. The highest acceptable level for total hardness (TH) is 500 mg/L. However, the preferred level is 100 mg/L [52]. Continued use of hard water in irrigation can increase soil pH, affecting plants that require acidic conditions. However, high TH in the water does not pose a health risk [7].

Regarding elements such as iron and manganese, WHO reports that in 2023, 0.30 mg/L of iron and 0.05 mg/L of manganese were recommended in drinking water. However, they did not exceed the recommended limits in all the sampled areas. Chloride levels in one sample (Sogoreng-BH; Mankweng) were higher than 100 mg/L. Again, a high amount of Fluoride levels was detected in one sample (Sogoreng-BH; Mankweng). According to WHO [53], high fluoride may cause various health problems. Fluorite, mica, and hornblende are the most common fluoride-bearing minerals found in thin sections of rock from western Bushveld areas of South Africa [54].

Fluoride in water can come from nature or humans. High levels cause fluorosis, affecting teeth and bones [55,56]. Based on WHO [42], the amount of fluoride above 1500 μg/l has a chronic effect, in which one sample in Mankweng was detected with 1730 μg/l. However, the remaining water samples were acceptable for drinking and irrigation purposes.

4.3. Bacterial Analysis

The microbial contamination in the water samples may be attributed to the depth of the boreholes. The minimum depth of the sampled borehole was 55 m in B51 (Dalmada). As groundwater moves through the first 30 m of saturated sand or unfissured rock, it effectively cleanses and purifies itself by filtering out surface microbial contamination [7]. According to Moyo [7], in the unsaturated zone, purifying groundwater may require no more than 3 m of filtration. Nevertheless, if there is a fractured aquifer, microbial contaminants can quickly pass through the unsaturated zone and reach the water table. In such cases, the efficiency of the purifying process is reduced. Studies carried out in Southern Africa have revealed that certain boreholes contain contaminated water [7,47]. Moyo [7] suggested that the borehole depth affected the water samples in Zimbabwe, as no significant amount of coliform bacteria were detected in the higher depth of borehole water samples. In contrast, water samples from Magoebaskloof River contain a high number of coliform bacteria, as indicated by Potgieter et al. [57]. This is due to the animal community of the Magoebaskloof, which uses the river as a natural resource, and the low depth of the river, for not filtered through the natural process. The presence of *E. coli* in water is typically an indicator of recent fecal contamination or poor hygienic practices. This signifies a potential risk of waterborne diseases due to the presence of harmful bacteria [58]. Fecal contamination, inadequate sanitation measures, and improper storage conditions can all contribute to the proliferation and persistence of *E. coli*, posing a public health concern. Therefore, effective control and prevention strategies are necessary to mitigate the impact of these factors on the prevalence of *E. coli* in water. Generally, in the studied areas of Limpopo Province, no potential risk of *E. coli* was detected. However, constant monitoring is necessary for a healthy water system.

All water samples were collected from the same aquifer (Figure S2) [59], albeit from varying depths (Table 2). This indicates that the aquifer can impact contamination levels, mineral composition, and overall groundwater quality. Although the water samples from Mankweng exhibited lower quality compared to those from Dalmada, they still demonstrated low microbial contamination. Continuous monitoring is crucial to detect any changes in water quality.

5. Environmental, Animal, and Human Effects on Analysis

The present study faced a few challenges, primarily related to accessing certain boreholes, which complicated the collection of water samples. Water samples were obtained from boreholes at depths ranging from 60 to 120 m (Table S1), with the distilled water serving as a control sample. The only environmental factor that could potentially influence the results was temperature; however, all samples were collected during the same season to mitigate this effect. As a result, the data gathered is not influenced by animal or human activities.

6. Conclusions

The physical and chemical properties of the borehole water showed that two water types are available, namely Na-Cl and Ca-Cl, neither of which conforms to the acceptable range of SSP (<50%) and KR (<1). Hence, it is crucial to constantly monitor the water sources of the studied areas for their chemical composition. Additionally, ion exchange can also be a source of sodium ions, while the Ca^{2+}/Mg^{2+} ratio indicated that calcium was more prevalent than magnesium. Based on the SAR values, 60% of the water samples showed excellent quality for drinking, and 40% showed good quality for irrigation. During the current study, it was observed that certain boreholes in Mankweng were inadequately maintained. Consequently, it is recommended that these boreholes receive proper maintenance. Despite the low presence of bacteria such as *E. coli* in the water samples from Mankweng and Dalmada, regular microbiological analysis of water sources is necessary to mitigate the negative impact on the health system of Limpopo Province. The limitation of the study was access to the borehole water. For future perspective, it is strongly recommended that the South African government, in collaboration with local municipalities in the Capricorn district of Limpopo Province, take proactive measures to implement sustainable water management practices. Key strategies should include the regulation of groundwater pumping rates to prevent over-extraction, the establishment of stringent guidelines to prevent water contamination, and the development of comprehensive educational campaigns aimed at raising public awareness about responsible water use. By fostering community involvement and promoting sustainable habits, this initiative can contribute significantly to the preservation of vital water resources for future generations.

Supplementary Materials: The following supporting information can be downloaded at https://www.mdpi.com/article/10.3390/environments12060174/s1, Table S1: The locations of the samples were used for the water analysis from Mankweng and Dalmada, Limpopo Province, South Africa; Table S2: Pearson correlation matrix for the water samples parameters, Figure S1: Location of sampling in Limpopo Province, South Africa [59], Figure S2. Sampling site of the water samples from Limpopo Province, South Africa (Source: Council for Geosciences; Johnson et al., 2006 [59]), where prominent aquifers are marked (red circle), Figure S3. Distribution of mineral across the Limpopo River basin (Source: https://limpopocommission.org/maps/maps-the-river-basin [60]).

Author Contributions: Conceptualization, E.S. and N.M.; methodology, E.S. and N.M.; software, E.S.; investigation, E.S. and N.M.; resources, E.S. and N.M.; writing—original draft preparation, E.S.; writing—review and editing, E.S. and N.M.; funding acquisition, E.S. and N.M. All authors have read and agreed to the published version of the manuscript.

Funding: This research was funded by RNA-UL-2022, provided by the University of Limpopo, South Africa.

Data Availability Statement: The original contributions presented in this study are included in the article/Supplementary Material. Further inquiries can be directed to the corresponding authors. All data provided in the paper or as a Supplementary File.

Acknowledgments: The authors acknowledge the University of Limpopo for the funds and facilities for water analysis.

Conflicts of Interest: The authors declare no conflicts of interest.

References

1. WHO. *Guidelines for Drinking Water Quality*, 4th ed.; WHO: Geneva, Switzerland, 2017.
2. Jakeman, A.J.; Barreteau, O.; Hunt, R.; Rinaudo, J.D.; Ross, A.; Arshad, M.; Hamilton, S. Integrated Groundwater Management: An Overview of Concepts and Challenges. In *Integrated Groundwater Management*; Jakeman, A.J., Barreteau, O., Hunt, R.J., Rinaudo, J.D., Ross, A., Eds.; Springer: Cham, Switzerland, 2016. [CrossRef]

3. Levallois, P.; Villanueva, C.M. Drinking Water Quality and Human Health: An Editorial. *Int. J. Environ. Res. Public Health* **2019**, *16*, 631. [CrossRef] [PubMed]
4. Verlicchi, P.; Grillini, V. Surface and Groundwater Quality in South African Area—Analysis of the Most Critical Pollutants for Drinking Purposes. *Proceedings* **2020**, *48*, 3. [CrossRef]
5. Gavrilescu, M. Water, Soil, and Plants Interactions in a Threatened Environment. *Water* **2021**, *13*, 2746. [CrossRef]
6. Pitchard, M.; Mkandawire, T.; Óneill, T.G. Assessment of groundwater quality in shallow wells within the southern districts of Malawi. In Proceedings of the 8th WaterNet/WARFSA/GWP-SA Annual Symposium, Lusaka, Zambia, 31 October – 2 November 2007.
7. Moyo, N. An analysis of the chemical and microbiological quality of ground water from boreholes and shallow wells in Zimbabwe. *Phys. Chem. Earth* **2013**, *66*, 27–32. [CrossRef]
8. Rietveld, L.C.; Haarhoff, J.; Jagals, P. A tool for technical assessment of rural water supply systems in South Africa. *Phys. Chem. Earth Parts A/B/C* **2009**, *34*, 43–49. [CrossRef]
9. Odiyo, J.O.; Makungo, R. Fluoride concentrations in groundwater and impact on human health in Siloam Village, Limpopo province, South Africa. *Water SA* **2012**, *38*, 731–736. [CrossRef]
10. Potgieter, N.; Mudau, L.S.; Maluleke, F.R.S. The microbiological quality of private and communal boreholes in the Tshitale-hlanganani region of the Limpopo province, South Africa. *Water Sci. Technol.* **2006**, *54*, 371–377. [CrossRef]
11. Vegter, J.R. *Hydrogeology of Groundwater Region 7 Polokwane/Pietersburg Plateau*; WRC Consultancy No. K8/466; Water Research Commission: Pretoria, South Africa, 2003; 53p.
12. Rivett, U.; Champanis, M.; Wilson-Jones, T. Monitoring drinking water quality in South Africa: Designing information systems for local needs. *Water SA* **2013**, *39*, 409–414. Available online: http://www.scielo.org.za/scielo.php?script=sci_arttext&pid=S1816-79502013000300010&lng=en (accessed on 6 March 2025). [CrossRef]
13. Misstear, B.; Vargas, C.R.; Lapworth, D.; Ouedraogo, I.; Podgorski, J. A global perspective on assessing groundwater quality. *Hydrogeol. J.* **2023**, *31*, 11–14. [CrossRef]
14. APHA. *Standard Methods for the Examination of the Water and Wastewater*, 20th ed.; American Public Health Association: Washington, DC, USA, 1998.
15. Hach. *Water Analysis Handbook*, 7th ed.; Hach Company: Loveland, CO, USA, 2012; 1796p.
16. US Salinity Laboratory. *Diagnosis and Improvement of Saline and Alkaline Soils*; Richards, L.A., Ed.; Handbook No. 60; US Department of Agriculture: Washington, DC, USA, 1954.
17. Todd, D.K. *Groundwater Hydrology*; Wiley: New York, NY, USA, 1980.
18. Raghunath, H.M. *Groundwater*; Wiley Eastern: New Delhi, India, 1987.
19. Kelley, W.P. Use of saline irrigation water. *Soil. Sci.* **1963**, *95*, 355–391. [CrossRef]
20. Waterloo Hydrogeologic. Water Quality Data Analysis and Reporting Software. 2021. Available online: https://www.waterloohydrogeologic.com/products/aquachem/ (accessed on 1 January 2023).
21. Addinsoft. *XLSTAT, Analyse de Données et Statistique Avec MS Excel*; Addinsoft: New York, NY, USA, 2007.
22. MacFaddin, J.F. *Biochemical Tests for Identification of Medical Bacteria*, 2nd ed.; Williams and Wilkons: Baltimore, MD, USA, 1980; 527p.
23. Bain, R.; Cronk, R.; Hossain, R.; Bonjour, S.; Onda, K.; Wright, J.; Yang, H.; Slaymaker, T.; Hunter, P.; Prüss-Ustün, A.; et al. Global assessment of exposure to faecal contamination through drinking water based on a systematic review. *Trop. Med. Int. Health* **2014**, *19*, 917–927. [CrossRef]
24. Raihan, F.; Alam, J.B. Assessment of groundwater quality in Sunamganj Bangladesh. *Iran. J. Environ. Health Sci. Eng.* **2008**, *6*, 155–166.
25. Mukonazwothe, M.; Munyai, L.F.; Mutoti, M.I. Groundwater quality evaluation for domestic and irrigation purposes for the Nwanedi Agricultural Community, Limpopo Province, South Africa. *Heliyon* **2022**, *8*, e09203. [CrossRef]
26. Monira, U.; Sattar, G.S.; Mostafa, M.G. Assessment of surface water quality using the Water Quality Index (WQI) and multivariate statistical analysis (MSA), around tannery industry effluent discharge areas. *H2Open J.* **2024**, *7*, 130–148. [CrossRef]
27. Wantasen, S.; Luntungan, J.N.; Tarore, A.E. Determination of the water quality of panasen river as a source of irrigation water. *IOP Conf. Series Earth Environ. Sci.* **2019**, *314*, 012034. [CrossRef]
28. Joshi, D.M.; Kuman, A.; Agrawal, N. Assessment of the irrigation water quality of River Ganga in Haridwar district. *Indian J. Chem.* **2009**, *2*, 285–292.
29. Bispham, N.Z.; Nowak, K.L. Drinking Water: The Saltier The Better? *J. Am. Heart Assoc.* **2019**, *8*, e012758. [CrossRef] [PubMed]
30. Dinka, M.O.; Loiskand, W.; Ndambuki, J.M. Hydrochemical characterization of various surface water and groundwater resources available in Matahara areas, Fantalle Woreda of Oromiya region. *J. Hydrol. Reg. Stud.* **2015**, *3*, 444–456. [CrossRef]
31. Van Reenen, D.D.; Roering, C.; Brand, G.; Smit, C.A.; Barton, J.M. The granulite-facies rocks of the Limpopo belt, Southern Africa. In *Granulites and Crustal Evolution*; Vielzeuf, D., Vidal, P., Eds.; NATO ASI Series; Springer: Dordrecht, The Netherlands, 1990; Volume 311. [CrossRef]

32. Ayers, R.S.; Westcot, D.W. *Water Quality for Agriculture, Irrigation and Drainage (Paper No. 29)*; FAO: Rome, Italy, 1985.
33. Tanvir Rahman, M.A.T.M.; Saadat, A.H.M.; Islam, M.S.; Al-Mansur, M.A.; Ahmed, S. Groundwater characterization and selection of suitable water type for irrigation in the western region of Bangladesh. *Appl. Water Sci.* **2017**, *7*, 233–243. [CrossRef]
34. Ali, A.A.H. Overview of the vital roles of macro minerals in the human body. *J. Trace Elem. Miner.* **2023**, *4*, 100076. [CrossRef]
35. Mohammed, S.; Arshad, S.; Bashir, B.; Vad, A.; Alsalman, A.; Harsányi, E. Machine learning driven forecasts of agricultural water quality from rainfall ionic characteristics in Central Europe. *Agric. Water Manag.* **2024**, *293*, 108690. [CrossRef]
36. Hakami, R.A.; Naser, R.S.; El-Bakkali, M.; Othman, M.D.M.; Yahya, M.S.; Raweh, S.; Belghyti, D. Groundwater quality deterioration evaluation for irrigation using several indices and geographic information systems: A case study. *Desalination Water Treat.* **2024**, *320*, 100645. [CrossRef]
37. Nur, A.; Ishaku, J.; Yusuf, S. Groundwater Flow Patterns and Hydrochemical Facies Distribution Using Geographical Information System (GIS) in Damaturu, Northeast Nigeria. *Int. J. Geosci.* **2012**, *3*, 1096–1106. [CrossRef]
38. Sengupta, P. Potential health impacts of hard water. *Int. J. Prev. Med.* **2013**, *4*, 866–875.
39. Council for Geoscience (CGS). Limestone and Dolomite Map. 2023. Available online: https://login.mdpi.com/login?_target_path=https://www.preprints.org/production/layout?authAll=true (accessed on 1 January 2023).
40. Bucher, K. Metamorphic Rocks. In *Petrogenesis of Metamorphic Rocks*; Springer Textbooks in Earth Sciences, Geography and Environment; Springer: Cham, Switzerland, 2023. [CrossRef]
41. Zabala, M.; Manzano, M.; Vives, L. The origin of groundwater composition in the Pampeano aquifer underlying the Del Azul Creek basin, Argentina. *Sci. Total Environ.* **2015**, *518*, 168–188. [CrossRef]
42. Viljoen, M. The Mpumalanga/Limpopo Escarpment: Geology and Fluvial Landforms. In *Landscapes and Landforms of South Africa*; Grab, S., Knight, J., Eds.; World Geomorphological Landscapes; Springer: Cham, Switzerland, 2015. [CrossRef]
43. Elango, L.; Kannan, R. Rock–water interaction and its control on chemical composition of groundwater. *Dev. Environ. Sci.* **2007**, *5*, 229–243.
44. Zou, Y.; Lou, S.; Zhang, Z.; Liu, S.; Zhou, X.; Zhou, F.; Radnaeva, L.D.; Nikitina, E.; Fedorova, I.V. Predictions of heavy metal concentrations by physiochemical water quality parameters in coastal areas of Yangtze river estuary. *Mar. Pollut. Bull.* **2024**, *199*, 115951. [CrossRef]
45. Zhao, S.; Zhao, Y.; Cui, Z.; Zhang, H.; Zhang, J. Effect of pH, Temperature, and Salinity Levels on Heavy Metal Fraction in Lake Sediments. *Toxics* **2024**, *12*, 494. [CrossRef]
46. Li, L.; He, Y.; Song, K.; Xie, F.; Li, H.; Sun, F. Derivation of water quality criteria of zinc to protect aquatic life in Taihu Lake and the associated risk assessment. *J. Environ. Manag.* **2021**, *296*, 113175. [CrossRef]
47. Bakare, B.F.; Adeyinka, G.C. Evaluating the Potential Health Risks of Selected Heavy Metals across Four Wastewater Treatment Water Works in Durban, South Africa. *Toxics* **2022**, *10*, 340. [CrossRef]
48. Malan, M.; Müller, F.; Cyster, L.; Raitt, L.; Aalbers, J. Heavy metals in the irrigation water, soils and vegetables in the Philippi horticultural area in the Western Cape Province of South Africa. *Environ. Monit. Assess.* **2014**, *185*, 4085. [CrossRef] [PubMed]
49. Ashton, P.; Love, D.; Mahachi, H.; Dirks, P. *An Overview of the Impact of Mining and Mineral Processing Operations on Water Resources and Water Quality in the Zambezi, Limpopo and Olifants Catchments in Southern Africa*; Report to Minerals, Mining and Sustainable Development Project, Southern Africa; MMSD: Birnam Park, South Africa, 2001; 338p.
50. Malakar, A.; Snow, D.D.; Ray, C. Irrigation Water Quality—A Contemporary Perspective. *Water* **2019**, *11*, 1482. [CrossRef]
51. Department of Water Affairs and Forestry (DWAF). *South African Water Quality Guidelines*, 2nd ed.; Holmes, S., Ed.; Department of Water Affairs and Forestry: Pretoria, South Africa, 1996; Volume 1.
52. WHO. *Guidelines for Drinking Water Quality*, 4th ed.; WHO: Geneva, Switzerland, 2011.
53. WHO. *Guidelines for Drinking Water Quality*, 4th ed.; WHO: Geneva, Switzerland, 2023.
54. McCaffrey, L.P. Distribution and origin of high fluoride groundwater in the Western Bushveld Areas. In *Fluoride and Fluorosis: The Status of South African Research*; University of Cape Town: Cape Town, South Africa, 1995; Volume 2.
55. Grobler, S.R.; Dreyer, A.G.; Blignaut, R.J. Drinking water in South Africa: Implications for fluoride supplementation. *J. South African Dent. Assoc.* **2001**, *56*, 557–559.
56. Shaji, E.; Bindu, J.V.; Thambi, D. High fluoride in groundwater of Palghat District, Kerala. *Curr. Sci.* **2007**, *92*, 240.
57. Potgieter, N.; Becker, P.J.; Ehlers, M.M. Evaluation of the CDC safe water-storage intervention to improve the microbiological quality of point-of-use drinking water in rural communities in South Africa. *Water SA* **2009**, *35*, 505–516. [CrossRef]
58. Odonkor, S.T.; Mahami, T. *Escherichia coli* as a tool for disease risk assessment of drinking water sources. *Int. J. Microbiol.* **2020**, *2020*, 2534130. [CrossRef]

59. Johnson, M.R.; Anhaeusser, C.R.; Thomas, R.J. (Eds.) *The Geology of South Africa*; The Geological Society of South Africa; Council for Geosciences: Pretoria, South Africa, 2006; 691p.
60. LIMCOM (Limpopo Watercourse Commission) Maps—The River Basin. Available online: https://limpopocommission.org/maps/maps-the-river-basin (accessed on 1 April 2025).

environments

MDPI

Article

Pipeline-Related Residential Benzene Exposure and Groundwater Natural Attenuation Capacity in the Eastern Niger Delta, Nigeria

Dogo Lawrence Aleku [1], Harald Biester [2] and Thomas Pichler [1,*]

[1] Institute of Geosciences, University of Bremen, 28359 Bremen, Germany; lawrence@uni-bremen.de
[2] Institute of Geoecology, Technical University Braunschweig, 38106 Braunschweig, Germany; h.biester@tu-braunschweig.de
* Correspondence: pichler@uni-bremen.de; Tel.: +49-421-2186-5100

Abstract: This study was conducted to investigate the presence of benzene in the ground and drinking water in the eastern Niger Delta, where multiple oil and gas production facilities are present. Samples from drinking water wells were collected for measurements of benzene, toluene, ethylbenzene, and xylenes (BTEX). Additionally, the dissolved organic carbon (DOC) concentration was determined for the first time to establish the groundwater's total hydrocarbon and non-hydrocarbon load. The groundwater BTEX and benzene levels were up to 3904 μg/L and 3500 μg/L, respectively. DOC concentrations were up to 49 mg/L. The highest benzene concentrations were detected in wells near an underground petroleum pipeline. However, the concentrations decreased with distance from the pipeline to levels less than 0.1 μg/L. Despite benzene contamination, the aquifer has shown promising aerobic attenuation potential, having up to a 7.5 (95%) mg/L DO level and 2.11 mg/L BTEX biodegradation capacity for DO. However, the high groundwater temperature of up to 32.5 °C may weaken attenuation. The benzene and BTEX point attenuation rates ranged from 0.128 to 0.693 day^{-1} and 0.086 to 0.556 day^{-1}, respectively. Hence, by natural attenuation alone, up to 66.5 and 85 years would be required to reach Nigeria's groundwater benzene and BTEX remediation goals, respectively.

Keywords: benzene; BTEX; DOC; groundwater; source; pipeline; petroleum; contamination; attenuation

1. Introduction

Using pipelines to transfer oil and gas has become a widespread global practice because they are considered the safest and most economical, particularly when transporting over long distances [1,2]. The Nigerian National Petroleum Corporation (NNPC) maintains a pipeline network of over 5000 km. Since the pipelines are constructed of carbon steel, leakage due to corrosion is a common problem; for example, see [3]. Additionally, problems of pipeline leakages resulting from oil theft/sabotage and vandalism (commonly called bunkering), aging infrastructure, equipment failures, and operational failures have been frequently observed in oil and gas infrastructures worldwide [1,4–7]. For instance, in Nigeria, Shell Nigeria reported 11,000 barrels per day of crude oil loss in their pipeline networks in 2018 due to leakages, an increment of about 550% compared to the previous year's data [2]. Such leaks are the main cause of oil losses and can lead to extreme environmental pollution, degradation, and economic losses [6–9]. Several studies have investigated the occurrence of pipeline leakages into groundwater (e.g., [10–17]), in the submarine environment (e.g., [18–20]), and in soils (e.g., [21]) at several locations worldwide. Also, investigations were reported for leakages in soils and groundwater near petroleum facilities and petroleum product spills (e.g., [22–26]), petrol stations, and other gasoline-containing storage facilities (e.g., [27–32]). As underground petroleum pipelines leak, the frequently co-occurring monocyclic aromatic compounds, i.e., benzene, toluene, ethylbenzene, and the three forms of xylenes (o, m, p), commonly referred to as BTEX, might be released into groundwater along with other hydrocarbon constituents, including several polycyclic aromatic hydrocarbons and organochlorine compounds [30,33–35]. These compounds, especially BTEX, can readily migrate through the soil [36] and contaminate groundwater due to their highly soluble nature. The aqueous solubilities of benzene, toluene, ethylbenzene, and xylenes are 1780 mg/L, 515 mg/L, 161 mg/L, and 204 mg/L, respectively [37–39]. To our knowledge, despite the substantial use of underground pipelines for crude oil transfer in the Niger Delta region [40], investigations on groundwater contamination from pipeline-related leakages are unavailable.

Of all the BTEX compounds, benzene is considered the most toxic (e.g., [41]). Prolonged human exposure to benzene can lead to various health issues, including an attack on the central nervous system, the immune system, and the hematopoietic system, cancer (e.g., acute myeloid leukemia, myelodysplastic syndrome, aplastic anemia, and pancytopenia), kidney conditions, liver damage, reduction in the size of women's ovaries, disruption of the menstrual cycle in women, and many other short-term health effects [42–44]. Exposure to TEX compounds can also be detrimental to human health (e.g., [45,46]). As a result, the WHO [42] set a drinking water guideline value of 10 µg/L for benzene. Since groundwater is Nigeria's primary drinking water source, National Environmental Regulations [47] set a remedial target value of 0.2 µg/L and an intervention value of 30 µg/L for benzene.

In groundwater, benzene undergoes several redox-controlled reactions. Under aerobic conditions, microbes can reduce benzene molecules into less toxic compounds, producing carbon dioxide and water: $C_6H_6 + 7.5O_2 \rightarrow 6CO_2 + 3H_2O$. Although benzene readily degrades under aerobic conditions [48–50], the extent of degradation depends on several factors, such as the benzene concentration, the available population of microbes, the availability of oxygen, the available nutrients, the prevailing pH conditions, and the temperature [36,49,51]. Degradation is slow if (1) the benzene concentration is very high and toxic to microbes [52] and (2) the conditions required to change the ring structure of the benzene to carbon dioxide gas molecules during benzene oxidation are not met (the process requires multiple electron transfer and substantial activation energies to remove electrons from the carbon atoms in the benzene ring) [53]. Nevertheless, the fate of benzene in contaminated groundwater systems seems to be primarily controlled by the availability of oxygen as a terminal electron acceptor [54]. Oxygen is rapidly depleted during benzene oxidation due to microbial respiration, creating anoxic conditions [55]. Therefore, to further biodegrade the benzene, an alternative electron acceptor, such as NO_3^-, Mn^{4+}, Fe^{3+}, or SO_4^{2-}, and other benzene-degrading microbes capable of using the alternative electron acceptor must be available [56–58]. This process is, however, usually slow and is often associated with long lag times [54,58], as benzene appears to be more persistent under anaerobic conditions than the TEX compounds [54].

Nevertheless, these natural processes, collectively referred to as natural attenuation (NA), can significantly reduce the concentration, mobility, bioavailability, and toxicity of benzene in groundwater. As a well-known clean-up measure for petroleum-contaminated groundwater, assessing the NA potential at contaminated sites is important. Although NA has been widely applied (e.g., [59–63]), studies of its effectiveness in attenuating the petroleum-contaminated groundwater in the eastern Niger Delta have not been conducted.

The hydrocarbon-contaminated areas in the Niger Delta provide a unique opportunity to investigate benzene contamination and improve our understanding of NA and the time required to reach Nigeria's groundwater clean-up goal by NA under aerobic conditions. To our knowledge, only a single study has been published on benzene occurrence in groundwater in the entire region [26]. That study reported concentrations that ranged from 0.155 to 48.2 µg/m^3 in indoor air and 161 to 9280 µg/L in groundwater. Benzene data for Alode and Okochiri does not exist. Total petroleum hydrocarbon concentration in groundwater ranged from 649 to 86,100 µg/L in Nsisioken Agbi Ogale and 1310 to 16,500 µg/L in Nkeleoken-Alode. Soil remediation was carried out in the affected community recently following UNEP recommendations, but published data on the current benzene level in the groundwater of Nsisioken Ogale and many other affected communities in the region are unavailable. Here, several investigations reported various disease symptoms associated with oil pollution, such as kidney diseases, respiratory problems, skin rashes, congenital disabilities, diabetes, headache, dizziness, throat irritation, and chest pain (e.g., [64–66]). Furthermore, Howard, Okpara [67] and Enuneku, Ogbeide [68] reported a high cancer risk in the region due to oil pollution. Particularly detrimental is benzene since, through dermal absorption alone, the risk of contracting leukemia can increase by 70% (e.g., [69]).

Since long-term exposure to benzene presents acute risks to human health [42], assessing its concentration and fate remains essential to ensure the safe use of groundwater as a source of drinking water. This paper presents (i) the current level and distribution of benzene, (ii) source information, and (iii) the potential and rates of the NA and biodegradation capacity for benzene-contaminated groundwater in the eastern Niger Delta, Nigeria.

2. Materials and Methods

2.1. Site Description, Geology, and Hydrogeology

The study site is located in the eastern part of the petroleum-rich Niger Delta Region of Nigeria (Latitude 4°44′57″ N to 4°47′42″ N and Longitude 7°5′26″ E to 7°6′52″ E), where several oil spills were reported by UNEP [26]. Groundwater samples were collected from the Okochiri in the Okrika Local Government Area, Alode, and Ogale in the Eleme Local Government Area of Rivers State. The sampling locations and the oil and gas production facilities are shown in Figure 1. Okochiri is situated between the Port Harcourt Refinery (PHR) and the Okochiri Creek at approximately 4 m above sea level. Groundwater is flowing to the SSW. The community is characterized by (1) the Nigerian National Petroleum Corporation Limited (NNPCL) underground petroleum pipeline, which runs from the refinery to Umu Nwa (through Alode and Ogale), and the Indorama Eleme Petrochemicals Plant (through Alode and Aleto), (2) the Pipelines and Products Marketing Company (PPMC) 12-joint surface pipeline, which runs from the refinery to the Nigerian Port Authority (Rivers Port Complex), and (3) the wastewater discharge channel stretching for over 1 km from the PHR (through the Okochiri) to Okochiri Creek. Alode is characterized by the presence of (1) the Shell Petroleum Development Corporation (SDPC)-owned underground pipeline, (2) the NNPCL-owned refined petroleum product pipeline, and (3) two petroleum truck parking stations situated on the edge of the PHR to the north, and the north-west. Soil remediation for oil spill clean-up was completed at this location in April 2021. It was overseen by the Hydrocarbon Pollution Remediation Project (HYPREP) (an agency established in July 2012 under the Nigerian Ministry of Petroleum to manage the implementation of recommendations in the UNEP [26] report).

Similarly, Ogale is characterized by (1) the presence of both SPDC- and NNPCL-owned petroleum crude and product pipelines, (2) the intersection of several petroleum pipelines, including the SPDC-owned 28-inch trunk line, which runs from Rumuekpe to Bomu, the 36-inch trunk line, which runs from Nkpoku to New Ebubu, and the abandoned 20-inch trunk line, which runs from the Rumuekpe manifold to the Bomu manifold [26], and (3) petroleum-contaminated drainage systems resulting from sales of petroleum products in plastic containers at roadside stalls. Although the pipelines run underground, visible signs indicate their routes and the width extent in the three communities. Nevertheless, residential houses have been built next to the pipelines, the closest within 5 m from the pipelines.

Geologically, the region comprises three (3) major lithostratigraphic units: the (1) Benin Formation, (2) Agbada Formation, and (3) Akata Formation [70]. The Benin Formation is Oligocene to Recent in age. It mainly comprises coarse-grained, sub-angular to well-rounded, and poorly sorted coastal plain sand and alluvial deposits at shallow horizons [71,72]. The formation, which extends to a depth of about 2 km [73], serves as a groundwater reservoir for the region [74]. It also has excellent water-yielding capacity, up to 6 to 9 L/s [75]. The groundwater resources can be tapped within 3 to 300 m depth within the formation [76]. Based on the hydrostratigraphic unit model developed by Akpokodje, Etu-Efeotor [77], the Benin Formation is considered homogeneous, and the groundwater flow direction is to the SSW. The Agbada Formation, Eocene to Recent, consists of a coastal sequence of alternating marine sands and shales [71,78]. It extends to a depth of up to 3.7 km [73]. Obaje [70] reported that the sand percentage variation within the formation ranges between 30 to 70%. The Paleocene—Recent Akata Formation consists of basal marine thick-shale units with <30% sand intercalations and little silt and clay [70,71]. The formation extends up to 7 km in depth [79]. The marine shales in the Agbada and the Upper Akata Formation are considered a source rock for the petroleum hydrocarbon in the region [72,80–82].

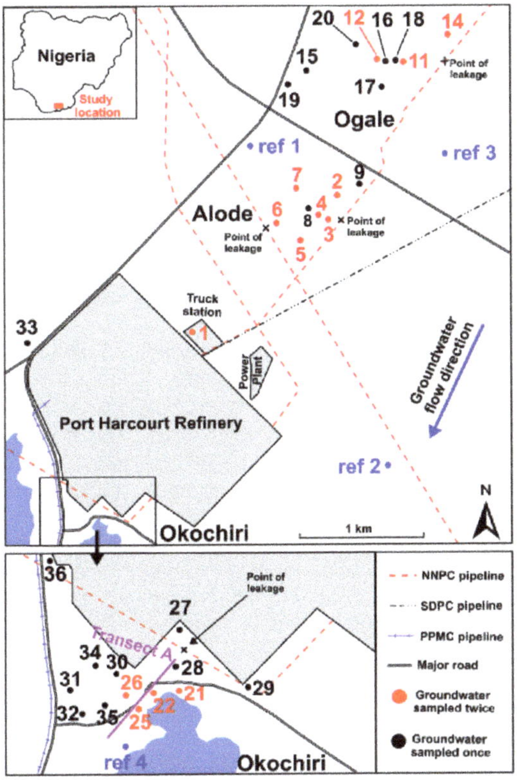

Figure 1. A site map showing the groundwater sample location and oil and gas production facilities. Transect A, indicated in purple, shows the groundwater flow path in Okochiri.

Both shallow and deeper wells tap the aquifers of the Benin Formation. They comprise clay units, unconsolidated sand, and sandy gravels of about 95 to 99% quartz grains [83]. The 2800 to 4000 mm/year precipitation rate, vast catchment area, geology, north–southwards groundwater flow, and rivers and streams contribute to the region's high perennial aquifer recharge [84,85]. Consequently, the increased precipitation, sea level rise, flood and coastal erosion, poverty, coastal location, and intensive oil and gas industrial activities have left the shallow groundwater vulnerable to pollution [86–89]. However, deep-seated aquifers are considered safe and less vulnerable due to the intercalations of clay units [85]. The protection of the Benin Formation's aquifers remains vital since the entire population in the region depends on them for their drinking water.

2.2. Groundwater Sampling

Groundwater samples were collected in May 2021, April 2022, and April 2023 (50 samples in 2021 and 53 samples in 2022 and 2023) for DOC, cations, anions, and trace metal analysis. Samples were filtered through 0.45 µm cellulose acetate (CA) membrane filters and collected into 25 mL glass vials for DOC, 30 mL brown HDPE vials for major cations, and 20 mL Zinnser vials for anions and trace metals. The sub-samples for DOC and major cations were preserved with 1% concentrated hydrochloric acid (HCl) and stored at 4 °C until transported to the laboratory for analysis. Samples were also collected for BTEX analysis during the 2022 and 2023 field campaigns.

The samples were collected from shallow wells (1 to 30 m) within the Benin Formation that serve as private supply wells (PSWs) and community supply wells (CSWs). Sampling locations were in three communities with suspected hydrocarbon contamination: Alode, Ogale, and Okochiri (Figure 1). In Alode, 17 samples were collected from PSWs within 350 m of the pipeline. In Ogale, 14 samples were collected from PSWs within 640 m of the pipeline. Similarly, 22 samples collected in both PSWs and CSWs in Okochiri were within 285 m from the pipeline. In addition, 4 samples (REF 1 to REF 4) were collected in areas considered unaffected by oil and gas activities in Alode, Ogale, and Okochiri (Figure 1). The groundwater sampling for BTEX measurement followed the US EPA method 5021A [90]. Duplicate samples were collected unfiltered in 22 mL headspace vials, preserved with one spatula tip of $CuSO_4/Na_2SO_4$, and stored at 4 °C until transported to the laboratory for analyses.

To ensure that "true or fresh" groundwater samples were collected, sampling was conducted during the early hours (between 6:00 and 8:00 a.m.) when residents were using the wells. Samples were collected by pumping where wells already had electric submersible pumps installed. First, groundwater was pumped into overhead storage tanks to purge the wells for 30 min before sampling directly from the wellhead. Manual sample collection was completed with a water bail made of polyvinyl chloride. The bail was rinsed with water from the well three times before collecting a sample. All samples were collected fresh after pumping.

The pH, conductivity, total dissolved solids (TDS), temperature, dissolved oxygen (DO), salinity, redox potential (ORP), and resistivity were immediately determined in situ using a Hanna instrument HI98494 multiparameter. The total alkalinity ($CaCO_3$) was determined in the field by colorimetric titration with 0.16 N H_2SO_4 in combination with a bromcresol green-methyl red indicator. The bromcresol green-methyl red indicator powder was added to 100 mL of the groundwater sample and titrated using a Hach digital titrator to a light pink color. The total alkalinity was reported as $CaCO_3$.

During the sampling, a strong petroleum odor was observed in most wells, especially in Alode. Samples with a strong petroleum odor also had free-phase petroleum hydrocarbon on the water surfaces, especially when left to sit for up to 10 min. Subsequently, a change in color, from colorless to reddish-brown, was observed in some samples collected within 5 m from the pipeline. Unfortunately, it was impossible to determine the water table and well depth for most wells because the wellheads were sealed with concrete slabs to prevent surface contamination and the theft of submersible pumps installed in the wells. Most well owners rejected unsealing the wells for depth measurement during sampling. In those wells where measurement was possible, the water table and well depth ranged from 1 to 11 m and 9.8 to 30 m, respectively.

2.3. Analytical Procedures

2.3.1. Determination of BTEX and DOC

The BTEX measurements were performed using the "DIN 38407-F 43: 2014-10" method. The method combines gas chromatography and mass spectrometry using the static headspace technique (HS-GC/MS). The instrument used for the measurement was the Shimadzu QP2020 with GC-2030 and HS-20 Trap. If necessary, the samples were diluted before measurement. Five MilliQ water (Sartorius Inc., Göttingen, Germany) samples and five reference groundwater samples were measured, along with the study samples, for quality control.

Dissolved organic carbon (DOC), a fraction of organic carbon that can pass through the 0.45 μm pore size, was determined using a Shimadzu TOC analyzer TOC-V CPN (Shimadzu Corporation, Kyoto, Japan). A certified Total Organic Carbon Standard of 50 mg/L (Aqua Solution Inc., Deer Park, TX, USA) was used to check the accuracy and precision of the method. The measurement error observed was <6%.

2.3.2. Anion and Cation Measurements

Major cations and trace metals were determined by inductively coupled plasma-optical emission spectrometry (ICP-OES) using a Perkin Elmer Optima 7300 DV instrument (Waltham, MA, USA). Measurement precision was checked using EnviroMAT Groundwater Low (ES-L-2) and High (ES-H-2) certified water from SCP Science, Baie-d'Urfé, QC, Canada, showing errors of <3% for all analytes. Also, the accuracy of the measurement was checked using an internal standard, and errors < 4% were observed. Anions were determined using a Metrohm 883 Basic IC plus instrument (Herisau, Switzerland) with a 5 μL injection loop and a Metrosep A Supp5 (150 × 4.0 mm; 5 μm) column. The accuracy and precision of the measurement were checked with an internal standard, and errors < 10% were recorded.

2.3.3. Expressed Biodegradation Capacity and Natural Attenuation Rate

Based on Wiedemeier, Rifai [59], the calculation of the expressed biodegradation capacity (EBC) was used to estimate the amount of BTEX degraded by a given terminal electron-accepting process (dissolved oxygen (DO) in this study):

$$EBC_{DO} = \frac{[C_B - C_P]}{F} \quad (1)$$

where EBC_{DO} (mg/L) = expressed biodegradation capacity for dissolved oxygen, C_B (mg/L) = average background concentration of the DO, C_P (mg/L) = the concentration of the DO within the plume, and F = the BTEX utilization factor using DO. Furthermore, point attenuation rates and half-lives were calculated for benzene and BTEX, following methods described by Newell [91], McAllister and Chiang [92], and Bockelmann, Zamfirescu [93], shown in Equation (2), which is transformed into Equation (3). The natural logarithm of benzene and BTEX concentrations were plotted against time to obtain the point attenuation.

$$C_t = C_0 e^{-kt} \quad (2)$$

$$\ln C_t = -kt \times \ln C_0 \quad (3)$$

where C_t = concentration of attenuated contaminant at time t (mg/L), C_0 = the initial concentration of contaminant (mg/L), t = time (days) after attenuation, and k = the first-order attenuation (decay) rate constant (day^{-1}). The half-life, $t_{1/2}$ (days) (i.e., the time required to attenuate the initial concentration of the contaminant by 50%) was computed using Equation (4):

$$t_{1/2} = \ln(2)/k \quad (4)$$

Furthermore, using Equation (5), the time required to reach the contaminant remediation goal can be estimated. Here, we used 1.3 mg/L, 0.620 mg/L, 1.0 mg/L, and 0.460 mg/L as $C_{initial}$ for benzene and 1.305 mg/L, 0.6225 mg/L, 1.1194 mg/L, and 0.4609 mg/L as $C_{initial}$ for BTEX in samples W–21, W–22, W–12, and W–1, respectively, to calculate the time required to reach Nigeria's groundwater remediation goal for benzene (i.e., 0.2 μg/L) and BTEX (i.e., 0.8 μg/L) [47].

$$t_{goal} = \left[-\ln\left(C_{goal}/C_{initial}\right)\right]/k \quad (5)$$

where t_{goal} = the time (days) required to reach goal concentrations C_{goal} (mg/L); $C_{initial}$ = the initial concentration of contaminant (mg/L)

3. Results and Discussion

3.1. Results

3.1.1. Field Measurements and Chemical Data

The groundwater's physicochemical and inorganic chemical composition showed slight variations in concentration between the three communities (Table 1). The pH was slightly acidic, in the range of 4.7 to 6.3, 4.1 to 6.6, and 4.5 to 6.1 in Alode, Ogale, and Okochiri, respectively. Overall, the mean pH in the groundwater across the three communities showed only slight differences in the order of Ogale > Alode > Okochiri; however, the pH values were mainly below the WHO [42] drinking water guideline range of 6.5 to 8.5, a likely result of the breakdown of petroleum hydrocarbon into organic acid by microbes. Lin, Chaocheng [94] observed a decrease in the pH when microbes degraded hydrocarbons in aqueous systems.

Table 1. Groundwater chemical measurements in the study area.

Parameter	Units	Okochiri (n = 15)	Ogale (n = 7)	Alode (n = 8)
pH		4.5–6.1 (5.4)	4.1–6.6 (5.2)	4.7–6.1 (5.1)
DO	mg/L	3.2–7.4 (6.3)	0.7–5.9 (4.2)	1.4–7.5 (5.5)
DO saturation	%	41.4–97 (85)	9–74.5 (55.4)	19–95 (76)
Eh	mV	238–801 (652)	118–561 (427)	113–596 (401)
EC	µS/cm	21–207 (59)	20–364 (67)	20–106 (49)
TDS	mg/L	11–105 (30)	10–183 (33)	10–53 (25)
Temperature	°C	26.6–32.5 (29.3)	25.12–30.7 (29.4)	26.45–31 (30.8)
Salinity	PSU	0.01–0.1 (0.03)	0.01–0.17 (0.03)	0.01–0.1 (0.02)
Alkalinity	mg/L	0–25 (1)	0–100 (0)	0–30 (3)
DIC	mg/L	0–31 (1.3)	0–122 (0)	0–37 (4)
F^-	mg/L	<0.01	<0.01–0.2 (<0.01)	<0.01–3 (0.1)
Cl^-	mg/L	1–6 (5)	3–17 (8)	<0.01–12 (4)
NO_2^-	mg/L	<0.01	<0.01–1 (<0.01)	<0.01–3 (<0.01)
NO_3^-	mg/L	<0.01–3 (2)	<0.01–39 (2)	<0.01–2 (<0.01)
SO_4^{2-}	mg/L	1–14 (9)	1–23 (6)	1–20 (6)
Ca	mg/L	0.5–1 (0.6)	0.2–13 (1)	0.5–12 (2)
Na	mg/L	0.6–15 (4)	1–14 (4)	0.6–10 (3)
K	mg/L	0.2–1 (0.4)	0.1–12 (1)	0.1–7 (1)
Mg	mg/L	0.04–0.4 (0.2)	0.02–4 (0.1)	0.1–1 (0.3)
Si	mg/L	0.5–5 (4)	0.3–4 (3)	1–4 (3.6)
Fe	mg/L	0.01–25 (0.2)	0.01–50 (2)	0.01–7 (1)
Mn	mg/L	0.01–0.3 (0.1)	0.01–0.2 (0.02)	0.01–0.2 (0.03)
Sr	mg/L	0.002–0.01 (0.003)	0.001–0.02 (0.003)	0.001–0.03 (0.01)
DOC	mg/L	3–33 (24)	9–49 (30)	16–47 (32)

Notes: DO: dissolved oxygen; DOC: dissolved organic carbon; DIC: Dissolved Inorganic Carbon; EC: electrical conductivity; TDS: total dissolved solids; NA: not analyzed; n: number of samples. The values in parentheses are the median.

The electrical conductivity (EC) values in the drinking water were relatively low across the three communities, ranging from 20 to 106 µS/cm, 20 to 364 µS/cm, and 21 to 207 µS/cm in Alode, Ogale, and Okochiri, respectively. The EC values showed slight differences in the order of Alode > Okochiri > Ogale. Similarly, total dissolved solids (TDS) in the groundwater ranged from 10 to 53 mg/L, 10 to 183 mg/L, and 11 to 105 mg/L in Alode, Ogale, and Okochiri, respectively. The wide range in EC and TDS values is controlled by the salinity ($r = 0.99$) and Cl^- ($r = 0.53$) distribution in the study area. While the source of the salinity and Cl^- values has not been established yet, the concentrations might be heavily influenced by anthropogenic activities, e.g., the indiscriminate disposal of household wastes into community drainages. Notably, the Cl^- concentrations were not influenced by the distance from the coast. The concentrations were lower in PSWs near the coast and higher in PSWs near the community drainages. Leachates from those drainages, which often contain contaminants, including Cl^-, infiltrate the groundwater and increase its Cl^- levels [95]. The EC, TDS, Cl^-, and salinity levels at the reference sites, which have comparatively less anthropogenic influence, had less variation, ranging from 19 to 24 µS/cm EC, 9 to 12 mg/L TDS, 4 to 6.6 mg/L Cl^-, and 0.01 PSU salinity. The EC, TDS, and Cl^- values were consistent with the findings by Eyankware, Akakuru [96], Nwankwoala and Walter [97], and Abam and Nwankwoala [85]. Although the study area is in a coastal region, there was no indication of saltwater intrusion. Groundwater salinity values ranged from 0.01 to 0.1 PSU (Alode), 0.01 to 0.17 PSU (Ogale), and 0.01 to 0.1 PSU (Okochiri).

The concentrations of the major ions and trace metals in the groundwater were generally within the acceptable drinking water guidelines set by the WHO [42]. However, other trace metals (e.g., As, Cr, Ni, Co, and Zn) were not detected in the groundwater. Since the aquifer comprises coastal plain sands of >95% quartz grains [83], limited ion exchange and minimal dissolution are expected, given that quartz is relatively inert under typical groundwater conditions. As such, the low TDS, EC, and major ion values are attributed to the dominance of quartz grains in the aquifer.

Based on the major ions, the hydrochemical facies of the groundwater were evaluated through the trilinear Piper diagram [98]. Two water types were identified: sodium–chloride (Na-Cl, 74%) and calcium–chloride (Ca-Cl, 26%) (Figure 2). In Alode, the facies were 62% Na-Cl and 38% Ca-Cl type. In Ogale, the facies were 75% Na-Cl and 25% Ca-Cl type; whereas, in Okochiri, closest to the coast, the facies were 83% Na-Cl and 17% Ca-Cl type. Despite the Na-Cl type's dominance, there was no incidence of saltwater intrusion in the groundwater (salinity: 0.01 to 0.17 PSU). Anthropogenic influence, rather than mixing freshwater with saltwater, was likely responsible for the Na-Cl water type in this study.

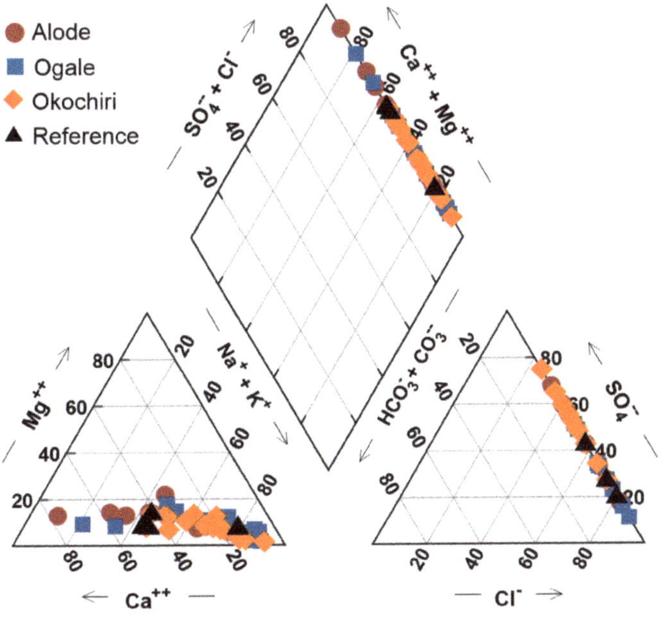

Figure 2. Piper diagram for groundwater of the study area.

3.1.2. Benzene and TEX Concentration

As shown in Figure 3, the concentration of benzene, toluene, ethylbenzene, xylenes, and trimethylbenzene varied between less than 0.1 and 3500 µg/L, less than 0.1 to 210 µg/L, less than 0.1 to 370 µg/L, less than 0.1 to 360 µg/L, and less than 0.1 to 220 µg/L. The total BTEX levels ranged from 0.1 to 3904 µg/L.

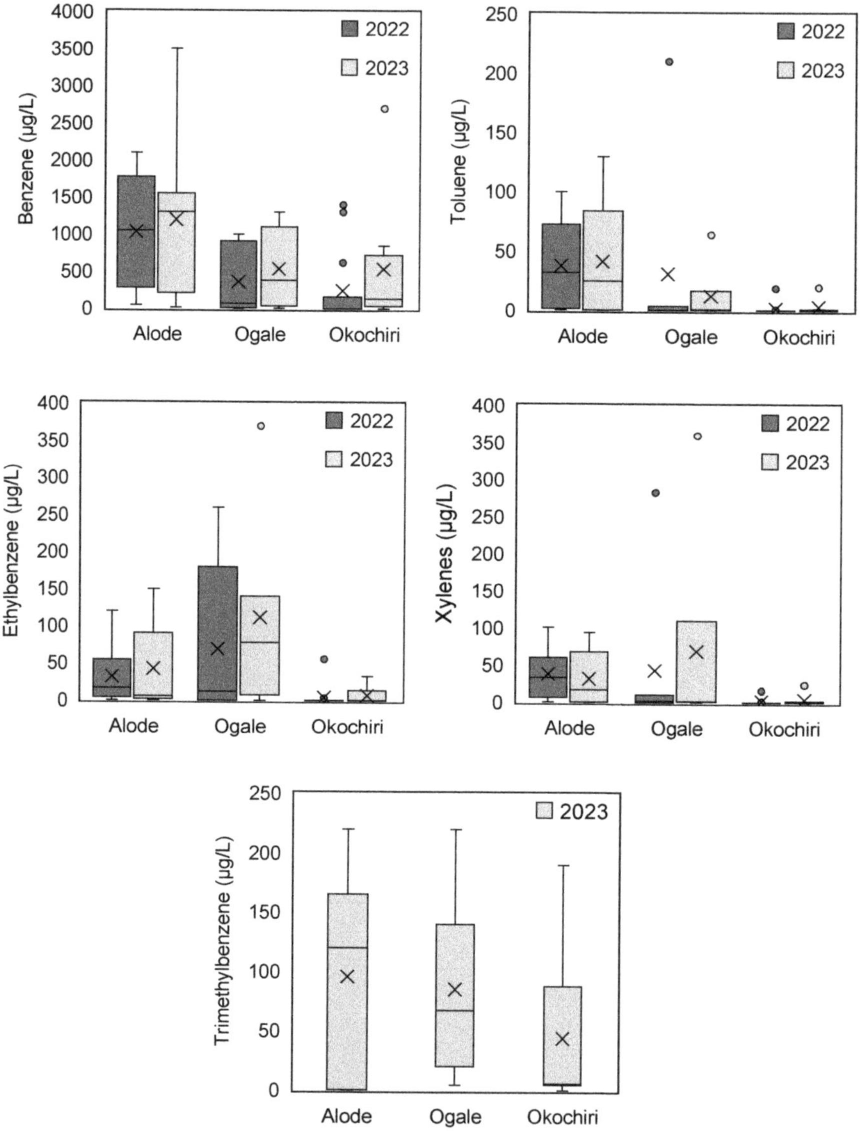

Figure 3. A box plot of benzene, toluene, ethylbenzene, xylenes, and trimethylbenzene concentrations in the study area. The edges of the box represent the 75th and 25th percentiles, respectively. The "x" sign in the box represents the mean value. The solid line represents the median value. The branch gives the range of the data, except for the outliers.

Although benzene, toluene, ethylbenzene, trimethylbenzene, and xylenes were released into the subsurface simultaneously, their concentrations vary significantly in the groundwater. The benzene concentration dominates in the wells, with a Pearson correlation coefficient of r = 0.97 in 2022 and 2023. Figures 3 and 4 show the percentage comparison between concentrations of benzene, toluene, ethylbenzene, xylenes, and trimethylbenzene in Alode, Ogale, and Okochiri, demonstrating the dominance of benzene in the study groundwaters. Overall, the order of abundance was benzene > trimethylbenzene > ethylbenzene > xylenes > toluene. In 2022, benzene constituted 91% in Alode, 72% in Ogale, and 97% in Okochiri. In 2023, those values were 85% in Alode, 66 % in Ogale, and 90% in Okochiri.

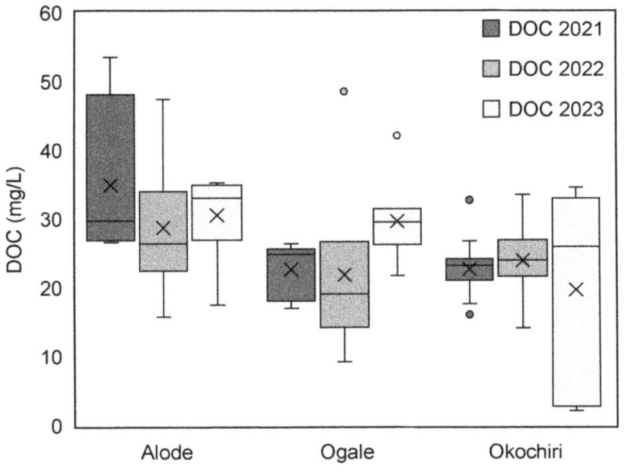

Figure 4. DOC concentrations in the study area. The box's edges represent the 75th and 25th percentiles, respectively. In the box, the "x" sign and the solid line represent the mean and median values, respectively. The branch gives the range of the data except for the outliers.

3.1.3. Dissolved Organic Carbon (DOC)

The DOC results are given in Figure 4 and Table 1. The concentrations varied between 2.7 and 54 mg/L. Overall, high concentrations were observed in wells with a strong hydrocarbon odor. Concentrations across the study sites were consistently within the same range throughout the sampling campaign, averaging 26 mg/L, 26 mg/L, and 28 mg/L for Okochiri, Ogale, and Alode, respectively.

3.2. Discussion

3.2.1. The Source, Transport, and Fate of the Benzene

Source

At Alode, the benzene concentration in the groundwater was up to 3500 μg/L. The elevated benzene concentrations in the groundwater were found along the NNPCL petroleum pipeline, which runs from the PHR to Umu Nwa, passing through Alode and Ogale. Samples from wells on the western side of the NNPCL pipeline had a strong hydrocarbon odor and free-phase hydrocarbon when left to sit for a few minutes. In contrast, the wells on the pipeline's eastern side do not show similar characteristics. As groundwater flows in the SSW direction, benzene is transported in that direction, contaminating wells on the pipeline's eastern side. Concentrations ranging from 220 to 3500 μg/L occur between 1 and 550 m, while lower concentrations occurred in PSWs further away from the pipeline. Nevertheless, all the sampled wells in 2022 and 2023 had elevated benzene, up to 350 times higher than the WHO-recommended drinking water value of 10 μg/L.

In Ogale, the benzene level was up to 1300 μg/L in the groundwater. Like the situation in Alode, wells on the western side of the NNPCL-owned petroleum pipeline had a strong hydrocarbon odor with free-phase hydrocarbon when left to sit for a few minutes. However, similar behavior was not observed in wells sited on the eastern side of the pipeline. Elevated benzene concentrations, up to 1300 μg/L, were observed at a distance of up to 700 m from the pipeline, along the groundwater flow direction. UNEP [26] observed a high benzene concentration in PSW 11 (labeled 001-005-BH-102 in their study), but the concentration was approximately eight times higher than our data in that well. This may be attributed to the NA effect due to DO availability and the soil remediation by HYPREP in Ogale in 2021.

In Okochiri (the host community of the PHR), the benzene concentration was up to 2700 µg/L in groundwater. The highest benzene concentrations were observed in PSW 21, 22, 27, and 28, close to the NNPCL underground petroleum pipeline, which runs from the PHR to Umu Nwa through Alode and Ogale. Based on the groundwater samples taken along transect A in Okochiri (Figure 1) along the groundwater flow direction, the benzene (and other TEX) concentration decreased significantly away from the pipeline leakage point (Figure 5). Benzene and other TEX compounds were undetected in wells more than 250 m from the pipeline.

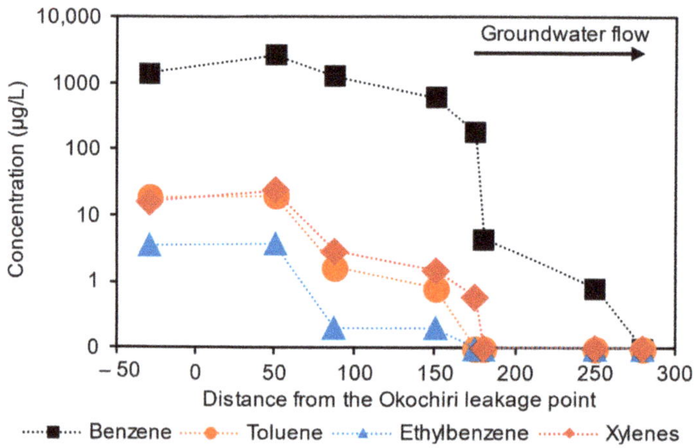

Figure 5. The relationship between groundwater benzene and other TEX concentrations and the distance from the petroleum pipeline leakage point in Okochiri for groundwater samples taken along the flow direction (transect A). The wells plotted are, from left to right, PSW 27, 28, 21, 22, 25, 26, 35, and REF 4.

The presence, level, and distribution of benzene in the groundwater of the three communities depend mainly on (1) the distance of PSWs and CSWs from the underground NNPCL petroleum pipeline, suggesting possible leakages of petroleum hydrocarbon into the corresponding groundwater, (2) the amount of hydrocarbon leakage from the NNPCL pipeline, and (3) the groundwater flow direction.

As the NNPCL petroleum pipeline leaks in the subsurface, benzene and other organic pollutants are released, contaminating groundwater. This occurs in three stages, as outlined by Zhang [99] and Freeze and Cherry [100]: (1) the seepage stage, where gravitational and capillary forces influence the lateral migration of benzene from the oil-wetting zone in the soil into the corresponding groundwater; (2) the seepage of the petroleum ceases when it reaches the water table and subsequently floats on the groundwater; and (3) the separate phase migration, where vaporization and slight dissolution of some of the benzene into the groundwater occurs. The dissolution further leads to benzene lateral transport with the groundwater flow [101]. However, concentrations usually decrease further away from the seepage source. Hence, in the study area, the high amount of precipitation facilitates benzene transport into the groundwater [102], while the groundwater flow direction controls the extent and movement of the contamination plume.

Table 2 compares groundwater BTEX levels of the current study with previous results from the same locations and published results from other countries with similar underground petroleum leakage issues (e.g., [10,26,30–32,103,104]). All other study locations, except Eleme (Ogale) in the Niger Delta of Nigeria [26] and Utah, USA [105], have lower groundwater benzene levels than the current study. The pipeline presence and the continuous transport of petroleum products from the PHR refinery to other parts of the country through the pipeline might be responsible for the high benzene levels observed in Ogale, Alode, and Okochiri groundwater compared to different locations. The groundwater benzene concentration was several orders higher than that of Brazil (Bragança), India (Bengaluru), China (a Petrochemical Site near the Yangtze River), and Nigeria (Bonny) groundwater, as shown in Table 2. However, the current study's toluene, ethylbenzene, xylenes, and trimethylbenzene concentrations are comparable to results published for other areas.

Table 2. A comparison of BTEX concentrations (μg/L) in the eastern Niger Delta region groundwater with other groundwater studies.

Location	Benzene	Toluene	Ethylbenzene	Xylenes	Trimethylbenzene	Source
Bragança, Brazil [1]	<0.1–0.6	<0.1–10.4	<0.1	<0.1–0.5	NA	UST at gas station
Bengaluru, India [2]	<0.1–485	<0.1–153	<0.1–80	<0.1–2620	NA	UST at gas station
China	<0.1–644	<0.1–16.7	<0.1–209	<0.1–181	NA	Petrochemical site
Bonny, Nigeria [3]	<0.1–660	<0.1–800	<0.1–250	<0.1–4200	NA	Petroleum spillage
Eleme, Nigeria [4]	161–9280	NA	NA	NA	NA	Petroleum spillage
Minnesota, USA [5]	<0.1–2550	<0.1–10.37	0.3–3.26	<0.1–1230.7	<0.1–678.23	Pipeline rupture
Utah, USA [6]	<0.1–5600	<0.1–5870	2–950	36–9050	2–650	Hydrocarbon storage facility
Eleme/Okrika, Nigeria [7]	<0.1–3500	<0.1–210	<0.1–370	<0.1–360	<0.1–220	Pipeline leakage

ND: Not detected; NA: not analyzed; UST: Underground Storage Tank; References: [1]: Gomes, Oliva [30]; [2]: Joshua [31]; [3]: Nwankwoala and Omofuophu [25]; [4]: UNEP [26]; [5]: Cozzarelli, Baedecker [10]; [6]: Weidemeier, Swanson [105]; [7]: this study.

Benzene Concentrations in Relation to Toluene, Ethylbenzene, Trimethylbenzene, and Xylenes

Being a soluble compound, and in the presence of the high BTEX concentration, the aerobic removal of the more easily degraded TEX compounds may be responsible for the abundance of benzene in the Ogale and Okochiri groundwater [106]. It has been shown that TEX compounds degrade at higher rates than benzene when specific benzene-degrading bacteria capable of oxidizing the aromatic ring are unavailable [107]. Although oxygen appears to be available for degradation in the study area, the groundwater benzene levels remain high. In an experiment, Eziuzor, Schmidt [108] collected sediments from the study area, i.e., Eleme (Alode and Ogale), as well as other locations in the region (i.e., Tai, Gokana, and Khana) to test the benzene degradation potential. The sediments were spiked with a mixture of benzene, ethylbenzene, and naphthalene dissolved in acetone and left to sit for one year to monitor the potential for hydrocarbon degradation in the sediment. The result suggests that the benzene degradation is too slow to detect by analyzing benzene removal or that the benzene degraders were absent in the sediments. Relatively, sediments collected from Khana showed slightly higher degradation potential than sediments collected from Eleme (the study area), Gokana, and Tai.

Few samples, however, have less dominance of benzene (e.g., PSW 11 had 55% in 2022 and 54% in 2023, and PSW 7 had 64% in 2022 and 75% in 2023). In contrast, sample PSW 17 had an ethylbenzene (11 μg/L) dominance, up to 91%, with 1% benzene, 1% toluene, and 7% total xylenes. The dominance of ethylbenzene in PSW 17, while benzene is dominant in all the other 52 samples, suggests the possible presence of distinct microbial communities that may preferentially degrade benzene, toluene, and xylenes over ethylbenzene. Since the groundwater is mainly oxygen-saturated, the aerobic microbial communities can break down the benzene, toluene, and xylenes using the available oxygen as an electron acceptor. This process likely depleted the DO level in PSW 17 to 1.53 mg/L (19%), making it the lowest in the 2022 data set. The role of DO in the natural degradation of benzene will be discussed in more detail in Section 3.2.2.

Dissolved Organic Carbon (DOC) in the Eastern Niger Delta

The residents of Ogale and Alode suggested that oil spills resulting from pipeline failure first occurred in 2005 in Ogale. As a result, HYPREP carried out soil remediation to clean up the affected areas. Although the remediation activities ended in 2021 in Ogale and Alode, most wells in those communities still have a mild to pungent hydrocarbon odor, and some well waters show free-phase hydrocarbon after sitting for about 10 min. Those wells, as well as a few other wells with little or no hydrocarbon odor, all have elevated DOC concentrations. The generally high DOC (>4 mg/L), unusual for natural groundwater, indicates groundwater contamination derived from anthropogenically released organic pollutants [109]. Most of the DOC is likely dominated by hydrocarbon degradation intermediates [15,110,111]. The DOC levels, therefore, reflect the total hydrocarbon and non-hydrocarbon load in the groundwater. The shallow and sandy nature of the aquifer and the amount of precipitation in the area facilitate the infiltration of DOC from these sources into the aquifer [112,113].

In oxic groundwater where aerobic respiration is present, DOC typically shows an inverse relationship with DO [114] that depends on the bioavailability of the DOC [115]. In Alode and Ogale, DOC poorly correlates with DO, indicating either the absence of aerobic respiration or that DOC is not bioavailable for microbial respiration. In Okochiri, however, there seems to be a moderate negative trend ($r = -0.43$), suggesting possible microbial respiration ongoing in the aquifer. Generally, the microbes in the study area utilized the available oxygen to metabolize and break down the DOC into CO_2 and HCO_3^-, leading to its slight decrease in concentration (DOC vs. HCO_3^- showed a moderate positive correlation, $r = 0.5$). In instances where a specific microbial population capable of degrading DOC is high in Alode, Ogale, and Okochiri aquifers, the DOC will be degraded, depleting the DO in the process. When the DO is wholly consumed, electron acceptors such as NO_3^-, Mn^{4+}, Fe^{3+}, and SO_4^{2-}, if available, will further oxidize the DOC [53].

Fe Contamination and Influence on the Fate of Benzene

While the major ions and trace metal concentrations were relatively low or undetected in some wells, Fe^{2+} concentrations in some samples were elevated. Alode, Ogale, and Okochiri concentrations varied between 0.01 to 7 mg/L, 0.01 to 50 mg/L, and 0.01 to 25 mg/L, respectively. The highest concentrations were observed in wells located near the NNPCL pipeline in Okochiri and Ogale. Wells situated further away have lower concentrations, usually between 0.01 to 0.3 mg/L of Fe, and in some cases, they are undetected. Along the NNPCL petroleum pipelines, the accumulation of Fe precipitates as stains (reddish-brown rust particles) were observed on surfaces of (1) polyvinyl chloride (PVC) overhead tanks used for the storage of drinking water (Figure 6) and (2) plumbing fixtures, as well as other domestic water containers. This phenomenon usually occurs under oxic conditions [116–118]. The reddish-brown rust particles ($Fe(OH)_3$) resulted from the reaction between Fe and DO in the groundwater, as shown by the following reaction: $4Fe^{2+} + O_2 + 10H_2O \rightarrow 4Fe(OH)_3 + 8H^+$. The precipitation was also observed in samples left sitting for about 10 min after abstraction, suggesting the release of Fe from the nearby underground pipelines.

pH plays a significant role in the oxidation process, as the solubility and precipitation of Fe are pH-dependent. In the study wells, while >95% of Fe concentrations are <4 mg/L, the elevated concentration shows a positive relationship with pH ($r = 0.75$ in 2022 and 0.54 in 2023, Figure 7), demonstrating Fe precipitation under slightly alkaline conditions. As groundwater becomes more alkaline, Fe ions become less soluble and tend to come out of the solution, forming visible solid particles (i.e., the reddish-brown rust particles). Since rusts primarily contain Fe oxides, and the groundwater samples were acidified to $pH < 2$

with 2% HNO_3 after filtration, the low pH condition favored the further dissolution of the rust, which was analyzed as Fe^{2+}.

Figure 6. Accumulated precipitates on PVC overhead drinking water storage tank situated directly above the NNPCL pipeline in Okochiri.

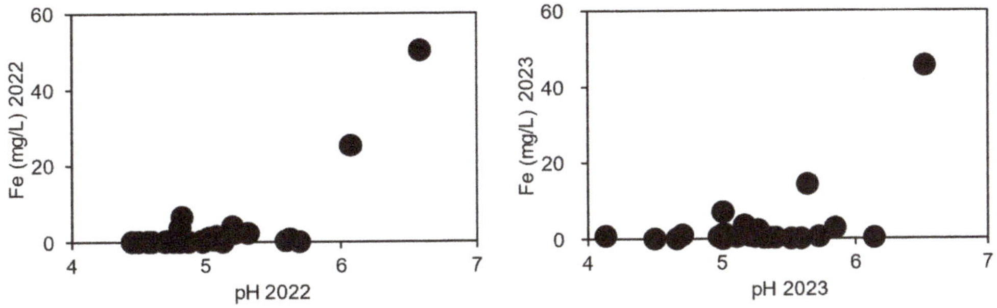

Figure 7. The relationship between Fe and pH in the groundwater in 2022 and 2023.

Being an electron acceptor, iron is significant in the anaerobic degradation of petroleum hydrocarbons [119]. van Leeuwen, Gerritse [120] and Botton and Parsons [121] showed that benzene was degraded under iron-reducing conditions in groundwater. From a thermodynamic perspective, such conditions occur when DO, NO_3^-, and Mn^{4+} are consumed [118]. In Ogale, for instance, iron-reducing conditions occurred in PSW-11. Here, the DO, NO_3^-, Mn^{4+}, and Fe^{3+} levels were 2.3 mg/L, 2 mg/L, 0.1 mg/L, and 50 mg/L in 2022 and 0.7 mg/L, less than 0.01 mg/L, 0.1 mg/L, and 46 mg/L in 2023, respectively. While the Fe^{3+} in the system decreased in 2023, the continuous release of benzene overwhelmed the iron-reducing capacity to attenuate benzene in the groundwater. Thus, the benzene level increased from 910 μg/L in 2022 to 1100 in 2023 μg/L. This suggests that benzene degradation by iron reduction may not have occurred. Furthermore, the SO_4^{2-} level was higher in 2023 (7 mg/L) than in 2022 (3 mg/L), indicating that SO_4^{2-} did not react with benzene.

3.2.2. Potential for Natural Attenuation of Benzene

Since DO is considered the primary electron acceptor for hydrocarbon degradation [122], the available amount of DO in the studied groundwater shows considerable potential for the NA of benzene contamination. DO, however, poorly correlates with benzene in the groundwater (Figure 8a). The lowest DO level (0.7 mg/L, 9%) was observed in well PSW 11, which is located 111 m away from the NNPC underground petroleum pipeline in Ogale but is contaminated with both benzene (1100 µg/L) and Fe (45.6 mg/L). The high Fe content contributed to the depletion of oxygen levels in the groundwater. The Fe contamination was discussed in detail earlier in Fe contamination and influence on the fate of benzene section.

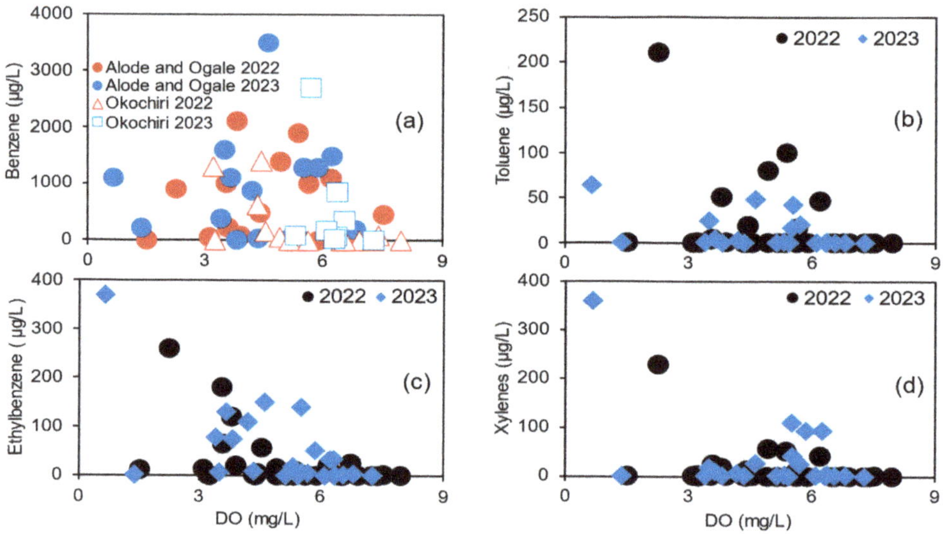

Figure 8. The relationship between (**a**) benzene vs. DO, (**b**) toluene vs. DO, (**c**) ethylbenzene vs. DO, and (**d**) total xylenes vs. DO in Alode, Ogale, and Okochiri groundwater.

Although DO and benzene have a weak negative correlation (r = −0.22 in 2022 and −0.11 in 2023, Figure 8a), DO is consistently higher in wells with undetected benzene than in the contaminated wells. For instance, DO levels in the five reference wells with undetected benzene ranged from 6.4 (88%) to 7.9 (102%) mg/L. In contrast, the DO levels at the contaminated sites were lower in most wells, ranging from 0.7 (9%) to 6.7 (86%) mg/L. Data from Okochiri indicated that the considerable decrease in benzene must be related to the amount of DO available in the groundwater (Figure 8a). For instance, the DO concentrations at PSW 21 were 3.2 (41.4%) mg/L and 6.4 (85%) mg/L, while the benzene concentrations were 1300 µg/L and 850 µg/L in 2022 and 2023, respectively. Similarly, at well PSW 22, the DO concentrations were 4.4 (57%) mg/L (620 µg/L of benzene) and 6.6 (88%) mg/L (350 µg/L of benzene) in 2022 and 2023, respectively.

In 2022 and 2023, higher DO levels and corresponding lower benzene levels were consistently observed downgradient of the NNPCL product pipeline in Okochiri. Similarly, benzene levels in the groundwater samples taken along transect A in the Okochiri (shown earlier) showed an inverse relationship with DO along an established flow path (Figure 9); DO levels increase as the benzene concentration decreases. In Alode and Ogale, however, the relationship between DO and benzene was inconsistent. This may be affected by several factors, such as multidimensional groundwater flow directions and other contaminants.

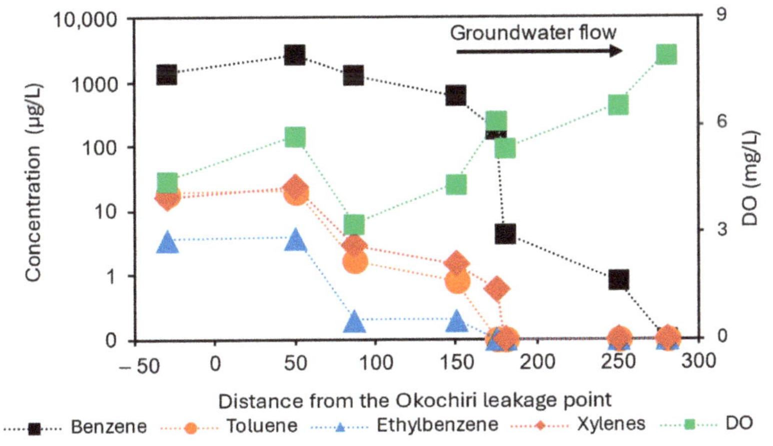

Figure 9. The relationship between DO, benzene, toluene, ethylbenzene, and xylenes concentrations along the groundwater flow direction (transect A) in Okochiri. From left to right, the wells plotted are PSW 27, 28, 21, 22, 25, 26, 35, and REF 4.

Despite the elevated benzene levels in the three communities, DO is still available due to constant recharge from precipitation. The precipitation, saturated with oxygen, continuously infiltrates the soils and the corresponding shallow and sandy aquifers in the study area. Since the area has permeable sandy soils, oxygen consumption in the soil zone is resupplied by gaseous oxygen transport through the soil, resulting in insignificant oxygen consumption (e.g., [118]) given that the depth to the aquifer is 11 m, 9 m, and 1 m in Ogale, Alode, and Okochiri, respectively. As a result, groundwater across the three communities remained saturated, up to 95%, with an average of 66%. Oxygen in the groundwater can initiate aerobic benzene degradation [123] under favorable conditions, such as appropriate pH levels (i.e., 6.5 to 8.5) [124,125] and specific microbial populations capable of degrading benzene. In this process, aerobic bacteria, fungi, or algae utilize DO for both ring activation and the cleavage of the aromatic nucleus and as an electron acceptor for the complete mineralization of the benzene, as presented in Equation (1) [107].

$$C_6H_6 + 7.5O_2 \rightarrow 6CO_2 + 3H_2O \quad (6)$$

According to Jindrová, Chocová [123], under ambient conditions, 8 to 12 mg/L of DO is sufficient to degrade 3000 to 4000 µg/L of benzene. However, such high DO levels were absent in the study area due to the high groundwater temperature, up to 32.5 °C.

Typically, 100% DO saturation at 25 °C is 8.45 mg/L and, at 32.5 °C, only 7.2 mg/L. The highest values for DO were 7.4 mg/L at Okochiri, 5.9 mg/L at Ogale, and 7.5 mg/L at Alode. Similarly, the highest values for temperature were 32.5 °C at Okochiri, 31 °C at Ogale, and 31 °C at Alode. The groundwater temperature, therefore, appears to be a limiting factor to the aquifer's benzene NA potential. The rate and efficiency of benzene degradation, however, do not only depend on the oxygen availability but also the discontinuous release of BTEX from the pollutant source, temperature, nutrient availability, and overall microbial activity in the groundwater environment [56,126–129]. In the case of Ogale, for instance, the benzene concentration in well PSW 11 decreased from 9280 µg/L [26] to 910 µg/L (this study) in 11 years but increased from 910 µg/L to 1100 µg/L in 1 year. Figure 10 shows a general comparison between benzene levels in 2022 and 2023. Besides two samples from Alode, which were significantly higher in 2023, benzene concentrations were similar for the two sampling campaigns. The lack of a consistent decline between 2022 and 2023 concentrations suggests that there may be a continuous release of benzene from the pipeline into the groundwater. Also, benzene concentrations in the three communities appear to exceed the capacity of microbial degradation. Therefore, the high initial concentration or continuous benzene release overwhelmed the NA capacity, slowing the degradation.

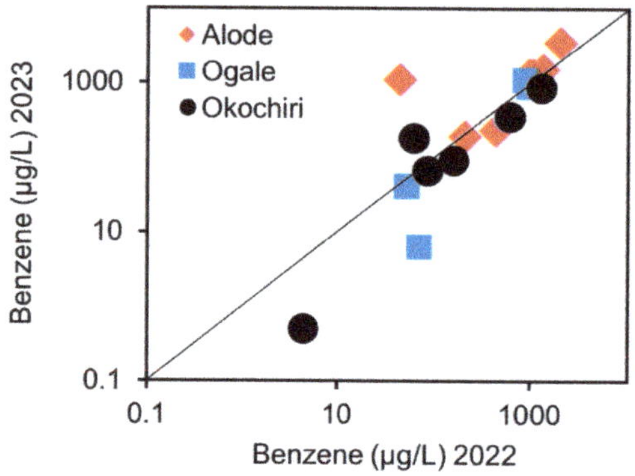

Figure 10. A comparison between benzene concentrations in 2022 and 2023.

Nevertheless, the EBC calculations shown in Table 3 further suggest that the DO levels could be sufficient to support aerobic benzene-degrading bacteria to naturally attenuate the benzene-contaminated groundwater of Alode, Ogale, and Okochiri. The EBC calculations were based on the lowest and the highest DO measured in the contaminated and reference groundwater in the 2022 and 2023 samples. The calculated EBC_{DO} for BTEX levels ranged from 0.62 to 2.11 mg/L, which should be sufficient for BTEX removal [63].

Table 3. The results of EBC calculations.

Year	Parameter	Alode	Ogale	Okochiri
2022	C_P (mg/L, DO)	3.6	1.53	3.23
	C_B (mg/L, DO)	8.89	8.17	8.44
	F	3.14	3.14	3.14
	EBC_{DO} (mg/L)	1.68	2.11	1.66
2023	C_P (mg/L, DO)	1.4	0.66	5.32
	C_B (mg/L, DO)	7.86	7.28	7.27
	F	3.14	3.14	3.14
	EBC_{DO} (mg/L)	2.06	2.11	0.62
Mean	EBC_{DO} (mg/L)	1.87	2.11	1.14

EBC: expressed biodegradation capacity; F: BTEX utilization factor using DO.

Furthermore, the point attenuation rate for benzene and BTEX were computed. Most wells, especially in Alode, showed fluctuated concentrations or insufficient data and were unsuitable for first-order decay calculations. Thus, only a few suitable samples were selected for the NA investigation. The computed point attenuation rates for benzene and BTEX are shown in Table 4. The BTEX point attenuation rates varied between 0.086 and 0.556 day^{-1}, equivalent to half-lives of 51.7 years and 4.5 years, respectively. These attenuation rates were comparable to the 0.19 day^{-1} and 0.038 day^{-1} values reported for the sandy aquifer in Florida [130] and Utah, USA [105], respectively. The time required to reach the Nigerian remediation goal (0.8 μg/L total BTEX) was 9.2 to 85 years. Also, the point attenuation rates for benzene were smaller, varying between 0.128 and 0.693 day^{-1}, with half-lives of 4.6 to 32.8 years. To reach the Nigeria benzene remediation goal of 0.2 μg/L by NA alone, 11.2 to 66.5 years would be required. Notably, most of the NA at this site may be attributed to dilution due to the high precipitation rate, up to 4000 mm/year, and dispersion enhanced by groundwater flow. Although the aquifer has a promising NA potential, active remediation measures would be required to reduce the time needed to reach the Nigerian benzene remediation goal. Due to the lack of sufficient data, it was impossible to estimate the study site's bulk attenuation rate.

Table 4. Estimated point attenuation rates for total BTEX and benzene.

Sample	Site	Contaminant	Point Attenuation (day^{-1})	Half-Life (yr)	Remediation Goal [a]	Remediation Time (yr) [b]
W-21	Okochiri	Benzene	0.425	7.1	0.0002	20.7
		BTEX	0.383	7.7	0.0008	19.3
W-22	Okochiri	Benzene	0.572	4.6	0.0002	14.1
		BTEX	0.556	4.5	0.0008	12
W-12	Ogale	Benzene	0.128	32.8	0.0002	66.5
		BTEX	0.086	51.7	0.0008	85
W-1	Alode	Benzene	0.693	3.5	0.0002	11.2
		BTEX	0.4609	4.8	0.0008	9.2

[a] Groundwater remediation goal of Nigeria [47]; [b] time required to reach the remediation goal.

4. Conclusions

The current study demonstrates that Alode, Ogale, and Okochiri groundwater is heavily contaminated by benzene. Other drinking water quality parameters are within the WHO-recommended values. The maximum benzene concentration (3500 µg/L) was recorded in Alode. Ogale and Okochiri had benzene concentrations up to 1300 µg/L and 2700 µg/L, respectively. Elevated values, however, were observed only in wells next to the NNPCL underground petroleum product pipeline, which runs from the PHR to Umu Nwa through Alode and Ogale. The concentration decreased with distance from the pipeline. While the benzene contamination is at a dangerous level, concentrations of the other co-occurring monocyclic aromatic compounds, i.e., toluene, ethylbenzene, trimethylbenzene, and the three forms of xylenes, are within the WHO-accepted drinking water limits (i.e., 210 µg/L, 370 µg/L, 220 µg/L, and 360 µg/L, respectively). Additionally, DOC levels are elevated in most samples, up to 47 mg/L, 49 mg/L, and 33 mg/L in Alode, Ogale, and Okochiri. The elevated DOC reflects the groundwater's total hydrocarbon and non-hydrocarbon load.

While both the benzene and DOC levels in the groundwater are high, the aquifer has shown promising potential for possible aerobic degradation with its available level of DO (up to 7.5 (95%) mg/L). The calculated BTEX biodegradation capacity for DO alone ranged from 0.62 to 2.11 mg/L. The point attenuation rates of benzene ranged from 0.128 to 0.693 day^{-1}. The time required to reach the groundwater benzene clean-up goal by NA alone was 11.2 to 66.5 years. Despite the promising potential for NA, the degradation rate appears to be too slow owing to (1) the elevated concentrations in the samples, (2) the continuous release of benzene into the groundwater, (3) the high groundwater temperature (up to 32.5 °C), or (4) the possible absence of a specific benzene-degrading microbial population to utilize the available oxygen. The analysis and identification of the microbial population were not within this study's scope. Hence, considerable uncertainties and distinct unknowns regarding the nature and population of the appropriate microbial population needed for such attenuation remain.

Immediate measures are required to save the health of the growing population residing in the affected communities. These could include (1) using granular activated carbon or charcoal filters for benzene removal or (2) discontinuing the use of the contaminated groundwater. Water from safe sources (e.g., bottled or sachet) should be used instead. In the long term, (1) the local authorities should provide alternative drinking water sources in the form of a centralized water supply system, where the quality of the drinking water can be controlled before being distributed to the affected communities, (2) the local authorities should implement measures to ensure a decline in benzene concentration, and (3) the NNCPL should carry out active remedial intervention on sediments and aquifers to facilitate the biodegradation and NA capacities in Alode, Ogale, and Okochiri. The benzene level in the affected communities already exceeded the National Environmental Regulations [47] drinking water target value at levels over 17,500 times and the intervention value at levels over 117 times.

Supplementary Materials: The following supporting information can be downloaded at: https://www.mdpi.com/article/10.3390/environments11100221/s1.

Author Contributions: Conceptualization, T.P.; methodology, T.P.; formal analysis, D.L.A.; investigation, D.L.A.; writing—original draft presentation, D.L.A.; writing—review and editing, H.B. and T.P.; visualization, D.L.A.; supervision, T.P.; project administration, T.P. and D.L.A.; funding acquisition, D.L.A. and T.P. All authors have read and agreed to the published version of the manuscript.

Funding: This work was partly funded by the Deutsche Forschungsgemeinschaft (DFG) grant PI 746/19-1 to Pichler and Petroleum Technology Development Fund (PTDF), Nigeria to Dogo.

Data Availability Statement: Data is contained within the article/Supplementary Materials.

Acknowledgments: The authors thank Kay Hammer (University of Bremen) for contributing to this work. We thank Obrike Ewoma Stephen (Nasarawa State University, Keffi), Faith Nyerma Aleku, and Clement Domgbara for assistance in the field. We thank Henning Fröllje, Christian Hansen, Tobias Himmler, and Janin Scheplitz (University of Bremen) for laboratory support. The Dr. Döring laboratory in Bremen is thanked for the excellent collaboration during the BTEX analyses.

Conflicts of Interest: The authors declare no conflicts of interest.

References

1. Feo, A.; Pinardi, R.; Scanferla, E.; Celico, F. How to Minimize the Environmental Contamination Caused by Hydrocarbon Releases by Onshore Pipelines: The Key Role of a Three-Dimensional Three-Phase Fluid Flow Numerical Model. *Water* **2023**, *15*, 1900. [CrossRef]
2. Ahmed, S.; Le Mouël, F.; Stouls, N.; Lipeme Kouyi, G. Development and Analysis of a Distributed Leak Detection and Localisation System for Crude Oil Pipelines. *Sensors* **2023**, *23*, 4298. [CrossRef] [PubMed]
3. Unueroh, U.; Omonria, G.; Efosa, O.; Awotunde, M. Pipeline corrosion control in oil and gas industry: A case study of NNPC/PPMC system 2A pipeline. *Niger. J. Technol.* **2016**, *35*, 317–320. [CrossRef]
4. Umar, H.; Abdul Khanan, M.; Magashi, S.; Ja'afar, M.; Sani, M. Fractal Analysis for Oil Spills Clustering in Ahoada Communities of the Niger Delta Region of Nigeria. *Int. Arch. Photogramm. Remote Sens. Spat. Inf. Sci.* **2023**, *48*, 371–377. [CrossRef]
5. PHMSA. *Pipeline Failure Causes*; Technical Report; U.S. Department of Transportation: Washington, DC, USA, 2019.
6. Ekong, A.P.; James, G.G.; Ohaeri, I. Oil and Gas Pipeline Leakage Detection using IoT and Deep Learning Algorithm. *J. Inf. Syst. Inform.* **2024**, *6*, 421–434. [CrossRef]
7. Ambituuni, A.; Hopkins, P.; Amezaga, J.; Werner, D.; Wood, J. Risk assessment of a petroleum product pipeline in Nigeria: The realities of managing problems of theft/sabotage. In *Safety and Security Engineering V*; WIT Press: Southampton, UK, 2015; pp. 49–60.
8. Lu, H.; Iseley, T.; Behbahani, S.; Fu, L. Leakage detection techniques for oil and gas pipelines: State-of-the-art. *Tunn. Undergr. Space Technol.* **2020**, *98*, 103249. [CrossRef]
9. Korlapati, N.V.S.; Khan, F.; Noor, Q.; Mirza, S.; Vaddiraju, S. Review and analysis of pipeline leak detection methods. *J. Pipeline Sci. Eng.* **2022**, *2*, 100074. [CrossRef]
10. Cozzarelli, I.M.; Baedecker, M.J.; Mumford, A.C.; Jaeschke, J.B.; Spencer, T.A. Understanding the Evolution of Groundwater-Contaminant Plume Chemistry Emanating from Legacy Contaminant Sources: An Example from a Long-Term Crude Oil Spill. *Groundw. Monit. Remediat.* **2022**, *42*, 30–42. [CrossRef]
11. McGuire, J.T.; Cozzarelli, I.M.; Bekins, B.A.; Link, H.; Martinović-Weigelt, D. Toxicity assessment of groundwater contaminated by petroleum hydrocarbons at a well-characterized, aged, crude oil release site. *Environ. Sci. Technol.* **2018**, *52*, 12172–12178. [CrossRef]
12. Baedecker, M.J.; Siegel, D.I.; Bennett, P.; Cozzarelli, I.M. The fate and effects of crude oil in a shallow aquifer. In *US Geological Survey Toxic Substances Hydrology Program, Proceedings of the Technical Meeting, Phoenix, AZ, USA, 26–30 September 1988*; Department of the Interior, US Geological Survey: Washington, DC, USA, 1989.
13. Baedecker, M.J.; Cozzarelli, I.; Siegel, D.; Bennett, P.; Egan-house, R. Crude oil in a shallow sand and gravel aquifer. III. Biochemical reactions and mass balance modeling in anoxic groundwater. *Appl. Geochem.* **1993**, *8*, 569586. [CrossRef]
14. Bennett, P.; Siegel, D.; Baedecker, M.J.; Hult, M. Crude oil in a shallow sand and gravel aquifer—I. Hydrogeology and inorganic geochemistry. *Appl. Geochem.* **1993**, *8*, 529–549. [CrossRef]
15. Bekins, B.A.; Brennan, J.C.; Tillitt, D.E.; Cozzarelli, I.M.; Illig, J.M.; Martinović-Weigelt, D. Biological effects of hydrocarbon degradation intermediates: Is the total petroleum hydrocarbon analytical method adequate for risk assessment? *Environ. Sci. Technol.* **2020**, *54*, 11396–11404. [CrossRef] [PubMed]

16. Essaid, H.I.; Bekins, B.A.; Godsy, E.M.; Warren, E.; Baedecker, M.J.; Cozzarelli, I.M. Simulation of aerobic and anaerobic biodegradation processes at a crude oil spill site. *Water Resour. Res.* **1995**, *31*, 3309–3327. [CrossRef]
17. Cozzarelli, I.M.; Eganhouse, R.P.; Baedecker, M.J. Transformation of monoaromatic hydrocarbons to organic acids in anoxic groundwater environment. *Environ. Geol. Water Sci.* **1990**, *16*, 135–141. [CrossRef]
18. Poursanidis, K.; Sharanik, J.; Hadjistassou, C. World's largest natural gas leak from nord stream pipeline estimated at 478,000 tonnes. *iScience* **2024**, *27*, 1108772. [CrossRef]
19. Zhao, H.; Zhang, D.; Lv, X.; Song, L.; Li, J.; Chen, F.; Xie, X. Numerical Simulation of Crude Oil Leakage from Damaged Submarine-Buried Pipeline Keywords: Submarine buried pipeline Crude oil leakage Oil spill Numerical simulation Multiphase flow. *J. Appl. Fluid Mech.* **2024**, *17*, 75–88.
20. Lu, H.; Xu, Z.-D.; Song, K.; Cheng, Y.; Dong, S.; Fang, H.; Peng, H.; Fu, Y.; Xi, D.; Han, Z.; et al. Greenhouse gas emissions from U.S. crude oil pipeline accidents: 1968 to 2020. *Sci. Data* **2023**, *10*, 563. [CrossRef] [PubMed]
21. Ismailov, N.; Nadjafova, S. Experience in Assessing Environmental Risks of Main Oil Pipelines in Azerbaijan through the Prism of Soil Biogeoresistance to Crude Oil Pollution. *Mosc. Univ. Soil Sci. Bull.* **2022**, *77*, 196–202. [CrossRef]
22. Adewuyi, G.; Olowu, R. Assessment of oil and grease, total petroleum hydrocarbons and some heavy metals in surface and groundwater NNPC oil depot, Apata, Ibadan, Nigeria. *Int. J. Aquat. Sci.* **2012**, *13*, 45–76.
23. Nambi, I.M.; Rajasekhar, B.; Loganathan, V.; RaviKrishna, R. An assessment of subsurface contamination of an urban coastal aquifer due to oil spill. *Environ. Monit. Assess.* **2017**, *189*, 148. [CrossRef]
24. Gross, S.A.; Avens, H.J.; Banducci, A.M.; Sahmel, J.; Panko, J.M.; Tvermoes, B.E. Analysis of BTEX groundwater concentrations from surface spills associated with hydraulic fracturing operations. *J. Air Waste Manag. Assoc.* **2013**, *63*, 424–432. [CrossRef] [PubMed]
25. Nwankwoala, H.O.; Omofuophu, E. Investigation of hydrocarbon contaminant levels and groundwater quality assessment in parts of Bonny Island, Rivers State of Nigeria. *Cent. Asian J. Environ. Sci. Technol. Innov.* **2020**, *1*, 61–70.
26. UNEP. *Environmental Assessment of Ogoniland, Nigeria*; United Nations Environment Programme: Nairobi, Kenya, 2011.
27. Sivasankar, V.; Gopalakrishna, G. Quantification of benzene in groundwater sources and risk analysis in a popular South Indian Pilgrimage City—A GIS based approach. *Arab. J. Chem.* **2017**, *10*, S2523–S2533.
28. doRego, E.C.P.; Pereira, N.A.D. PAHs and BTEX in groundwater of gasoline stations from Rio de Janeiro City, Brazil. *Bull. Environ. Contam. Toxicol.* **2007**, *79*, 660–664. [CrossRef]
29. Rao, S.M.; Joshua, R.E.; Arkenadan, L. BTEX contamination of Bengaluru aquifers, Karnataka, India. *J. Environ. Eng. Sci.* **2017**, *12*, 56–61. [CrossRef]
30. Gomes, K.J.M.; Oliva, P.A.C.; da Rocha, H.O.; de Alcantara Mendes, R.; da Costa, A.C.G.; dos Santos Miranda, C.; de Oliveira Almeida, N. Evaluation of the contamination of the subsurface and groundwater by monoaromatic hydrocarbons in an eastern Amazonian town in northern Brazil. *Environ. Earth Sci.* **2023**, *82*, 23. [CrossRef]
31. Joshua, R.E. Evaluation of Btex Contamination in Bengaluru Groundwater and Remediation of Contaminated Water Samples. Ph.D. Dissertation, Indian Institute of Science, Bangalore, India, 2020.
32. Chen, X.; Zhang, S.; Yi, L.; Liu, Z.; Ye, X.; Yu, B.; Shi, S.; Lu, X. Evaluation of Biodegradation of BTEX in the Subsurface of a Petrochemical Site near the Yangtze River, China. *Int. J. Environ. Res. Public Health* **2022**, *19*, 16449. [CrossRef]
33. Belpaire, C.; Goemans, G. Eels: Contaminant cocktails pinpointing environmental contamination. *ICES J. Mar. Sci.* **2007**, *64*, 1423–1436. [CrossRef]
34. An, Y.-J. Toxicity of benzene, toluene, ethylbenzene, and xylene (BTEX) mixtures to Sorghum bicolor and Cucumis sativus. *Bull. Environ. Contam. Toxicol.* **2004**, *72*, 1006–1011. [CrossRef]
35. Yang, Y.; Li, J.; Lv, N.; Wang, H.; Zhang, H. Multiphase migration and transformation of BTEX on groundwater table fluctuation in riparian petrochemical sites. *Environ. Sci. Pollut. Res.* **2023**, *30*, 55756–55767. [CrossRef]
36. Kaur, G.; Lecka, J.; Krol, M.; Brar, S.K. Novel BTEX-degrading strains from subsurface soil: Isolation, identification and growth evaluation. *Environ. Pollut.* **2023**, *335*, 122303. [CrossRef] [PubMed]
37. Alexander, M. *Biodegradation and Bioremediation*; Academic Press: San Diego, CA, USA, 1999.
38. López, E.; Schuhmacher, M.; Domingo, J.L. Human health risks of petroleum-contaminated groundwater. *Environ. Sci. Pollut. Res.* **2008**, *15*, 278–288. [CrossRef]
39. Mohammadi, L.; Rahdar, A.; Bazrafshan, E.; Dahmardeh, H.; Susan, M.A.B.H.; Kyzas, G.Z. Petroleum hydrocarbon removal from wastewaters: A review. *Processes* **2020**, *8*, 447. [CrossRef]
40. Umar, H.; Abdul Khanan, M.; Ogbonnaya, C.; Shiru, M.; Ahmad, A.; Baba, A. Environmental and socioeconomic impacts of pipeline transport interdiction in Niger Delta, Nigeria. *Heliyon* **2021**, *7*, e06999. [CrossRef]
41. Behnami, A.; Jafari, N.; Benis, K.Z.; Fanaei, F.; Abdolahnejad, A. Spatio-temporal variations, ozone and secondary organic aerosol formation potential, and health risk assessment of BTEX compounds in east of Azerbaijan Province, Iran. *Urban Clim.* **2023**, *47*, 101360. [CrossRef]
42. WHO. *Guidelines for Drinking-Water Quality: First Addendum to the Fourth Edition*; WHO: Geneva, Switzerland, 2017.
43. WHO. *Exposure to Benzene: A Major Public Health Concern*; WHO: Geneva, Switzerland, 2019.
44. IARC. *IARC Monographs on the Evaluation of Carcinogenic Risks to Humans*; Volume 120: Benzene; International Agency for Research on Cancer: Lyon, France, 2018; Volume 120.

45. Kuranchie, F.A.; Angnunavuri, P.N.; Attiogbe, F.; Nerquaye-Tetteh, E.N. Occupational exposure of benzene, toluene, ethylbenzene and xylene (BTEX) to pump attendants in Ghana: Implications for policy guidance. *Cogent Environ. Sci.* **2019**, *5*, 1603418. [CrossRef]
46. Zoleikha, S.; Mirzaei, R.; Roksana, M. Exposure to chemical hazards in petrol pumps stations in Ahvaz City, Iran. *Arch. Environ. Occup. Health* **2017**, *72*, 3–9. [CrossRef] [PubMed]
47. National Environmental Regulations. *National Environmental (Surface and Ground Water Quality Control) Regulations of Nigeria*; National Environmental Regulations: Lagos, Nigeria, 2011.
48. Billersjö, S. *In-Situ Remediation of Benzene-Contaminated Groundwater—A Bench-Scale Study*; TRITA–LWR Degree Project 13:19; Royal Institute of Technology: Stockholm, Sweden, 2013; 37p.
49. Bedics, A.; Táncsics, A.; Tóth, E.; Banerjee, S.; Harkai, P.; Kovács, B.; Bóka, K.; Kriszt, B. Microaerobic enrichment of benzene-degrading bacteria and description of *Ideonella benzenivorans* sp. nov., capable of degrading benzene, toluene and ethylbenzene under microaerobic conditions. *Antonie Leeuwenhoek* **2022**, *115*, 1113–1128. [CrossRef]
50. Sohrabi, T.; Shakiba, M.; Mirzaei, F.; Pourbabaee, A. BTEX biodegradation using *Bacillus* sp. in a synthetic hypoxic aquatic environment: Optimization by Taguchi-based design of experiments. *Int. J. Environ. Sci. Technol.* **2022**, *19*, 5571–5578. [CrossRef]
51. ATSDR. *Toxicological Profile for Benzene*; Agency for Toxic Substances and Disease Registry: Atlanta, GA, USA, 2007.
52. Vaishnav, D.; Babeu, L. *Comparison of Occurrence and Rates of Chemical Biodegradation in Natural Waters*; Center for Lake Superior Environmental: Superior, WI, USA, 1987.
53. Christensen, T.H.; Bjerg, P.L.; Banwart, S.A.; Jakobsen, R.; Heron, G.; Albrechtsen, H.-J. Characterization of redox conditions in groundwater contaminant plumes. *J. Contam. Hydrol.* **2000**, *45*, 165–241. [CrossRef]
54. Vogt, C.; Kleinsteuber, S.; Richnow, H.H. Anaerobic benzene degradation by bacteria. *Microb. Biotechnol.* **2011**, *4*, 710–724. [CrossRef] [PubMed]
55. Melkonian, C.; Fillinger, L.; Atashgahi, S.; da Rocha, U.N.; Kuiper, E.; Olivier, B.; Braster, M.; Gottstein, W.; Helmus, R.; Parsons, J.R. High biodiversity in a benzene-degrading nitrate-reducing culture is sustained by a few primary consumers. *Commun. Biol.* **2021**, *4*, 530.
56. Wu, J.; Bian, J.; Wang, Q.; Ruan, D. Degradation of benzene in anaerobic groundwater in the typical cold industrial region: Identification, interactions, and optimization of nitrate-/sulfate-reducing assemblages. *Biochem. Eng. J.* **2023**, *192*, 108833. [CrossRef]
57. Atashgahi, S.; Hornung, B.; van der Waals, M.J.; da Rocha, U.N.; Hugenholtz, F.; Nijsse, B.; Molenaar, D.; van Spanning, R.; Stams, A.J.M.; Gerritse, J.; et al. A benzene-degrading nitrate-reducing microbial consortium displays aerobic and anaerobic benzene degradation pathways. *Sci. Rep.* **2018**, *8*, 4490. [CrossRef]
58. Toth, C.R.A.; Luo, F.; Bawa, N.; Webb, J.; Guo, S.; Dworatzek, S.; Edwards, E.A. Anaerobic Benzene Biodegradation Linked to the Growth of Highly Specific Bacterial Clades. *Environ. Sci. Technol.* **2021**, *55*, 7970–7980. [CrossRef] [PubMed]
59. Wiedemeier, T.H.; Rifai, H.S.; Newell, C.J.; Wilson, J.T. *Natural Attenuation of Fuels and Chlorinated Solvents in the Subsurface*; John Wiley & Sons: Hoboken, NJ, USA, 1999.
60. Kao, C.; Prosser, J. Evaluation of natural attenuation rate at a gasoline spill site. *J. Hazard. Mater.* **2001**, *82*, 275–289. [CrossRef]
61. Scow, K.M.; Hicks, K.A. Natural attenuation and enhanced bioremediation of organic contaminants in groundwater. *Curr. Opin. Biotechnol.* **2005**, *16*, 246–253. [CrossRef] [PubMed]
62. Cozzarelli, I.M.; Bekins, B.A.; Eganhouse, R.P.; Warren, E.; Essaid, H.I. In situ measurements of volatile aromatic hydrocarbon biodegradation rates in groundwater. *J. Contam. Hydrol.* **2010**, *111*, 48–64. [CrossRef]
63. Choi, H.-M.; Lee, J.-Y. Groundwater contamination and natural attenuation capacity at a petroleum spilled facility in Korea. *J. Environ. Sci.* **2011**, *23*, 1650–1659. [CrossRef]
64. Ordinioha, B.; Brisibe, S. The human health implications of crude oil spills in the Niger delta, Nigeria: An interpretation of published studies. *Niger. Med. J.* **2013**, *54*, 10–16. [CrossRef]
65. Nriagu, J.; Udofia, E.A.; Ekong, I.; Ebuk, G. Health risks associated with oil pollution in the Niger Delta, Nigeria. *Int. J. Environ. Res. Public Health* **2016**, *13*, 346. [CrossRef] [PubMed]
66. Nanadeinboemi, O.A.; Uju, M.L.; Christopher, C.N.; Hakeem, O.O.; David, D.S. Environmental and Health Influences of Crude Oil Spills in Niger Delta, Nigeria: Case Study Oporoma Community. *J. Health Environ. Res.* **2024**, *8*, 29–40. [CrossRef]
67. Howard, I.C.; Okpara, K.E.; Techato, K. Toxicity and risks assessment of polycyclic aromatic hydrocarbons in river bed sediments of an artisanal crude oil refining area in the Niger Delta, Nigeria. *Water* **2021**, *13*, 3295. [CrossRef]
68. Enuneku, A.; Ogbeide, O.; Okpara, B.; Kubeyinje, B.F.; Job, O.; Asemota, C.O.; Imoobe, T.; Ezemonye, L.I. Ingestion and dermal cancer risk via exposure to polycyclic aromatic hydrocarbon–contaminated soils in an oil-producing community, Niger Delta, Nigeria. *Environ. Toxicol. Chem.* **2021**, *40*, 261–271. [CrossRef]
69. Kalnas, J.; Teitelbaum, D.T. Dermal Absorption of Benzene: Implications for Work Practices and Regulations. *Int. J. Occup. Environ. Health* **2000**, *6*, 114–121. [CrossRef]
70. Obaje, N.G. *Geology and Mineral Resources of Nigeria*; Springer: Berlin/Heidelberg, Germany, 2009; Volume 120.
71. Short, K.; Stäuble, A. Outline of geology of Niger Delta. *AAPG Bull.* **1967**, *51*, 761–779.
72. Avbovbo, A.A. Tertiary lithostratigraphy of Niger Delta. *AAPG Bull.* **1978**, *62*, 295–300.
73. Tuttle, M.L.; Charpentier, R.R.; Brownfield, M.E. *The Niger Delta Petroleum System: Niger Delta Province, Nigeria, Cameroon, and Equatorial Guinea, Africa*; US Department of the Interior, US Geological Survey: Washington, DC, USA, 1999.

74. Ajaegwu, N.; Ozumba, B.; Anomneze, D.; Ugwueze, C. The impact of global drawdown of sea-level on the Messinian deposits of shallow offshore, Niger Delta, Nigeria. *Basic Phys. Res.* **2017**, *7*, 68–86.
75. Akujieze, C.N.; Coker, S.; Oteze, G. Groundwater in Nigeria–a millennium experience–distribution, practice, problems and solutions. *Hydrogeol. J.* **2003**, *11*, 259–274. [CrossRef]
76. Adelana, S.M.A.; Olasehinde, P.I.; Bale, R.B.; Vrbka, P.; Edet, A.E.; Goni, I.B. *An Overview of the Geology and Hydrogeology of Nigeria*; Taylor & Francis Group eBooks: London, UK, 2008; pp. 171–197.
77. Akpokodje, E.; Etu-Efeotor, J.; Mbeledogu, I. *A Study of Environmental Effects of Deep Subsurface Injection of Drilling Waste on Water Resources of the Niger Delta*; CORDEC, University of Port Harcourt: Port Harcourt, Nigeria, 1996.
78. Ogbe, O.; Opatola, O.; Idjerhe, W.; Ocheli, A. Reservoir quality evaluation of sand bodies of K-field, onshore Niger Delta, using wireline logs. *Int. J. Sci. Emerg. Technol. Latest Trends* **2013**, *13*, 46–64.
79. Adagunodo, T.A.; Sunmonu, L.A.; Adabanija, M.A. Reservoir characterization and seal integrity of Jemir field in Niger Delta, Nigeria. *J. Afr. Earth Sci.* **2017**, *129*, 779–791. [CrossRef]
80. Diab, A.I.; Sanuade, O.; Radwan, A.E. An integrated source rock potential, sequence stratigraphy, and petroleum geology of (Agbada-Akata) sediment succession, Niger Delta: Application of well logs aided by 3D seismic and basin modeling. *J. Pet. Explor. Prod. Technol.* **2023**, *13*, 237–257. [CrossRef]
81. Adiela, U.; Odiri, N. Depositional Enviroment and Reservoir Characterization of the Z10 Reservoir sand, Niger Delta, Nigeria. *Int. J. Pure Appl. Sci. Technol.* **2018**, *38*, 008–012.
82. Ogbe, O.B. Sequence stratigraphic controls on reservoir characterization and architectural analysis: A case study of Tovo field, coastal swamp depobelt, Niger Delta Basin, Nigeria. *Mar. Pet. Geol.* **2020**, *121*, 104579. [CrossRef]
83. Onyeagocha, A. Petrography and depositional environment of the Benin Formation. *J. Min. Geol.* **1980**, *17*, 147–150.
84. Ohwoghere-Asuma, O.; Oteng, F.M.; Ophori, D. Simulation of Saltwater Intrusion into Coastal Aquifer of the Western Niger Delta. In *Recent Research on Hydrogeology, Geoecology and Atmospheric Sciences*; Springer Nature Switzerland: Cham, Switzerland, 2023.
85. Abam, T.; Nwankwoala, H. Hydrogeology of Eastern Niger Delta: A Review. *J. Water Resour. Prot.* **2020**, *12*, 741–777. [CrossRef]
86. Richard, G.; Izah, S.C.; Morufu, O.R.; Austin-Asomeji, I. Public and environmental health implications of artisanal petroleum refining and risk reduction strategies in the Niger Delta region of Nigeria. *Bio-Research* **2023**, *21*, 1896–1910. [CrossRef]
87. Ewim, D.R.E.; Orikpete, O.F.; Scott, T.O.; Onyebuchi, C.N.; Onukogu, A.O.; Uzougbo, C.G.; Onunka, C. Survey of wastewater issues due to oil spills and pollution in the Niger Delta area of Nigeria: A secondary data analysis. *Bull. Natl. Res. Cent.* **2023**, *47*, 116. [CrossRef]
88. Sam, K.S.; Onyena, A.P.; Erieegha, O.J.; Eze, F. Water quality evaluation using water quality index and pollution model in selected communities in Gbaramatu Kingdom, Niger Delta, Nigeria. *Afr. J. Environ. Sci. Technol.* **2023**, *17*, 118–134.
89. Adeniran, M.A.; Oladunjoye, M.A.; Doro, K.O. Soil and groundwater contamination by crude oil spillage: A review and implications for remediation projects in Nigeria. *Front. Environ. Sci.* **2023**, *11*, 1137496. [CrossRef]
90. USEPA. *Method 5021A: Volatile Organic Compounds in Various Sample Matrices Using Equilibrium Headspace Analysis*; Revision 1; US Environmental Protection Agency: Washington, DC, USA, 2003.
91. Newell, C.J. *Calculation and Use of First-Order Rate Constants for Monitored Natural Attenuation Studies*; US Environmental Protection Agency, National Risk Management Research Laboratory: San Francisco, CA, USA, 2002.
92. McAllister, P.M.; Chiang, C.Y. A practical approach to evaluating natural attenuation of contaminants in ground water. *Groundw. Monit. Remediat.* **1994**, *14*, 161–173. [CrossRef]
93. Bockelmann, A.; Zamfirescu, D.; Ptak, T.; Grathwohl, P.; Teutsch, G. Quantification of mass fluxes and natural attenuation rates at an industrial site with a limited monitoring network: A case study. *J. Contam. Hydrol.* **2003**, *60*, 97–121. [CrossRef] [PubMed]
94. Lin, L.; Chaocheng, Z.; Qiyou, L.; Yunbo, Z.; Chunshuang, L.; Jianliang, X. Optimization for microbial degradation of dibenzothiophene by Pseudomonas sp. Lky-5 using response surface methodology. *China Pet. Process. Petrochem. Technol.* **2014**, *16*, 19–26.
95. Aweto, K.; Ohwohere-Asuma, O.; Ovwamuedo, G.; Atiti, P. Hydro-geochemical characterization and Groundwater modelling of the subsurface around Ughelli West Engineered Dumpsite in the Western Niger Delta, Nigeria. *Niger. J. Technol. Dev.* **2023**, *20*, 62–72. [CrossRef]
96. Eyankware, M.; Akakuru, O.; Ulakpa, R.; Eyankware, E. Hydrogeochemical approach in the assessment of coastal aquifer for domestic, industrial, and agricultural utilities in Port Harcourt urban, southern Nigeria. *Int. J. Energy Water Resour.* **2023**, *7*, 401–419. [CrossRef]
97. Nwankwoala, H.O.; Walter, I. Assessment of groundwater quality in shallow coastal aquifers of Okrika Island, Eastern Niger Delta, Nigeria. *Ife J. Sci.* **2012**, *14*, 297–304.
98. Piper, A.M. A graphic procedure in the geochemical interpretation of water-analyses. *Eos Trans. Am. Geophys. Union* **1944**, *25*, 914–928.
99. Zhang, P.; Aagaard, P.; Gottschalk, L. Probability method used in predicting contaminant risk in groundwater adjacent to airport. *Water Air Soil Pollut.* **2010**, *211*, 323–339. [CrossRef]
100. Freeze, R.; Cherry, J. *Groundwater*; Prentice-Hall, Inc.: Englewood Cliffs, NJ, USA, 1979.
101. Pantazidou, M.; Sitar, N. Emplacement of nonaqueous liquids in the vadose zone. *Water Resour. Res.* **1993**, *29*, 705–722. [CrossRef]
102. Wadge, A.; Salisbury, J. *Benzene, National Environmental Health Forum Monographs*; Air Series No. 2; South Australian Health Commission, Rundle Mall: Adelaide, SA, Australia, 1997.

103. Baedecker, M.J.; Eganhouse, R.P.; Bekins, B.A.; Delin, G.N. Loss of volatile hydrocarbons from an LNAPL oil source. *J. Contam. Hydrol.* **2011**, *126*, 140–152. [CrossRef]
104. Baedecker, M.J.; Eganhouse, R.P.; Qi, H.; Cozzarelli, I.M.; Trost, J.J.; Bekins, B.A. Weathering of oil in a surficial aquifer. *Groundwater* **2018**, *56*, 797–809. [CrossRef] [PubMed]
105. Weidemeier, T.; Swanson, M.; Wilson, J.; Kampbell, D.; Miller, R.; Hansen, J. *Approximation of Biodegradation Rate Constants for Monoaromatic Hydrocarbons (BTEX) in Ground Water*; US Environmental Protection Agency Papers; U.S. Environmental Protection Agency: Washington, DC, USA, 1996; p. 26.
106. Johnson, S.J.; Woolhouse, K.J.; Prommer, H.; Barry, D.A.; Christofi, N. Contribution of anaerobic microbial activity to natural attenuation of benzene in groundwater. *Eng. Geol.* **2003**, *70*, 343–349. [CrossRef]
107. El-Naas, M.H.; Acio, J.A.; El Telib, A.E. Aerobic biodegradation of BTEX: Progresses and prospects. *J. Environ. Chem. Eng.* **2014**, *2*, 1104–1122. [CrossRef]
108. Eziuzor, S.; Schmidt, M.; Vogt, C. Anaerobic benzene mineralization by natural microbial communities from Niger Delta. *Biodegradation* **2021**, *32*, 37–52. [CrossRef] [PubMed]
109. Regan, S.; Hynds, P.; Flynn, R. An overview of dissolved organic carbon in groundwater and implications for drinking water safety. *Hydrogeol. J.* **2017**, *25*, 959. [CrossRef]
110. Bekins, B.A.; Cozzarelli, I.M.; Erickson, M.L.; Steenson, R.A.; Thorn, K.A. Crude oil metabolites in groundwater at two spill sites. *Groundwater* **2016**, *54*, 681–691. [CrossRef]
111. Podgorski, D.C.; Zito, P.; McGuire, J.T.; Martinovic-Weigelt, D.; Cozzarelli, I.M.; Bekins, B.A.; Spencer, R.G. Examining natural attenuation and acute toxicity of petroleum-derived dissolved organic matter with optical spectroscopy. *Environ. Sci. Technol.* **2018**, *52*, 6157–6166. [CrossRef]
112. Chinago, A. Analysis of rainfall trend, fluctuation and pattern over Port Harcourt, Niger Delta coastal environment of Nigeria. *Biodivers. Int. J.* **2020**, *4*, 1–8. [CrossRef]
113. McDonough, L.K.; Santos, I.R.; Andersen, M.S.; O'Carroll, D.M.; Rutlidge, H.; Meredith, K.; Oudone, P.; Bridgeman, J.; Gooddy, D.C.; Sorensen, J.P. Changes in global groundwater organic carbon driven by climate change and urbanization. *Nat. Commun.* **2020**, *11*, 1279. [CrossRef]
114. Rajendiran, T.; Sabarathinam, C.; Panda, B.; Elumalai, V. Influence of Dissolved Oxygen, Water Level and Temperature on Dissolved Organic Carbon in Coastal Groundwater. *Hydrology* **2023**, *10*, 85. [CrossRef]
115. Chapelle, F.H.; Bradley, P.M.; McMahon, P.B.; Kaiser, K.; Benner, R. Dissolved oxygen as an indicator of bioavailable dissolved organic carbon in groundwater. *Groundwater* **2012**, *50*, 230–241. [CrossRef] [PubMed]
116. Ryan, J.N.; Gschwend, P.M. Effect of iron diagenesis on the transport of colloidal clay in an unconfined sand aquifer. *Geochim. Cosmochim. Acta* **1992**, *56*, 1507–1521. [CrossRef]
117. White, A.F. Heterogeneous electrochemical reactions associated with oxidation of ferrous oxide and silicate surfaces. *Rev. Mineral. Geochem.* **1990**, *23*, 467–509.
118. Appelo, C.A.J.; Postma, D. *Geochemistry, Groundwater and Pollution*; CRC Press: Boca Raton, FL, USA, 2005.
119. Castro, A.R.; Martins, G.; Salvador, A.F.; Cavaleiro, A.J. Iron Compounds in anaerobic degradation of petroleum hydrocarbons: A review. *Microorganisms* **2022**, *10*, 2142. [CrossRef]
120. Van Leeuwen, J.A.; Gerritse, J.; Hartog, N.; Ertl, S.; Parsons, J.R.; Hassanizadeh, S.M. Anaerobic degradation of benzene and other aromatic hydrocarbons in a tar-derived plume: Nitrate versus iron reducing conditions. *J. Contam. Hydrol.* **2022**, *248*, 104006. [CrossRef]
121. Botton, S.; Parsons, J. Degradation of BTEX compounds under iron-reducing conditions in contaminated aquifer microcosms. *Environ. Toxicol. Chem.* **2006**, *25*, 2630–2638. [CrossRef]
122. Marić, N.; Štrbački, J.; Polk, J.; Slavković Beškoski, L.; Avdalović, J.; Lješević, M.; Joksimović, K.; Žerađanin, A.; Beškoski, V.P. Spatial–temporal assessment of hydrocarbon biodegradation mechanisms at a contaminated groundwater site in Serbia. *Chem. Ecol.* **2022**, *38*, 95–107. [CrossRef]
123. Jindrová, E.; Chocová, M.; Demnerová, K.; Brenner, V. Bacterial aerobic degradation of benzene, toluene, ethylbenzene and xylene. *Folia Microbiol.* **2002**, *47*, 83–93. [CrossRef] [PubMed]
124. NRC. *Alternatives for Managing the Nation's Complex Contaminated Groundwater Sites*; National Academies Press: Washington, DC, USA, 2013.
125. Finneran, K.T.; Housewright, M.E. Enhanced anaerobic bioremediation of BTEX and TCE in groundwater. In *Situ and On-Site Bioremediation*; Battelle Press: Columbus, OH, USA, 2001; Volume 7.
126. Singh, R.; Celin, S.M. Biodegradation of BTEX (benzene, toluene, ethyl benzene and xylene) compounds by bacterial strain under aerobic conditions. *J. Ecobiotechnol.* **2010**, *2*, 27–32.
127. Seeger, E.M. Treatment of Groundwater Contaminated with Benzene, MTBE, and Ammonium by Constructed Wetlands. Ph.D. Thesis, Universität Tübingen, Tübingen, Germany, 2013.
128. Ali, M.; Song, X.; Wang, Q.; Zhang, Z.; Zhang, M.; Chen, X.; Tang, Z.; Liu, X. Thermally enhanced biodegradation of benzo[a]pyrene and benzene co-contaminated soil: Bioavailability and generation of ROS. *J. Hazard. Mater.* **2023**, *455*, 131494. [CrossRef]

129. Eze, M.O. Metagenome analysis of a hydrocarbon-degrading bacterial consortium reveals the specific roles of BTEX biodegraders. *Genes* **2021**, *12*, 98. [CrossRef] [PubMed]
130. Wilson, J.T.; Pfeffer, F.M.; Weaver, J.W.; Kampbell, D.H.; Wiedemeier, T.H.; Hansen, J.E.; Miller, R.N. Intrinsic bioremediation of JP-4 jet fuel. In Proceedings of the Symposium on Intrinsic Bioremediation of Ground Water, EPA/540/R-94/515, Denver, CO, USA, 30 August–1 September 1994.

MDPI AG
Grosspeteranlage 5
4052 Basel
Switzerland
Tel.: +41 61 683 77 34

Environments Editorial Office
E-mail: environments@mdpi.com
www.mdpi.com/journal/environments

www.ingramcontent.com/pod-product-compliance
Lightning Source LLC
Chambersburg PA
CBHW061552210926
51850CB00009B/2080